Impedance Source Matrix Converters and Control

IEEE Press
445 Hoes Lane
Piscataway, NJ 08854

Impedance Source Matrix Converters and Control

Yushan Liu
Beihang University, Beijing, China

Xiao Li
Beihang University, Beijing, China

Baoming Ge
Ford Motor Company, Michigan, USA

Haitham Abu-Rub
Hamad bin Khalifa University, Doha, Qatar

Frede Blaabjerg
Aalborg University, Aalborg East, Denmark

IEEEPress
WILEY

Published by John Wiley & Sons, Inc., Hoboken, New Jersey.
Published simultaneously in Canada.

Library of Congress Cataloging-in-Publication Data Applied for:

Hardback ISBN: 9781119906896

Cover Design: Wiley
Cover Images: © Justin Paget/Getty Images, Courtesy of Baoming Ge

Set in 9.5/12.5pt STIXTwoText by Straive, Chennai, India

Contents

About the Authors

Yushan Liu received the BSc degree in automation from Beijing Institute of Technology, Beijing, China, in 2008, and the PhD degree in electrical engineering from the School of Electrical Engineering, Beijing Jiaotong University, Beijing, China, in 2014. She has been a postdoctoral fellow and an assistant research scientist in the Department of Electrical and Computer Engineering, Texas A&M University at Qatar, Doha, Qatar, from 2014 to 2017. She is currently an associate professor in the School of Automation Science and Electrical Engineering, Beihang University, Beijing, China. She has published over 100 journal and conference papers, one book, and one book chapter in her area of expertise. Her research interests include, but not limited to, power electronics, impedance source inverters, cascade multilevel converters, photovoltaic power integration, and model predictive control. Dr. Liu received the "Beijing Science and Technology Nova" from Beijing Municipal Science and Technology Commission, the "Research Fellow Excellence Award" from Texas A&M University at Qatar, and the "Excellent Doctoral Dissertations" prize from Beijing Jiaotong University. She is an associate editor of the *IEEE Transactions on Industrial Electronics* and the *IEEE Open Journal of Industrial Electronics Society*. She served the Power Electronics Devices and Components Committee of IEEE Industry Applications Society as the secretary and vice chair in 2020–2023.

Xiao Li received the BSc degree in automation from Harbin Institute of Technology, China, in 2012, and the PhD degree in electrical engineering from Texas A&M University, College Station, TX, USA, in 2017. From 2017 to 2020, he has worked with Intersil Co., participating in multiple technology programs, including power module design and application of wide-bandgap power devices. After that, he joined Efficient Power Conversion Co. as a senior application engineer, focusing on application of gallium nitride power devices into power electronics systems. Since 2021, he has been with Beihang University, Beijing, China, and now the Associate Professor. Dr. Li's research interests include power electronics, modeling and application of wide-bandgap power devices, and electric energy conversion and control technology.

Baoming Ge received the PhD degree in electrical engineering from Zhejiang University, Hangzhou. He was a postdoctoral researcher in Tsinghua University, a research associate professor in Michigan State University, and a professor in Beijing Jiaotong University. Also, he worked in Texas A&M University and the University of Coimbra. He is the technical expert of power electronics in Ford Motor Company, MI. His main research interests include power electronics, power converters and inverters, electric drives, hybrid/electric vehicles, renewable energy generation, and energy storage systems. He has published more than 230 journal and conference papers, three books, and two book chapters, and he has been granted more than 60 patents.

Haitham Abu-Rub received the MSc degree from Gdynia Maritime Academy, Gdynia, Poland, in 1990; the PhD degree from the Technical University of Gdansk, Gdansk, Poland, in 1995, both in electrical engineering; and the PhD degree in humanities from Gdansk University, Gdansk, in 2004. Since 2006, he has been with Texas A&M University at Qatar, Doha, Qatar, where he is currently with Hamad bin Khalifa University. He has co-authored more than 600 journal and conference papers, seven books, and seven book chapters. His main research interests include energy conversion systems, smart grid, renewable energy systems, electric drives, and power electronic converters. Dr. Abu-Rub is the recipient of many prestigious national and international awards and recognitions. He is fellow of the IEEE.

Frede Blaabjerg was with ABB-Scandia, Randers, Denmark, from 1987 to 1988. From 1988 to 1992, he got the PhD degree in electrical engineering at Aalborg University in 1995. He became an assistant professor in 1992, an associate professor in 1996, and a full professor of power electronics and drives in 1998 at AAU Energy. From 2017 he became a Villum Investigator. He is *honoris causa* at University Politehnica Timisoara (UPT), Romania, in 2017, and Tallinn Technical University (TTU), Estonia, in 2018.

His current research interests include power electronics and its applications such as in wind turbines, Photovoltaic (PV) systems, reliability, Power-2-X, power quality, and adjustable speed drives. He has published more than 600 journal papers in the field of power electronics and its applications. He is the co-author of eight monographs and editor of fourteen books in power electronics and its applications, e.g., the series (four volumes) Control of Power Electronic Converters and Systems published by Academic Press/Elsevier.

He has received 38 IEEE Prize Paper Awards, the IEEE PELS Distinguished Service Award in 2009, the EPE-PEMC Council Award in 2010, the IEEE William E. Newell Power Electronics Award 2014, the Villum Kann Rasmussen Research Award 2014, the Global Energy Prize in 2019, and the 2020 IEEE Edison Medal. He was the editor-in-chief of the *IEEE transactions on power electronics* from 2006 to 2012. He has been distinguished lecturer for the IEEE Power Electronics Society from 2005 to 2007 and for the IEEE Industry Applications Society from 2010 to 2011 as well as 2017 to 2018. In 2019–2020, he served as the president of the IEEE Power Electronics Society. He has been vice president of the Danish Academy of Technical Sciences. He was nominated in 2014–2021 by Thomson Reuters to be the most 250 cited researchers in Engineering in the world.

Preface

A significant challenge to current utility grids is to integrate various energy sources, particularly renewable energy sources and energy storage elements into the network by using power electronics devices. A matrix converter directly connects an AC source to an AC load/source without using intermediate energy storage. It has attracted the attention of academia and industry because of the simple and compact power circuit without dc-link capacitor, output voltage with variable amplitude and variable frequency, sinusoidal input/output currents, and unity power factor operation at input side. Design and improvement in reliability, efficiency, compact volume, power quality, and sustainability of the electricity supply are new challenges. The use of matrix converter topologies with wide voltage gain capability will easily make the system reaching such goals.

The impedance source matrix converters are promising candidates for future generations of the three-three-phase and three-one-phase power conversions. They fulfill the wide voltage gain capability in the compact topologies and overcome the limits of traditional matrix converters. Hence, the high reliability, high efficiency, and low cost can be achieved when using proper topologies and controls. In addition, impedance source matrix converters can be seen as the second type, besides the other type of traditional matrix converters, and these two types build the complete matrix converter family. The recently derived impedance networks with improved voltage gain will be discussed in this book with highlighting the future trends and various applications.

This book brings together the state-of-the-art knowledge and the edge techniques in various stages of research related to the impedance source matrix converters. The book illustrates comprehensive analysis, operating principles, and modulation methods of typical and various impedance source matrix converter topologies developed by the book authors. The book addresses and compares different kinds of typical topologies, in terms of voltage gain analysis, filtering function assessment, parameters design, etc. The innovative design approach and low-frequency power compensation control are detailed for the derived three-one-phase impedance source matrix converter. The book also discusses the model predictive control solution of impedance source matrix converters, overcoming the complexity of duty cycles calculation and composition of traditional modulation. The book also presents applications of impedance source matrix converters to motor drives. It includes the configuration, operation, and results of the discussed topologies and control. Future trends of the research and development in this area are also discussed.

The book encompasses a thorough understanding of the concepts, design, control, and applications of the impedance source matrix converters. Researchers, senior undergraduate and graduate students, and professional engineers investigating vital topics related to power electronic converters will find a high value in the content of the book. They will be able to use the presented design approaches in this book to develop future generation of efficient and reliable power electronics converters/inverters.

Acknowledgment

We would like to take this opportunity to express our sincere appreciation to all the people who directly or indirectly helped in making this book a reality. Our special thanks go to Beihang University for the support provided to realize this effort. Further appreciations go to Ford Motor Company, Hamad bin Khalifa University, and Aalborg University.

We are indebted to our family members for their continuous support, patience, and encouragement without which this book would not have been completed.

Sincerely,
Yushan Liu, Xiao Li, Baoming Ge,
Haitham Abu-Rub, and Frede Blaabjerg

1

Background

1.1 Power Electronics Converter Topologies and Applications in Modern Power Systems

1.1.1 Introduction

In modern society, electrical energy is the most convenient and widely available form of energy, making it the most crucial energy source. However, in recent years, with rapid economic development, global electricity consumption has surged, leading to prominent issues of energy scarcity and environmental pollution. On one hand, electrical energy cannot meet the demands of industrial production and people's daily lives. On the other hand, extensive reliance on traditional fossil fuels for electricity generation has caused severe environmental problems and inefficient utilization of electrical energy [1].

According to statistics from the International Energy Agency in 2014, from 1973 to 2012, the proportion of coal and oil in global terminal energy consumption decreased by 3.6% and 7.5%, respectively. In contrast, the share of electricity consumption increased from 9.4% to 18.1%, ranking second only to oil, as shown in Figure 1.1. It is projected that by 2030, electricity will constitute 25% of global terminal energy consumption, and by 2050, this share is expected to surpass 50%, as depicted in Figure 1.2 [2–5].

Power electronics technology, serving as the vital link for energy conversion and a necessary means to address environmental pollution in the context of new energy sources, has permeated various aspects of electrical applications. This includes applications in power systems, industry, transportation, aerospace, information technology, and telecommunications, as depicted in Figure 1.3 [6]. It has directly or indirectly generated significant economic and societal benefits. In the future, approximately 90% of electrical energy will need to be processed through power electronics technology to enhance energy efficiency and production efficiency, thereby maximizing the utilization of renewable energy sources [7].

AC variable frequency drive technology is a significant application of power electronics in energy-efficient and high-capacity AC transmission control systems. Within this technology, AC converters play a crucial role as integral components of AC speed control systems. Currently, AC converters are extensively employed in high-power AC motor drive systems and power systems [8]. The classification of converters can be seen in Figure 1.4 [9].

The frequency converter, known as a thyristor-based AC/AC converter circuit, directly converts AC power of a certain frequency into adjustable-frequency AC power. As it lacks a direct current (DC) stage, it falls into the category of direct-frequency conversion circuits. However, this type

Impedance Source Matrix Converters and Control, First Edition.
Yushan Liu, Xiao Li, Baoming Ge, Haitham Abu-Rub, and Frede Blaabjerg.

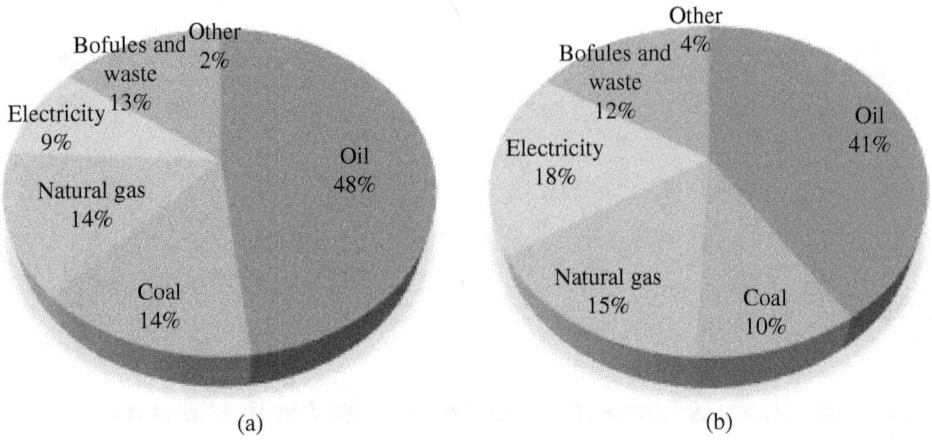

Figure 1.1 (a) Comparison of energy consumption structure between 1973 and 2012; (b) Global terminal energy consumption structure from 2010 to 2050.

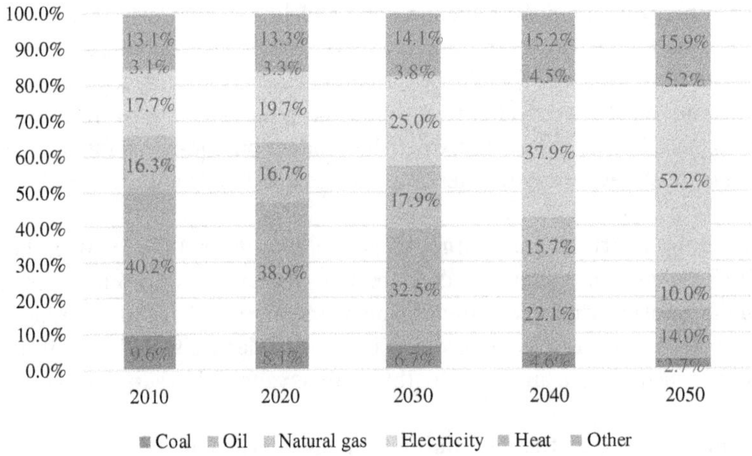

Figure 1.2 Global terminal energy consumption structure from 2010 to 2050.

of converter has notable drawbacks, with its output upper-frequency limit not exceeding 1/3 to 1/2 of the grid frequency. For single-phase AC circuits, two sets of converters are needed, while three-phase circuits require six sets, resulting in numerous components and highly complex control systems.

AC/DC/AC converter is presently one of the most widely used AC/AC frequency conversion circuits. This converter first rectifies AC power into DC power and then inverts DC power back into AC power. Due to the presence of a DC stage, this circuit falls under the category of indirect-frequency conversion circuits. Depending on whether the intermediate DC stage is composed of capacitors or inductors, it can be classified into voltage-source indirect AC/DC/AC converters and current-source indirect AC/DC/AC converters [8]. Among them, the voltage-source AC/DC/AC converter can be further divided into non-controlled rectifier + inverter (Figure 1.5a), which lacks boosting capability and generates high-input current harmonics, resulting in severe grid pollution. The controlled rectifier + inverter (Figure 1.5b) utilizes a boosting rectifier at the input stage, requiring the addition of an inductor. To mitigate harmonic pollution to the grid,

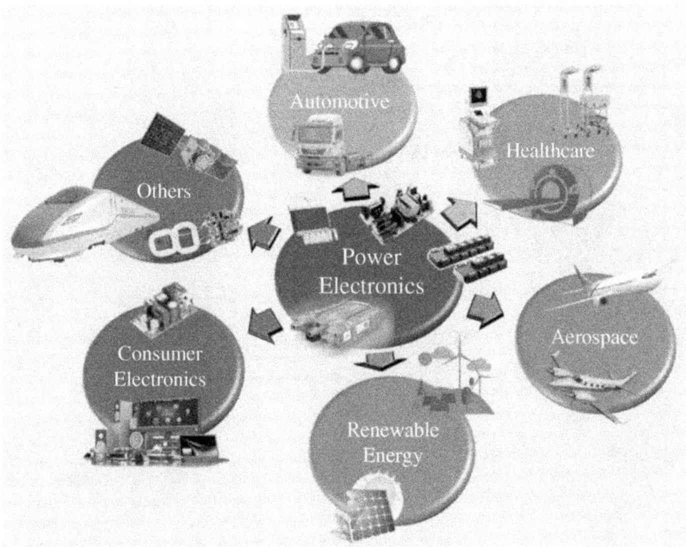

Figure 1.3 Application fields of power electronics [6].

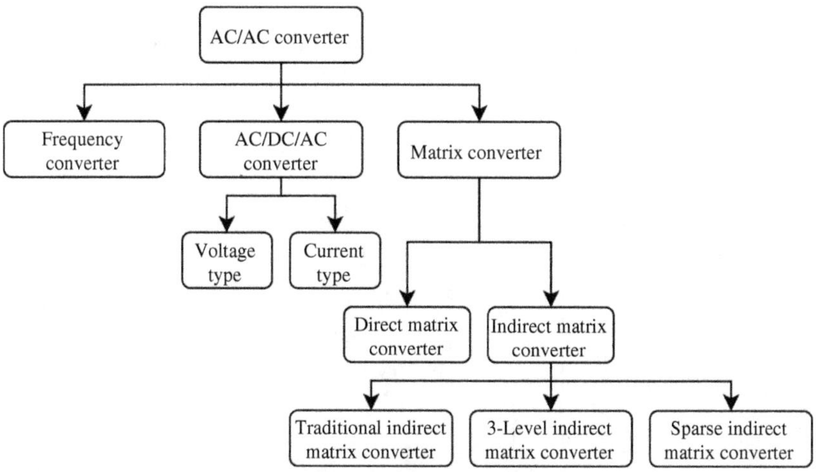

Figure 1.4 AC frequency converter classification.

inductor–capacitor (LC) or inductor–capacitor–inductor (LCL) filters need to be designed at the input stage. The primary drawback of both types of converters lies in the intermediate energy storage components, which not only have large volume and high mass but are also challenging to maintain, leading to lower power density in power converters.

The current-source AC/DC/AC converter Figure 1.6 introduces challenges related to the need for large-capacity flat-wave reactors and issues like current distortion and oscillations caused by AC-side LC filter. In comparison to voltage-source converters, it is more costly and complex to control, thereby limiting its application and research. However, with the advancement of superconducting technology, the current-source converter has found successful applications in superconducting energy storage. Furthermore, it has garnered significant attention in medium-voltage high-power wind power generation and motor drive applications [10, 11].

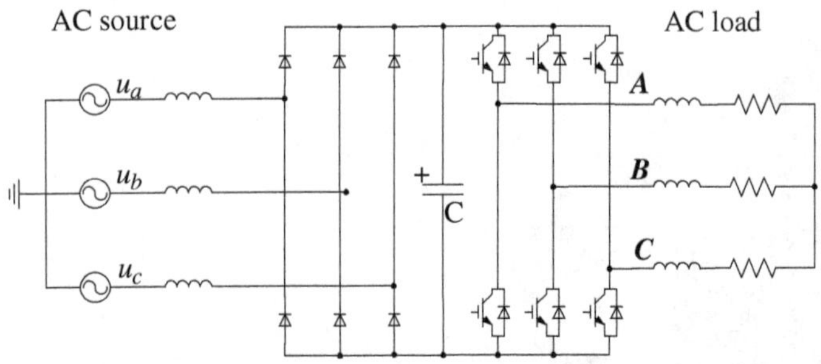

(a)

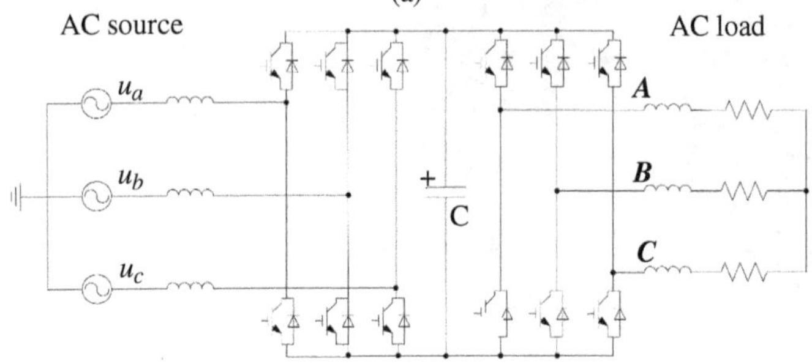

(b)

Figure 1.5 AC/DC/AC voltage-source converter (a) uncontrolled rectifier with inverter, (b) controlled rectifier with inverter.

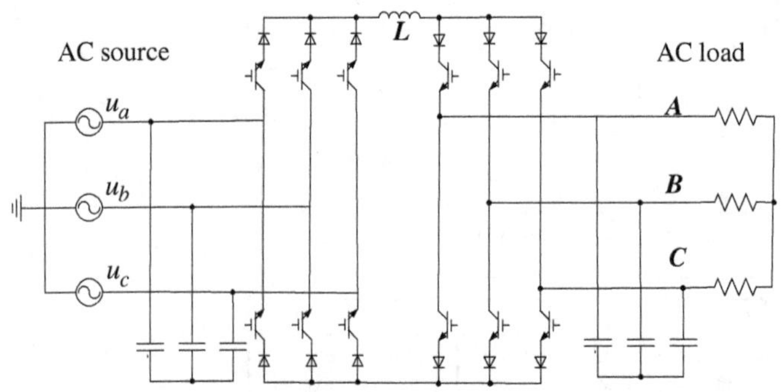

Figure 1.6 AC/DC/AC current-source converter.

To overcome the drawbacks associated with converters featuring intermediate energy storage components and to enhance the power density and reliability of AC/AC converters, researchers began to explore the possibility of AC/AC converters without the use of DC energy storage elements. It was at this juncture that matrix converter (MC) emerged. MC is an electrical conversion device based on bidirectional switches and utilizes pulse-width modulation to generate the desired output voltage. Among various novel AC power converters, MC has gained significant attention from researchers worldwide due to its simple structure and full silicon integration, among other excellent performance attributes [9]. Depending on their structural characteristics, MCs can be classified into two categories: direct matrix converters (DMCs) and indirect matrix converters (IMCs). IMCs not only inherit the advantageous features of DMC but also possess the advantage of zero-current switching at the rectifier stage, significantly reducing control complexity, making them one of the most promising types of AC power converters. IMCs have further led to the development of three-level MCs and generalized sparse IMCs.

1.1.2 Matrix Converter

MCs have been in development for over 40 years, and substantial progress has been made in key areas such as topology design, modulation strategies, control theory, and device development [12–14].

1.1.2.1 Direct Matrix Converter

The concept of DMCs and bidirectional switches was first proposed by Gugi and Pelly [15]. In 1980, Venturini and Alesina introduced the idea of using transistors to construct bidirectional switches for implementing MCs. They developed a prototype based on this concept and presented a series of attractive results. The topology of a DMC is shown in Figure 1.7. This topology employs nine bidirectional switches to interconnect each input phase with every output phase, allowing for the synthesis of the desired output and input currents through a single-stage transformation. Since each bidirectional switch consists of two antiparallel insulated gate bipolar transistors (IGBTs), a DMC requires a total of 18 IGBT power devices [16].

The advantages of a DMC include: (i) bidirectional energy flow, achieving four-quadrant operation; (ii) both input and output currents are sinusoidal; (iii) power factor at the input side can be unity for any load; and (iv) no need for a DC energy storage stage, resulting in a compact circuit structure and high integration level [17].

Despite over 40 years of development, MC technology still faces challenges preventing widespread industrial adoption [18]. These challenges include: (i) maximum boost ratio limited to 0.866; (ii) a relatively high number of power devices, leading to complex commutation control; (iii) difficulty in control under abnormal grid voltage conditions due to the absence of an intermediate DC stage, impacting system performance; (iv) interference on the load side directly affects input-side performance, leading to suboptimal electromagnetic compatibility with the grid; and (v) complex protection circuits, large physical footprint, and higher cost.

1.1.2.2 Indirect Matrix Converter

In pursuit of simplifying the structure of DMCs, reducing the count of power switching components, minimizing system energy losses, and alleviating control intricacies, scholars have introduced a category of IMCs, as depicted in Figure 1.8. In this topology, the input-side rectification employs bidirectional switches, while the inversion stage relies on unidirectional switches,

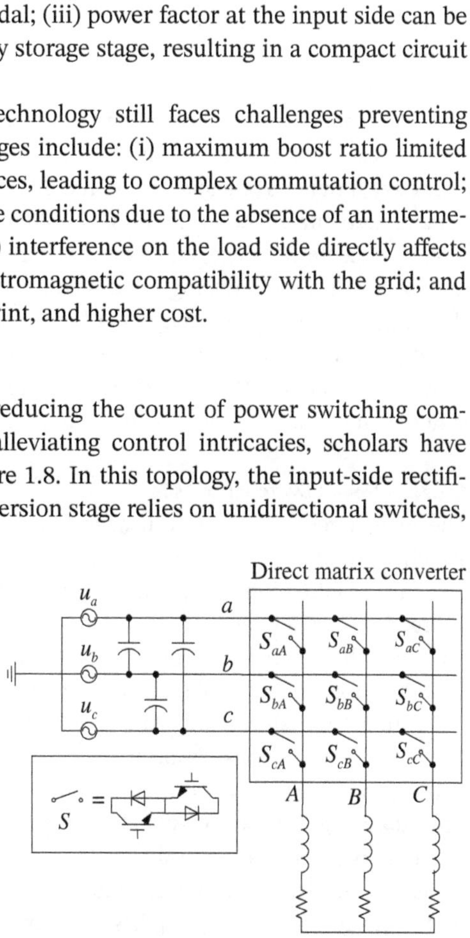

Figure 1.7 DMC circuit diagram.

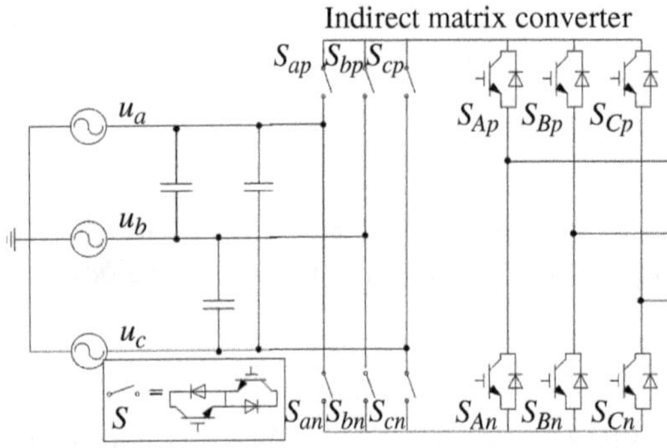

Figure 1.8 IMC.

necessitating a total of 18 IGBT power devices. The initial conceptualization of this topology was attributed to Wei at University of Wisconsin-Madison [19], USA. Subsequently, researchers, led by Kolar at wiss Federal Institute of Technology in Zurich, expanded and refined this concept, introducing various topological variants such as sparse matrix converters (SMCs), very sparse matrix converters (VSMCs), and ultra sparse matrix converters (USMCs) [20, 21], as illustrated in Figure 1.9. A comprehensive analysis of modulation techniques, commutation strategies, voltage transfer ratios, and switch losses for multiple MCs can be found in Table 1.1, as extensively discussed in [22].

IMCs have the potential to overcome the shortcomings of traditional AC-DC-AC PWM inverters and DMCs, making them a promising new category of AC-AC converters. Their advantages primarily manifest in the following ways: (i) they eliminate the need for large energy storage components like bulky inductors or capacitors in the intermediate DC stage; (ii) the rectifier stage switches can achieve zero-current commutation, simplifying the commutation control of the system; (iii) with an intermediate DC stage, mature PWM control methods can be separately applied to the rectifier and inverter stages, reducing control complexity; (iv) under certain constraints, they can reduce the number of switching devices; (v) by using the intermediate DC link as a common bus, they can facilitate multiple inverter outputs, supplying power to multiple AC motor loads [23–26].

In comparison to DMCs, IMCs offer more advantages. However, similar to DMCs, the limited system voltage gain is a significant hindrance to their industrialization.

1.1.2.3 Power Switches of MCs

The main circuit of a MC must employ bidirectional switches, also known as four-quadrant switches. To enable safe commutation and bidirectional power flow, these controllable switches can carry bidirectional current and block bidirectional voltage [27]. Since fully controllable bidirectional switch devices are not yet commercially available, the bidirectional switches used in MCs must be constructed from combinations of unidirectional switches. There are four common structures for bidirectional switches [12, 28, 29], as depicted in Figure 1.10.

Figure 1.10a,b shows antiparallel bidirectional switches with common emitter and common collector configurations, respectively. These bidirectional switches can be composed using two IGBTs (or MOSFETs) with internally integrated antiparallel diodes. These two antiparallel configurations possess the soft-switching capability and are commonly employed. By implementing a four-step commutation strategy [29], these bidirectional switches can control the switching sequence effectively, preventing input short circuits and providing a conduction path for inductive loads, thus avoiding voltage spikes. Each of these bidirectional switch configurations has distinct

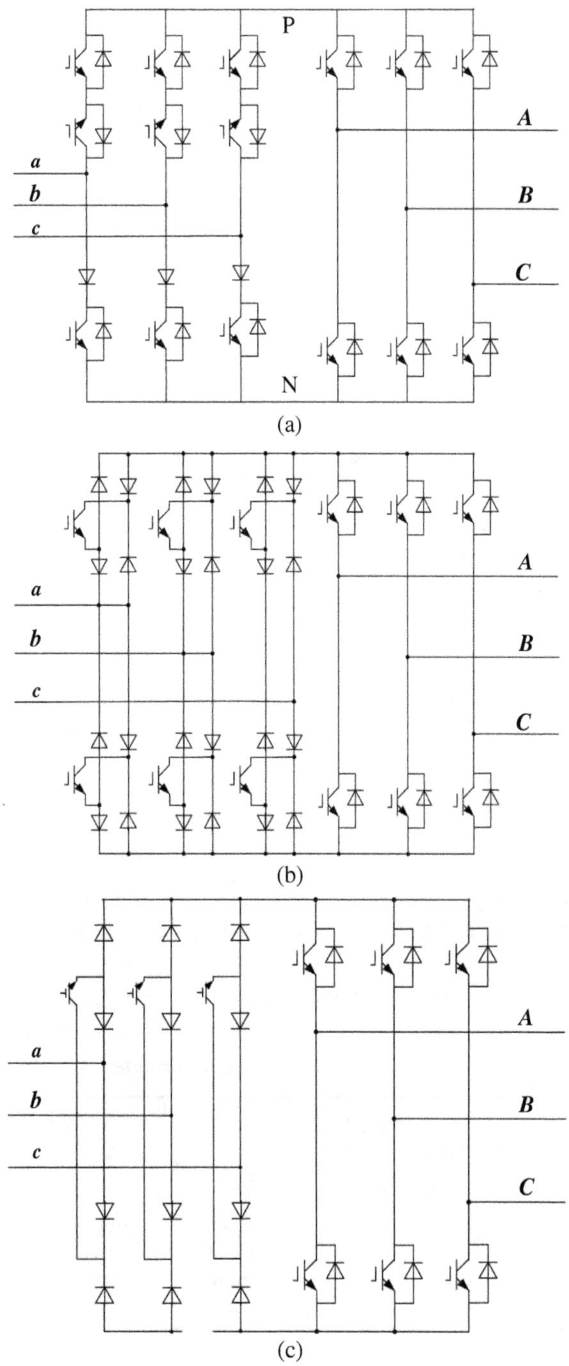

Figure 1.9 Other IMC circuit diagrams: (a) SMC; (b) VSMC; (c) USMC.

Table 1.1 Topologies comparison of MCs.

Topology	No. of switches	No. of diodes	Minimum No. of isolated drive power supplies
DMC	18	18	6
IMC	18	18	8
SMC	15	18	7
VSMC	12	30	10
USMC	9	18	7

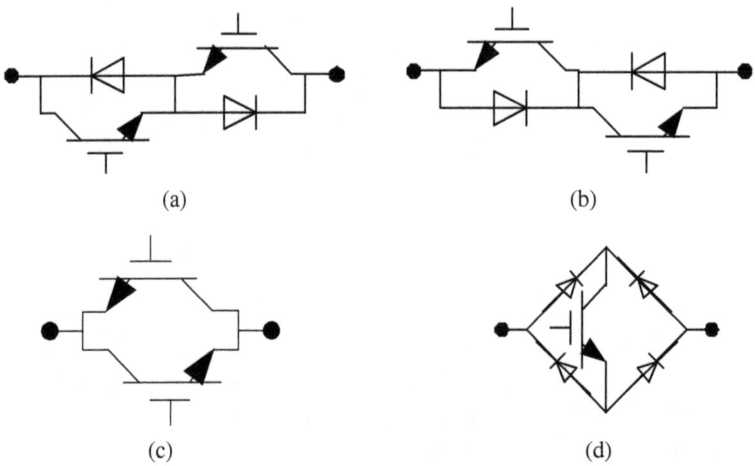

Figure 1.10 Different bi-directional switch configuration of MCs, (a) Common emitter IGBT, (b) Common collector IGBT, (c) RB-IGBT, (d) Diode bridge switch cell.

Table 1.2 Characteristics of bi-direction configurations.

	Common emitter	Common collector
Isolated power supply of gate drive circuitry	9	6
Detect all IGBT terminals	Yes	Cannot detect collector
Advantages	Overcurrent protection and reliable drive can be achieved by monitoring the collector–emitter voltage	Input and output voltage and current transformers can be powered by gate drive circuitry

characteristics in their main circuit operation [12]. Table 1.2 provides a simple comparison of these two structural configurations.

Figure 1.10c represents antiparallel bidirectional switches featuring Reverse Blocking IGBT (RB-IGBT) [30–32]. Compared to other bidirectional switches, RB-IGBTs exhibit symmetric voltage blocking capability, low switch conduction loss, high power density, and reduced switch device size, enhancing overall system efficiency. However, their driver circuit design is intricate, and they suffer from relatively high switching losses [33], which has limited their current usage.

The bidirectional switches shown in Figure 1.10d consist of a bridge circuit composed of diodes and IGBTs. This configuration offers the lowest cost and does not encounter the four-step commutation issue. Its driver circuitry is relatively simple. However, it has the drawback of higher conduction losses due to the presence of two diodes and one IGBT in the conduction path, resulting in increased conduction losses.

1.1.2.4 Research Status of MCs

Based on the relevant literatures, researchers all over the world have primarily focused their efforts on various aspects of MCs:

1) Research on Novel Circuit Topologies:
   To simplify MC structures and reduce control complexity, researchers have proposed several derivative topologies, including: (i) single-switch-based matrix AC/AC converters [34]; (ii) SMC series, such as SMC, VSMC, and USMC, which vary based on the number of single-phase switches used in IMCs [25, 35]; and (iii) three-phase four-wire MCs designed for unbalanced loads [36].
2) Introduction of Innovative Modulation Methods:
   To improve output waveform quality and reduce input-side harmonic content, researchers have introduced a series of novel modulation methods, including: (i) phase voltage-phase voltage direct transformation switching function modulation algorithms [37]; (ii) dual-space vector modulation (SVM) algorithms based on input line voltage–output line voltage direct transformation [26, 38]; (iii) dual-line voltage modulation algorithms [39, 40]; and (iv) output current hysteresis current modulation methods aimed at tracking output circuit currents [41, 42].
3) Research on Control Methods under Abnormal Operating Conditions:
   MCs lack a DC energy storage element, making the input-side disturbances directly affect the output side. In response, researchers have proposed methods such as using auxiliary diode clamp circuits to buffer energy storage capacitors [43] and improved modulation methods for unbalanced power supply conditions [44, 45].
4) Application of Advanced Control Algorithms:
   In recent years, scholars all over the world have explored the application of advanced control theories in MCs, including: (i) robust control for MC systems [46, 47]; (ii) sliding-mode control for MCs [48, 49]; and (iii) predictive control for MCs [50–53].
5) Application Research on New Power Electronic Devices:
   To further reduce converter switching loss, numerous companies in Europe, the United States, Japan, and elsewhere have developed Reverse Blocking IGBTs (RB-IGBTs). The systems built using these devices have effectively reduced overall system losses [54–56]. Additionally, the emergence of power electronic switching devices based on new materials like SiC and GaN has significantly increased switching frequencies, and lowered the system losses [57, 58].
6) Developed Prototypes:
   As power electronic devices become more integrated and MC technology matures, researchers worldwide have developed a variety of prototypes [53]: (i) Siemens in Germany introduced an MC solution for industrial drives in 2001 [23]. (ii) Aalborg University in Denmark developed a 4 kW MC for driving asynchronous motor speed control systems in 2002 [59]. (iii) Fuji Electric in Japan created an RB-IGBT module in 2003, using it to build a 22 kW MC prototype in 2004 [60]. (iv) In collaboration with the military, University of Nottingham in the UK released

a 150 kVA MC prototype for military vehicle transmission systems in 2004 [61]. (v) In 2003, University of Nottingham in the UK developed a jet aircraft wing electrohydraulic control system based on an MC [62]. (vi) Several prototypes have emerged at various universities. These include the first domestic MC prototype built using discrete IGBTs by Chen and Lu at Shanghai University in 1998 [63], as well as a constant-frequency sampling current tracking control MC prototype developed by Tang and Fang at Fuzhou University in 1999 [64]. In 2000, Xiangtan University began researching AC/AC MCs and produced an experimental prototype [65]. In 2006, Huang and Sun at Tsinghua University developed a 3.6 kW MC prototype based on RB-IGBT modules [66]. In 2009, she at Huazhong University of Science and Technology developed a 5.5 kW prototype driving an induction motor using discrete IGBTs [67]. In 2010, Li and Mei at North China University of Technology developed an IMC prototype based on dual IGBT modules [68]. To date, researchers worldwide have publicly disclosed MC prototypes, as shown in Table 1.3.

Table 1.3 MC prototype research and development.

Year	Affiliation	Power	Power module	Application
1988	Westinghouse [28]	22 kW	Bridge Type	Induction Motor Drive
1995	Virginia Tech [69]	2 kW	Anti-parallel MOSFET	Algorithm Verification
1999	Shanghai University [63]	2 kW	Discrete IGBT	Algorithm Verification
2001	University of Bologna [70]	7 kW	IGBT Module	Induction Motor Drive
2002	Aalborg University [59]	4 kW	IGBT Module	IM Speed Regulation
2002	University of Nottingham [71]	10 kW	IGBT Module	Induction Motor Drive
2002	University of Nottingham [72]	30 kVA	IGBT Module	Ground level power supply
2002	University of Karlsruhe [23]	7.5 kW	IGBT Module	Algorithm Verification
2003	University of Nottingham [73]	20 kW	IGBT Module	Aircraft aileron control
2004	University of Nottingham [74]	10 kW	Discrete IGBT	Diesel engine generator power supply
2004	University of Sheffield [75]	0.7 kW	Anti-tandem MOSFET	PMSM Drive
2005	University of Nottingham [61]	150 kVA	Discrete IGBT	Electric chariots
2005	Aalborg University [76]	3 kW	IGBT Module	Induction Motor Drive
2005	University of Nottingham [77]	30 kVA	Discrete IGBT	IM Speed Regulation
2005	University of Bologna [78]	10 kW	IGBT Module	Induction Motor Drive
2005	Nagaoka University of Technology and Science [24]	22 kW	RB-IGBT Module	Induction Motor Drive
2006	Tsinghua University [66]	3.6 kW	RB-IGBT Module	Induction Motor Drive
2008	University of Stuttgart [79]	18 kW	Discrete IGBT	Induction Motor Drive
2009	Huazhong University of Science and Technology [67]	5.5 kW	Discrete IGBT	Induction Motor Drive
2010	North China University of Technology [68]	2.7 kVA	Two-in-one IGBT Module	Algorithm Verification

7) Research on Expanding the Gains of MC Systems:
Given the characteristics of MCs, their most significant limitation is that the maximum voltage boost ratio is only 0.866. Many scholars are actively researching methods to improve this aspect. Current research focused on enhancing IMC gain primarily involves improving modulation strategies and adding auxiliary circuits. These approaches include overmodulation methods and combined modulation strategies [80–82]. While the gain improvement is not significant with overmodulation methods, the process is overly complex. On the other hand, adding auxiliary circuits to the intermediate DC bus, such as Boost circuits [83, 84], or Z-source/quasi-Z-source (ZS/QZS) circuits before and after MC, has been rapidly gaining attention in the field of voltage boost due to their structural symmetry and simplified control.

1.2 ZS/QZS Converters

ZS converter was introduced by Peng et al. in 2002 [85], as depicted in Figure 1.11a. By incorporating a ZS circuit between DC source and the inverter, it offers an alternative to traditional voltage-source and current-source inverters. ZS circuit consists of a diode D, two capacitors C_1 and C_2, and two inductors L_1 and L_2 interconnected. The distinguishing feature of ZS converter is its ability to achieve both voltage boost/buck and inversion in a single-stage power conversion, resulting in significant cost savings. Moreover, it permits a direct shoot through of the same bridge arm of the inverter, eliminating the need for a dead time, which greatly enhances the system's immunity to disturbances [86, 87].

In 2008, QZS, an improved topology of ZS converter, was proposed by Peng et al. [88], as shown in Figure 1.11b. This circuit possesses all the characteristics of ZS converter and offers the additional benefit of achieving continuous input current. Furthermore, the QZS network has a shared negative pole for input and output, which aids in suppressing electromagnetic interference (EMI). Due to the advantages of QZS network converter, various impedance network structures based on this architecture have been proposed, primarily aimed at maximizing voltage gain [89–91].

ZS/QZS converters were initially applied to DC/DC and DC/AC converters. Currently, ZS inverters (ZSIs) and QZS inverters (QZSIs) have garnered widespread research interest, especially in applications requiring voltage boosting, leading to significant breakthroughs. Based on available literature, scholars worldwide have primarily focused their research in the following directions:

1) Improvement of Topological Structures:
Researchers have proposed enhanced ZS and QZS structures, such as extended ZS and QZS [89], switched-inductor-type ZSI structures [90], L–C coupled inductor-based current-fed ZSI

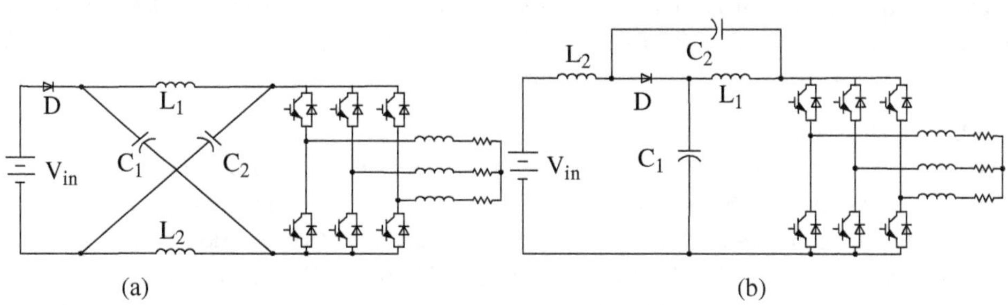

Figure 1.11 Classical topology of (a) ZSI; (b) QZSI.

(LCCT-ZSI) structures [92], and trans-ZS structures [93]. They have investigated the voltage boost ratios and characteristics of different structures.

2) Modulation Algorithm Research:
This includes investigations into modulation techniques like sinusoidal pulse-width modulation (SPWM) [94, 95] and SVM [96, 97].
3) Model Establishment and Analysis:
By developing state-space average models, scholars have created small-signal models for ZSI/QZSI [98–100], providing mathematical foundations for the study of converter dynamic responses and controller design.
4) Introduction of Advanced Control Algorithms:
Building upon model analysis, scholars have designed corresponding controllers, including sliding-mode variable structure control [101, 102] and model predictive control [101, 103, 104].
5) Multi-Level Expansion:
In the field of photovoltaic power generation systems, an extensive research is the ZS multi-level technology. This includes the development and study of cascaded multi-level inverters based on QZS and energy storage-type cascaded multi-level inverters [105–108].

1.3 Advantages of ZS/QZS MCs Compared to Existing Technology

The characteristics of MCs and ZS/QZS converters demonstrate that their integration can effectively enhance the performance of MCs. Currently, through the efforts of researchers, the combination of ZS/QZS with MCs has resulted in several topological structures, primarily including ZS/QZS placed before DMC, referred to as ZS/QZS-DMC converters, ZS/QZS circuits positioned on the DC side of IMCs, known as DC-ZS/QZS-IMC converters, and those located before or on grid side (GS) of IMCs, denoted as GS-ZS/QZS-IMC converters [109–111].

Introducing ZS and QZS networks in front of the DMC, as illustrated in Figure 1.12a,b, respectively, overcomes the issue of low-voltage gain in traditional DMC systems. Furthermore, the ZS network allows the subsequent bridge arms to operate in a direct shoot-through mode, simplifying the commutation process of DMC [109]. However, ZS-DMC exhibits discontinuous input current. On the other hand, QZS-DMC does not suffer from phase-shift issues, possesses a high system voltage gain, lower voltage and current stresses on the switches, and, through structural adjustments, ensures continuous input current, as depicted in Figure 1.12c.

The combination of ZS or QZS with IMC results in ZS/QZS-IMCs that effectively inherit the advantages of IMC. Additionally, by utilizing the voltage boost/buck capability and robustness of ZS/QZS converters, these hybrid systems can overcome the drawbacks of traditional IMC, such as low-voltage gain and poor disturbance rejection. ZS/QZS-IMCs with ZS/QZS networks introduced into IMC can be divided into two categories [110, 111]: the first category is DC-ZS/QZS-IMC, and the second category is GS-ZS/QZS-IMC, as illustrated in Figures 1.13 and 1.14.

1) DC-Link ZS/QZS-IMCs
The first category of ZS/QZS-IMCs, represented by DC-ZS-IMC and DC-QZS-IMC, is depicted in Figure 1.13. Aside from the distinction in ZS and QZS networks, these two variants are functionally identical, possessing the characteristics of both ZS/QZS networks and IMCs. However, the incorporation of ZS/QZS circuits on the DC side of the IMC disrupts the all-silicon structure of traditional MCs, resulting in a bulkier system. The primary drawbacks of this topology include: (i) discontinuous input current; (ii) the need for additional LC filters on GS,

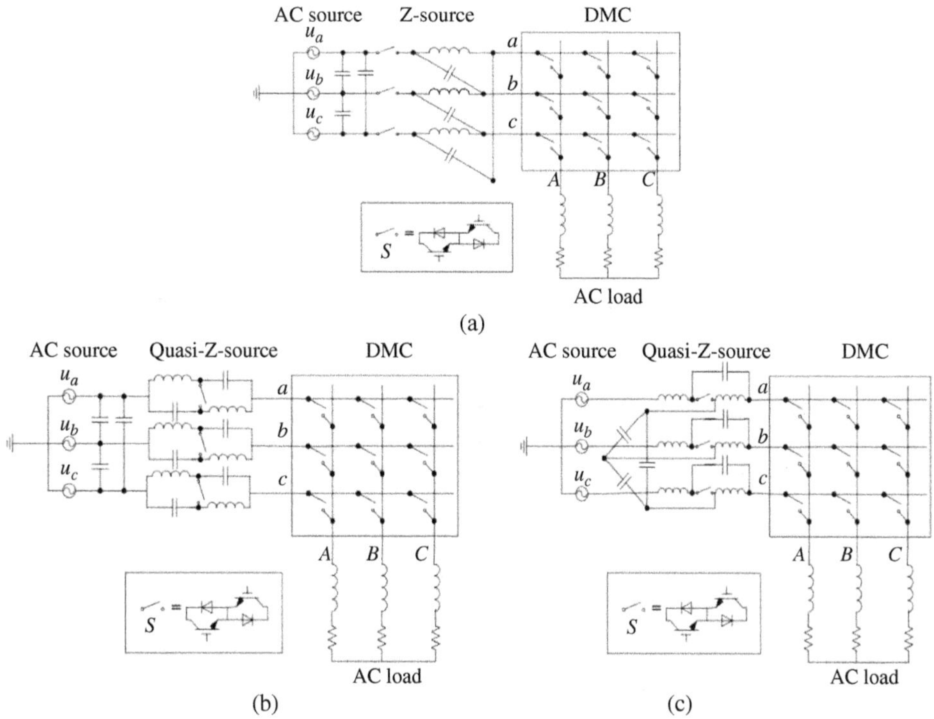

Figure 1.12 ZS/QZS-IMCs. (a) ZS DMC; (b) DCM-QZS DMC; (c) CCM-QZS DMC.

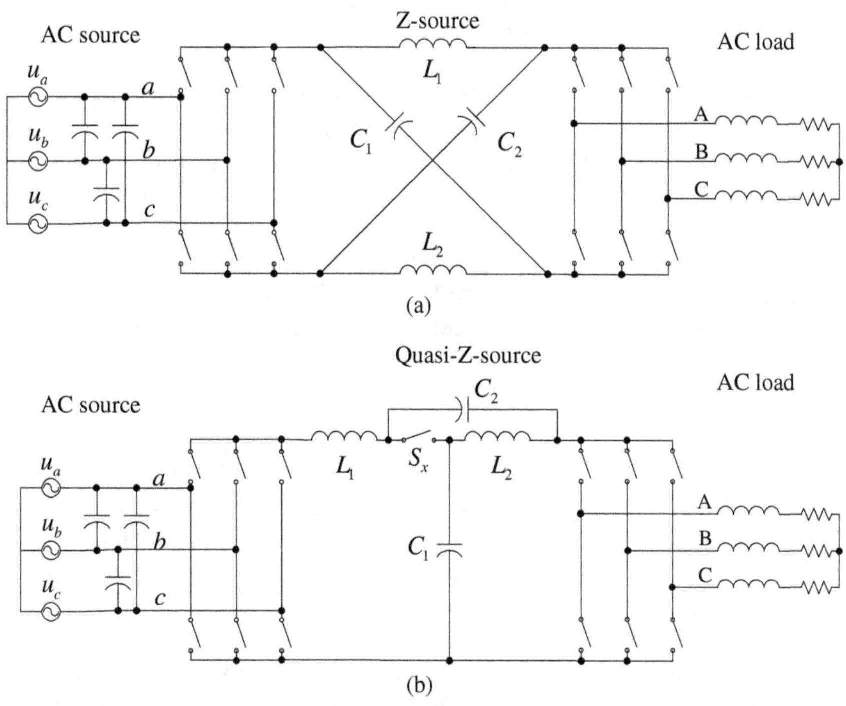

Figure 1.13 DC link ZS/QZS-IMC. (a) DC link ZS IMC; (b) DC link QZS-IMC.

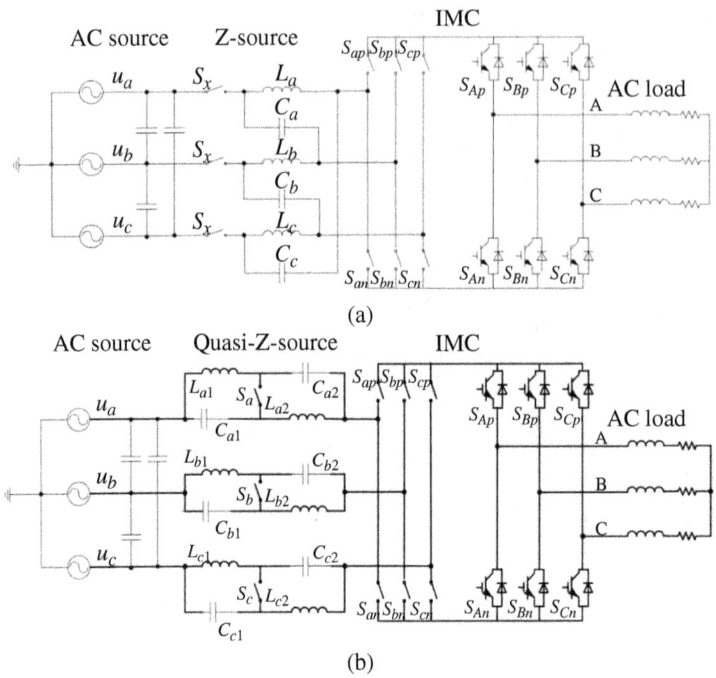

Figure 1.14 GS ZS/QZS-IMC. (a) GS ZS IMC; (b) GS QZS-IMC.

leading to system complexity, higher cost, increased volume, and greater loss; (iii) the addition of ZS/QZS circuits on the high-voltage DC side leads to higher current (voltage) stress on the inductors (capacitors), contributing to increased system cost and loss [110].

2) GS ZS/QZS-IMCs
   The second category of ZS/QZS-IMCs involves placing ZS and QZS networks on the grid side of the IMC, resulting in GS-ZS-IMC and GS-QZS-IMC configurations, as shown in Figure 1.14. These topologies also retain the characteristics of both ZS/QZS networks and IMCs. By locating ZS and QZS networks on the grid side, the full silicon characteristics of the IMC remain intact, facilitating integrated system design. However, each of these two topologies has its own characteristics. For the GS-ZS-IMC, the phase-shifting nature of ZS network limits the maximum voltage boost ratio to 1.15, and the input current becomes discontinuous. In contrast, the GS-QZS-IMC topology achieves a high-voltage boost ratio. Nevertheless, both configurations require the addition of extra filters at ZS/Quasi-ZS front end, resulting in increased system volume, weight, and cost [111].

Considering the drawbacks of traditional MCs, which include: (i) the complexity arising from a large number of power switching devices; (ii) low-voltage gain, with the maximum output voltage amplitude reaching only 86.6% of the input voltage amplitude; (iii) vulnerability to abnormal input voltages due to the absence of an intermediate buffer circuit; (iv) the complexity of bidirectional switch commutation, necessitating real-time monitoring of input voltage and output current for complex system control; and (v) discontinuous input current, requiring additional filters on the input side.

Taking into account the advantages of ZS/QZS converters, ZS/QZS MCs studied in this book incorporate ZS/QZS networks into traditional DMCs or IMCs. These converters possess the following characteristics: (i) adjustable voltage boost and buck capabilities, allowing the system voltage gain to exceed 1, thus overcoming the deficiencies of traditional MCs; (ii) strong immunity to disturbances, featuring the ability to traverse through drops in grid voltage, a capability not presented in traditional MCs; and (iii) elimination of the need for four-step commutation, simplifying the commutation process and ensuring straightforward and reliable system control.

1.4 Current Status and Future Trends

Currently, the research on ZS/QZS MCs primarily focuses on the following areas: (i) derivation and analysis of system voltage gain, by analyzing the circuit's operational behavior and characteristics to derive the system gain and conducting comparative analyses among converters of the same kind; (ii) analysis of voltage and current stress on switching devices, considering the impact of the shoot-through ratio on device stress, in combination with the characteristics of ZS/QZS circuits; (iii) improvement of modulation techniques, particularly in the rectification and inversion stages of ZS/QZS-IMCs. Insertion of a direct-zero vector during inversion enhances the system's voltage gain [112]; (iv) stable control of output voltage, achieved by controlling the capacitor voltage in ZS/QZS circuits to obtain a stable output voltage, ensuring the system stability even under input voltage disturbances [113]; (v) reduction in the number of switching devices in ZS MCs, thereby reducing system complexity and enhancing reliability, such as reducing the number of switches in IMCs from 18 to 12 [114]; and (vi) application of ZS MCs to the motor speed control and wind power generation [115, 116].

Additionally, as previously mentioned, the QZS-IMC offers the advantage of high-voltage gain. However, as mentioned in section 1.3, GS-QZS-IMC exhibits discontinuous input current and requires additional filtering. Therefore, an integrated LC filter QZS-IMC was introduced in 2013 [117], as shown in Figure 1.15. This structure is based on the GS-QZS-IMC but integrates QZS network with an LC filter, simultaneously achieving voltage boosting and filtering functions.

The integrated LC filter QZS-IMC exhibits the following characteristics: (i) QZS network is integrated into the input side of IMC, combining it with LC filter. This integration imparts 1) filtering capability to the QZS circuit, eliminating the need for additional LC filters on the input side, significantly simplifying the hardware of the system; (ii) exceptional immunity to disturbances, with the ability to withstand drops in grid voltage, a feature not present in traditional MCs; (iii) overcoming the drawback of discontinuous current in traditional ZS MCs; (iv) with the adoption of an IMC, the system control is simplified due to the two-step commutation process; and (v) inheritance of

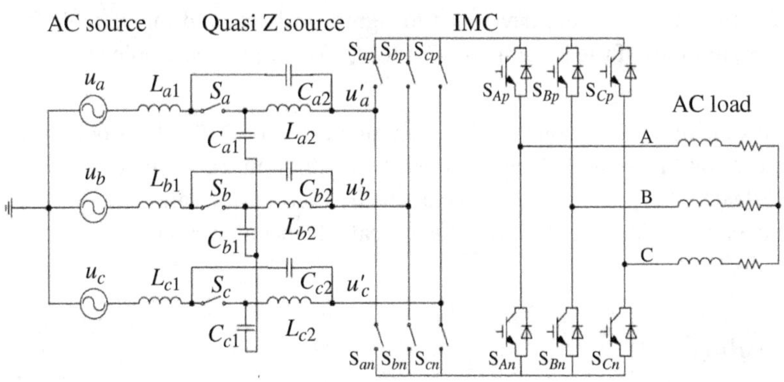

Figure 1.15 LC filter-Integrated QZS-IMC [117, 118, 121].

the full silicon characteristics of traditional MCs, resulting in a significant reduction in the size and weight of capacitors and inductors, leading to an improved efficiency [117, 118].

Given the numerous advantages of the integrated LC filter QZS-IMC, it has become a highly promising AC converter. Currently, researchers in the field are primarily focusing on the following aspects:

1) Analysis of the working principles and mathematical modeling of integrated LC filter QZS-IMCs.
   This includes inserting ZS/QZS circuits into the input side of IMC, analyzing the newly generated topology, mathematical modeling for small-signal analysis, non-minimum phase analysis, and filtering characteristics analysis [118, 119].
2) Research on modulation methods and common-mode voltage suppression methods for integrated LC filter QZS-IMCs.
   This includes the insertion of zero vectors during rectification, which significantly increases the system's switching frequency, analyzing the impact of different modulation methods on system switching frequency and input harmonics, and strategies for common-mode voltage suppression [120].
3) Research on the parameter design of the topology and circuit of integrated LC filter QZS-IMCs.
   This involves analyzing the stress on passive components and power switching devices in ZS/QZS circuits, methods for loss calculation, and parameter design standards [118, 121].
4) Research on three-phase-to-single-phase QZS MC.
   This includes an expansion of the integrated LC filter to three-phase-to-single-phase QZS MCs and research into methods to suppress double-frequency ripple for single-phase output [122].
5) Research on application of integrated LC filter QZS IMCs.
   It was applied to motor speed control systems, conducting dynamic and steady-state characteristic tests. [117, 123].
6) Research on integrated LC filter QZS-IMC systems based on new types of devices.
   This includes comparative studies of parameters, volumes, and losses between the systems composed of Si-IGBT, mixed SiC-IGBT, and SiC-MOSFET devices [124].
7) Optimization operation control.
   This mainly focuses on the operation curve optimization of the integrated LC filter QZS-IMC system based on constrained optimization theory, aimed at reducing losses and decreasing device stress [125].
8) Combining the D-minimization control laws of the integrated LC filter QZS-IMC with AC motor control.
   The comprehensive control methods were developed to significantly expand the variable frequency speed control capabilities of the integrated LC filter QZS-IMC, increase efficiency, and reduce cost [126].

In conclusion, due to its outstanding advantages, the integrated LC filter QZS-IMC is poised for further research and application in the field of electric power transmission. For instance, in modular dual-three-level QZS-IMCs [127], this technology can ensure high-performance operation of high-power motors, reduce harmonics, and promote theoretical advancement and innovation in variable frequency technology.

1.5 Contents Overview

Chapter 1 offers a brief introduction to the concept, classification, advantages, current status, and future trends in the field. The subsequent chapters in this book provide detailed and

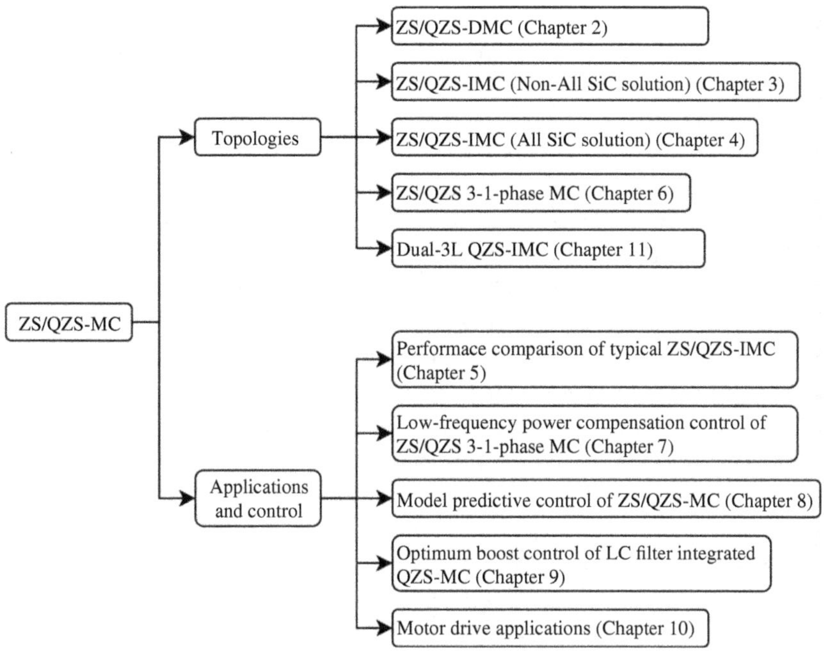

Figure 1.16 Summary of contents.

systematic content. Figure 1.16 illustrates the book's structure. In Chapter 2, ZS/QZS-DMC is introduced, covering its topology, operating principles, modulation methods, and demonstration results. Chapters 3 and 4 focus on ZS/QZS-IMC, with Chapter 3 addressing non-all SiC solutions and Chapter 4 exploring all SiC solutions. Chapter 5 conducts a comprehensive comparison of typical ZS/QZS MCs, including voltage gain analysis, filtering function assessment, parameters design, and verification results. Chapter 6 discusses ZS/QZS 3-1-phase MCs, while Chapter 7 presents a control method for low-frequency power compensation within this context. Chapter 8 details the model predictive control of ZS/QZS MCs. Chapter 9 demonstrates the optimal boost control in LC filter integrated QZS-IMC, and Chapter 10 explores its applications in motor drives. Finally, Chapter 11 outlines the future prospects of this promising topology.

References

1 P. Rajan and S. Jeevananthan, "Toward energy sustainability—domestic power station, tariff acquiescent EMS, and procedure to rejuvenate petrol scooter into electric scooter for accelerated participation of rural consumers in energy demand management," *IEEE J. Emerg. Select. Top. Power Electron.*, vol. 11, no. 4, pp. 4442–4452, Aug. 2023.

2 IEA, "World energy outlook 2014," 2014. Accessed Oct. 16, 2024 [Online]. Available: https://www.iea.org/reports/world-energy-outlook-2014

3 IEA, "Key world energy statistics 2014," 2014. Accessed Oct. 16, 2024 [Online]. Available: https://www.iea.org/reports/key-world-energy-statistics-2014

4 M. Guo, "Research on quasi-Z source indirect matrix converter-based AC motor drive systems," Ph.D. dissertation, School of Electrical Engineering, Beijing Jiaotong Univ., Beijing, 2019. (In Chinese)

5 Y. Goude, R. Nedellec, and N. Kong, "Local short and middle term electricity load forecasting with semi-parametric additive models," *IEEE Trans. Smart Grid*, vol. 5, no. 1, pp. 440–446, Jan. 2014.

6 B. S. P. Emandi, "Improving EMI performance through simulation," 2018. [Online]. Available: https://www.eetimes.com/improving-emi-performance-through-simulation/

7 Z. Tang, Y. Yang and F. Blaabjerg, "Power electronics: The enabling technology for renewable energy integration," *CSEE J. Power Energy Syst.*, vol. 8, no. 1, pp. 39–52, Jan. 2022.

8 Z. Pan and R. A. Bkayrat, "Modular motor/converter system topology with redundancy for high-speed, high-power motor applications," *IEEE Trans. Power Electron.*, vol. 25, no. 2, pp. 408–416, Feb. 2010, doi:https://doi.org/10.1109/TPEL.2009.2025948.

9 S. Liu, "Research on Z-source/quasi-Z-source indirect matrix converter," Ph.D. dissertation, School of Electrical Engineering, Beijing Jiaotong Univ., 2014. (In Chinese)

10 Z. -C. Zhang and B.-T. Ooi, "Multimodular current-source SPWM converters for a superconducting magnetic energy storage system," *IEEE Trans. Power Electron.*, vol. 8, no. 3, pp. 250–256, July 1993.

11 J. Dai, D. Xu and B. Wu, "A novel control scheme for current-source-converter-based PMSG wind energy conversion systems," *IEEE Trans. Power Electron.*, vol. 24, no. 4, pp. 963–972, Apr. 2009.

12 P. W. Wheeler, J. Rodriguez, J. C. Clare, L. Empringham and A. Weinstein, "Matrix converters: a technology review," *IEEE Trans. Ind. Electron.*, vol. 49, no. 2, pp. 276–288, Apr. 2002.

13 D. G. Holmes, "A unified modulation algorithm for voltage and current source inverters based on AC-AC matrix converter theory," *IEEE Trans. Ind. Appl.*, vol. 28, no. 1, pp. 31–40, Jan.-Feb. 1992.

14 M. Milanovic and B. Dobaj, "Unity input displacement factor correction principle for direct AC to AC matrix converters based on modulation strategy," *IEEE Trans. Circ. Syst. I*, vol. 47, no. 2, pp. 221–230, Feb. 2000.

15 L. Gyugi and B. Pelly, "Static power frequency changers," in *Theory, Performance and Applications*, New York, NY, USA: Wiley, 1976.

16 A. K. Sahoo, J. Meenakshi, S. S. Dash, and T. Thyagarajan, "Analysis and simulation of matrix converter using PSIM," in *7th Int. Conf. Power Electronics*, Daegu, Korea (South), 2007, pp. 414–419.

17 J. Zhang, L. Li and D. G. Dorrell, "Control and applications of direct matrix converters: A review," *Chin. J. Electr. Eng.*, vol. 4, no. 2, pp. 18–27, June 2018.

18 W. Deng, H. Liu, J. Zhu, and Z. Dai, "Research on double closed-loop control strategy based on two-stage matrix converter," in *Chinese Contr. Decision Conf.*, Yantai, Shandong, 2008, pp. 3599–3604.

19 L. Wei and T. A. Lipo, "A novel matrix converter topology with simple commutation," in *Conf. Rec. 2001 IEEE Industry Applications Conf. 36th IAS Annu. Meeting (Cat. No.01CH37248)*, Chicago, IL, USA, 2001, pp. 1749–1754, vol. 3.

20 C. Klumpner and F. Blaabjerg, "Two stage direct power converters: an alternative to the matrix converter," in *IEE Seminar on Matrix Converters (Digest No. 2003/10100)*, 2003, pp. 7/1–7/9.

21 C. Klumpner, P. Wheeler, and F. Blaabjerg, "Control of a two-stage direct power converter with a single voltage sensor mounted in the intermediary circuit," in *IEEE 35th Annu. Power Electronics Specialists Conf. (IEEE Cat. No.04CH37551)*, Aachen, Germany, 2004, pp. 2386–2392, vol. 3.

22 J. W. Kolar, M. Baumann, F. Schafmeister, and H. Ertl, “Novel three-phase AC-DC-AC sparse matrix converter,” in *APEC. 17th Annu. IEEE Applied Power Electronics Conf. Expo. (Cat. No.02CH37335)*, Dallas, TX, USA, 2002, pp. 777–791, vol. 2.

23 O. Simon, J. Mahlein, M. N. Muenzer, and M. Bruckmarm, “Modern solutions for industrial matrix-converter applications,” *IEEE Trans. Ind. Electron.*, vol. 49, no. 2, pp. 401–406, Apr. 2002.

24 J.-I. Itoh, I. Sato, A. Odaka, H. Ohguchi, H. Kodachi, and N. Eguchi, “A novel approach to practical matrix converter motor drive system with reverse blocking IGBT,” *IEEE Trans. Power Electron.*, vol. 20, no. 6, pp. 1356–1363, Nov. 2005.

25 J. W. Kolar, F. Schafmeister, S. D. Round, and H. Ertl, “Novel three-phase AC–AC sparse matrix converters,” *IEEE Trans. Power Electron.*, vol. 22, no. 5, pp. 1649–1661, Sept. 2007.

26 T. D. Nguyen and H.-H. Lee, “A new SVM method for an indirect matrix converter with common-mode voltage reduction,” *IEEE Trans. Industr. Inform.*, vol. 10, no. 1, pp. 61–72, Feb. 2014.

27 N. Burany, “Safe control of four-quadrant switches,” in *Conf. Rec. IEEE Industry Applications Soc. Annual Meeting*, San Diego, CA, USA, 1989, pp. 1190–1194, vol. 1.

28 C. L. Neft and C. D. Schauder, “Theory and design of a 30-hp matrix converter,” *IEEE Trans. Ind. Appl.*, vol. 28, no. 3, pp. 546–551, 1992.

29 S. Jia, K. J. Tseng, and X. Wang, “Study on reverse recovery characteristics of reverse-blocking IGBT applied in matrix converter,” in *20th Annu. IEEE Appl. Power Electr. Conf. Exposit. APEC 2005*, Austin, TX, USA, 2005, pp. 1917–1921, vol. 3.

30 S. Bala and G. Venkataramanan, “Matrix converter BLDC drive using reverse-blocking IGBTs,” in *21st Annu. IEEE Applied Power Electronics Conf. Expos., APEC’06*, Dallas, TX, USA, 2006, pp. 660–666.

31 T. Friedli, M. L. Heldwein, F. Giezendanner, and J. W. Kolar, “A high efficiency indirect matrix converter utifizing RB-IGBTs,” in *37th IEEE Power Electr. Special. Conf.*, Jeju, Korea (South), 2006, pp. 1–7.

32 E. R. Motto, J. F. Donlon, M. Tabata, H. Takahashi, Y. Yu, and G. Majumdar, “Application characteristics of an experimental RB-IGBT (reverse blocking IGBT) module,” in *Conf. Rec. 2004 IEEE Industry Applications Conf. 39th IAS Annu. Meeting*, Seattle, WA, USA, 2004, pp. 1540–1544, vol. 3.

33 J. Chang, A. Wang, and T. Sun, “VF-input and high-frequency matrix converter - recent development and evaluation,” in *IECON’03. 29th Annu. Conf. IEEE Industrial Electronics Society (IEEE Cat. No.03CH37468)*, Roanoke, VA, USA, 2003, pp. 2066–2071, vol. 3.

34 S. Kim, S.-K. Sul, and T. A. Lipo, “AC/AC power conversion based on matrix converter topology with unidirectional switches,” *IEEE Trans. Ind. Appl.*, vol. 36, no. 1, pp. 139–145, Jan.–Feb. 2000.

35 O. Kuseso, T. Shishaye, W. Xiaohua, and H. Ali, “Unidirectional step-up ultra sparse matrix converter for integration of wind energy resources to microgrids,” in *AFRICON 2015*, Addis Ababa, Ethiopia, 2015, pp. 1–5.

36 F. Yue, P. W. Wheeler, and J. C. Clare, “A novel four-leg matrix converter,” in *IECON 2006 – 32nd Annu. Conf. IEEE Industrial Electronics*, Paris, France, 2006, pp. 2694–2699.

37 S. Kwak, “Four-leg-based fault-tolerant matrix converter schemes based on switching function and space vector methods,” *IEEE Trans. Ind. Electron.*, vol. 59, no. 1, pp. 235–243, Jan. 2012.

38 J. Haruna, J. Tsuchiya, H. Ueda, and H. Funato, “An optimized switching pattern for reducing input current distortion of matrix converter using space vector modulation,” in *IEEE 3rd Int.*

Future Energy Electronics Conf. ECCE Asia (IFEEC 2017 – ECCE Asia), Kaohsiung, Taiwan, 2017, pp. 68–74.

39 X. Ma, Z. Zhang, D. Xu, and K. Wang, "On closed - loop control of matrix converter with double voltage control," in *IEEE 7th Data Driven Contr. Learn. Syst. Conf. (DDCLS)*, Enshi, China, 2018, pp. 229–234.

40 T. Shi, Y. Yan, H. An, M. Li, and C. Xia, "Improved double line voltage synthesis strategies of matrix converter for input/output quality enhancement," *IEEE Trans. Ind. Electron.*, vol. 60, no. 8, pp. 3034–3046, Aug. 2013.

41 B. Wang, D. -Q. Gao, D. Chen, C. Ning and G. Ling, "On a novel current control method for matrix converter," in *29th Chinese Contr. Decision Conf. (CCDC)*, Chongqing, China, 2017, pp. 5087–5092.

42 G. Hongjuan, Z. Shao, and Z. Bo, "Implementation of DSP-based matrix converter-permanent magnetic synchronous motor closed-loop control system," in *Int. Conf. Electr. Mach. Syst.*, Nanjing, China, 2005, pp. 362–366, vol. 1.

43 C. Klumpner and F. Blaabjerg, "Experimental evaluation of ride-through capabilities for a matrix converter under short power interruptions," *IEEE Trans. Ind. Electron.*, vol. 49, no. 2, pp. 315–324, Apr. 2002.

44 X. Yang, P. Ye, X. Liu, X. Yang, J. Wang, and L. Zhang, "Simulation analysis on current SVM algorithm of matrix rectifier," in *CES/IEEE 5th Int. Power Electronics and Motion Control Conf.*, Shanghai, China, 2006, pp. 1–7.

45 X. Wang, H. Lin, H. She, and B. Feng, "A research on space vector modulation strategy for matrix converter under abnormal input-voltage conditions," *IEEE Trans. Ind. Electron.*, vol. 59, no. 1, pp. 93–104, Jan. 2012.

46 Z. Chen and W. Deng, "Study on current control strategy for two-stage matrix converter (TSMC) fed induction motor using complex vectors," in *Int. Conf. on Computer Modeling and Simulation*, Macau, China, 2009, pp. 265–269.

47 Y. Mei and L. Huang, "A second-order auto disturbance rejection controller for matrix converter fed induction motor drive," in *IEEE 6th Int. Power Electronics and Motion Control Conf.*, Wuhan, China, 2009, pp. 1964–1967.

48 C. Der-fa and Y. Kai-chao, "A novel sliding-mode controller design for a matrix converter drive system," in *9th Int. Conf. Hybrid Intelligent Systems*, Shenyang, China, 2009, pp. 134–137.

49 S. T. Behrooz, M. Bekrani, and M. Heydari, "Fuzzy type-2 sliding mode control of matrix converter using indirect space vector modulation," in *8th Power Electronics, Drive Systems & Technol. Conf. (PEDSTC)*, Mashhad, Iran, 2017, pp. 371–376.

50 M. Rivera, J. Rodriguez, J. R. Espinoza, and H. Abu-Rub, "Instantaneous reactive power minimization and current control for an indirect matrix converter under a distorted AC supply," *IEEE Trans. Industr. Inform.*, vol. 8, no. 3, pp. 482–490, Aug. 2012.

51 C. F. Garcia, M. E. Rivera, J. R. Rodríguez, P. W. Wheeler, and R. S. Peña, "Predictive current control with instantaneous reactive power minimization for a four-leg indirect matrix converter," *IEEE Trans. Ind. Electron.*, vol. 64, no. 2, pp. 922–929, Feb. 2017.

52 W. Song, Y. Yang, W. Qin, and P. Wheeler, "Switching state selection for model predictive control based on genetic algorithm solution in an indirect matrix converter," *IEEE Trans. Transp. Electrif.*, vol. 8, no. 4, pp. 4496–4508, Dec. 2022.

53 Z. Gong, J. Li, P. Dai, D. Su, and X. Wu, "Design and evaluation of a virtual vector based modulated model predictive control for the indirect matrix converters with improved performance," *IEEE Trans. Ind. Electron.*, vol. 69, no. 12, pp. 12019–12029, Dec. 2022.

54 C. Klumpner and F. Blaabjerg, "Using reverse-blocking IGBTs in power converters for adjustable-speed drives," *IEEE Trans. Ind. Appl.*, vol. 42, no. 3, pp. 807–816, May-June 2006.

55 Takahashi, Kaneda, and Minato, "1200V class reverse blocking IGBT (RB-IGBT) for AC matrix converter," in *Proc. 16th Int. Symp. Power Semiconductor Devices and ICs*, Kitakyushu, Japan.

56 K. Zhou *et al.*, "Characterization and performance evaluation of the superjunction RB-IGBT in matrix converter," *IEEE Trans. Power Electron.*, vol. 33, no. 4, pp. 3289–3301, Apr. 2018.

57 F. Schafmeister, S. Herold, and J. W. Kolar, "Evaluation of 1200 V-Si-IGBTs and 1300 V-SiC-JFETs for application in three-phase very sparse matrix AC-AC converter systems," *18th Annu. IEEE Appl. Power Electronics Conf. Expos. APEC '03*, Miami Beach, FL, USA, 2003, pp. 241–255, vol.1.

58 A. Trentin, L. Empringham, L. de Lillo, P. Zanchetta, P. Wheeler, and J. Clare, "Experimental efficiency comparison between a direct matrix converter and an indirect matrix converter using both Si IGBTs and SiC mosfets," *IEEE Trans. Ind. Appl.*, vol. 52, no. 5, pp. 4135–4145, Sept.-Oct. 2016.

59 C. Klumpner, P. Nielsen, I. Boldea, and F. Blaabjerg, "A new matrix converter motor (MCM) for industry applications," *IEEE Trans. Ind. Electron.*, vol. 49, no. 2, pp. 325–335, Apr. 2002.

60 M. Takei, T. Naito, and K. Ueno, "The reverse blocking IGBT for matrix converter with ultra-thin wafer technology," in *IEEE 15th Int. Symp. Power Semiconductor Devices and ICs. Proc.*, Cambridge, UK, 2003, pp. 156–159.

61 T. F. Podlesak, D. C. Katsis, P. W. Wheeler, J. C. Clare, L. Empringham, and M. Bland, "A 150-kVA vector-controlled matrix converter induction motor drive," *IEEE Trans. Ind. Appl.*, vol. 41, no. 3, pp. 841–847, May-June 2005.

62 P. W. Wheeler, J. C. Clare, M. Apap, L. Empringham, C. Whitley, and G. Towers, "Power supply loss ride-through and device voltage drop compensation in a matrix converter permanent magnet motor drive for an aircraft actuator," *IEEE 35th Annu. Power Electronics Specialists Conf.*, Aachen, Germany, 2004, pp. 149–154, vol. 1.

63 B. Chen and H. Lu, "Matrix AC-AC converter and control," *Power Electr.*, vol. 33, no. 1, pp. 8–11, 1999. (In Chinese)

64 N. Tang and X. Fang, "Design and application of constant-frequency sampling current tracking control for matrix converter," *J. Fuzhou Univ. (Nat. Sci. Ed.)*, vol. 28, no. 1, pp. 38–43, 2000. (In Chinese)

65 W. Liu et al, "Control strategy of multi-drive system based on AC-DC-AC matrix converter," *Proc. CSEE*, vol. 26, no. 6, pp. 111–115, 2006. (In Chinese)

66 K. Sun, L. Huang, K. Matsuse, and T. Ishida, "A combined controller for induction motor fed by matrix converter," in *The 5th Int. Conf. Power Electronics and Drive Systems, PEDS 2003*, Singapore, 2003, pp. 189–192, vol. 1.

67 H. She, H. Lin, B. He, X. Wang, L. Yue, and X. An, "Implementation of voltage-based commutation in space-vector-modulated matrix converter," *IEEE Trans. Ind. Electron.*, vol. 59, no. 1, pp. 154–166, Jan. 2012.

68 Y. Mei and Z. Li, "A fault-tolerant mechanism for multi-drive system based on indirect matrix converter," in *37th Annu. Conf. IEEE Industrial Electronics Society*, Melbourne, VIC, Australia, 2011, pp. 1036–1040.

69 L. Huber and D. Borojevic, "Space vector modulated three-phase to three-phase matrix converter with input power factor correction," *IEEE Trans. Ind. Appl.*, vol. 31, no. 6, pp. 1234–1246, Nov.–Dec. 1995.

70 D. Casadei, G. Serra, and A. Tani, "The use of matrix converters in direct torque control of induction machines," *IEEE Trans. Ind. Electron.*, vol. 48, no. 6, pp. 1057–1064, Dec. 2001.

71 P. W. Wheeler, J. C. Clare, and L. Empringham, "A vector controlled MCT matrix converter induction motor drive with minimized commutation times and enhanced waveform quality," in *Conf. Rec. 2002 IEEE Industry Applications Conf. 37th IAS Annu. Meeting*, Pittsburgh, PA, USA, 2002, pp. 466–472, vol. 1.

72 S. Large *et al.*, "Matrix converter solution for aircraft starting," *IEE Sem. Mat. Conver.*, 2003, pp. 5/1–5/18.

73 P. W. Wheeler *et al.*, "A matrix converter based permanent magnet motor drive for an electro-hydrostatic aircraft actuator," in *IECON'03. 29th Annu. Conf. IEEE Industrial Electronics Society*, Roanoke, VA, USA, 2003, pp. 2072–2077, vol. 3.

74 P. Zanchetta, M. Sumner, J. C. Clare, and P. W. Wheeler, "Control of matrix converters for AC power supplies using genetic algorithms," in *IEEE Int. Symp. Industrial Electronics*, Ajaccio, France, 2004, pp. 1429–1433, vol. 2.

75 P. Snary, B. Bhangu, C. M. Bingham, and D. A. Stone, "Matrix converters for sensorless control of PMSMs and other auxiliaries on deep-sea ROVs," in *2nd Int. Conf. Power Electronics, Machines and Drives (PEMD 2004)*, Edinburgh, UK, 2004, pp. 703–708, vol. 2.

76 K.-B. Lee and F. Blaabjerg, "Reduced-order extended luenberger observer based sensorless vector control driven by matrix converter with nonlinearity compensation," *IEEE Trans. Ind. Electron.*, vol. 53, no. 1, pp. 66–75, Feb. 2006.

77 P. W. Wheeler *et al.*, "An integrated 30 kW matrix converter based induction motor drive," in *IEEE 36th Power Electronics Specialists Conf.*, Dresden, Germany, 2005, pp. 2390–2395.

78 D. Casadei, G. Serra, A. Trentin, L. Zarri, and M. Calvini, "Experimental analysis of a matrix converter prototype based on new IGBT modules," in *Proc. IEEE Int. Symp. Industrial Electronics. ISIE 2005*, Dubrovnik, Croatia, 2005, pp. 559–564, vol. 2.

79 S. Muller, U. Ammann, and S. Rees, "New time-discrete modulation scheme for matrix converters," *IEEE Trans. Ind. Electron.*, vol. 52, no. 6, pp. 1607–1615, Dec. 2005.

80 G. T. Chiang and J. Itoh, "Comparison of two overmodulation strategies in an indirect matrix converter," *IEEE Trans. Ind. Electron.*, vol. 60, no. 1, pp. 43–53, Jan. 2013.

81 Y. Xia, X. Zhang, M. Qiao, F. Yu, Y. Wei, and P. Zhu, "Research on a new indirect space-vector overmodulation strategy in matrix converter," *IEEE Trans. Ind. Electron.*, vol. 63, no. 2, pp. 1130–1141, Feb. 2016.

82 T. D. Nguyen and H. -H. Lee, "Development of a three-to-five-phase indirect matrix converter with carrier-based PWM based on space-vector modulation analysis," *IEEE Trans. Ind. Electron.*, vol. 63, no. 1, pp. 13–24, Jan. 2016.

83 G. T. Chiang and J.-I. Itoh, "Performance optimization of a square wave operation in an indirect matrix converter with a reactor free DC boost converter," in *Proc. 14th European Conf. Power Electronics and Applications*, Birmingham, UK, 2011, pp. 1–10.

84 G. T. Chiang and J.-I. Itoh, "DC/DC boost converter functionality in a three-phase indirect matrix converter," *IEEE Trans. Power Electron.*, vol. 26, no. 5, pp. 1599–1607, May 2011.

85 F. Z. Peng, "Z-source inverter," *IEEE Trans. Ind. Appl.*, vol. 39, no. 2, pp. 504–510, Mar.-Apr. 2003.

86 M. Shen, A. Joseph, J. Wang, F. Z. Peng, and D. J. Adams, "Comparison of traditional inverters and Z -source inverter for fuel cell vehicles," *IEEE Trans. Power Electron.*, vol. 22, no. 4, pp. 1453–1463, July 2007.

87 Y. P. Siwakoti, F. Z. Peng, F. Blaabjerg, P. C. Loh, and G. E. Town, "Impedance-source networks for electric power conversion part I: A topological review," *IEEE Trans. Power Electron.*, vol. 30, no. 2, pp. 699–716, Feb. 2015.

88 J. Anderson and F. Z. Peng, "Four quasi-Z-Source inverters," in *IEEE Power Electronics Specialists Conf.*, Rhodes, Greece, 2008, pp. 2743–2749.

89 C. J. Gajanayake, F. L. Luo, H. B. Gooi, P. L. So, and L. K. Siow, "Extended-boost Z-source inverters," *IEEE Trans. Power Electron.*, vol. 25, no. 10, pp. 2642–2652, Oct. 2010.

90 M.-K. Nguyen, Y.-C. Lim, and G.-B. Cho, "Switched-inductor quasi-Z-source inverter," *IEEE Trans. Power Electron.*, vol. 26, no. 11, pp. 3183–3191, Nov. 2011.

91 Y. P. Siwakoti, F. Blaabjerg, P. C. Loh, G.E. Town, "High-voltage boost quasi-Z-source isolated DC/DC converter," *IET Power Electron.*, vol.7, no.9, pp.2387–2395, Sep. 2014.

92 M. Adamowicz, J. Guzinski, R. Strzelecki, F. Z. Peng, and H. Abu-Rub, "High step-up continuous input current LCCT-Z-source inverters for fuel cells," in *IEEE Energy Conversion Congress and Expos.*, Phoenix, AZ, USA, 2011, pp. 2276–2282.

93 W. Qian, F. Z. Peng and H. Cha, "Trans-Z-source inverters," *IEEE Trans. Power Electron.*, vol. 26, no. 12, pp. 3453–3463, Dec. 2011.

94 F. Z. Peng, M. Shen, and Z. Qian, "Maximum boost control of the Z-source inverter," *IEEE Trans. Power Electron.*, vol. 20, no. 4, pp. 833–838, July 2005.

95 M. Shen, J. Wang, A. Joseph, F. Z. Peng, L. M. Tolbert, and D. J. Adams, "Maximum constant boost control of the Z-source inverter," in *Conf. Rec. 2004 IEEE Industry Applications Conf., 2004. 39th IAS Annu. Meeting*, Seattle, WA, USA, 2004, p. 147.

96 Y. Liu, B. Ge, H. Abu-Rub, and F. Z. Peng, "Overview of space vector modulations for three-phase Z-source / quasi-Z-source inverters," *IEEE Trans. Power Electron.*, vol.29, no.4, pp.2098–2108, Apr. 2014.

97 Y. Liu, H. Abu-Rub, and B. Ge, "Z-source/quasi-Z-source inverters – Derived networks, modulations, controls, and emerging applications to photovoltaic conversion," *IEEE Ind. Electron. Mag.*, vol. 8, no. 4, pp. 32–44, Dec. 2014.

98 C. J. Gajanayake, D. M. Vilathgamuwa, and P. C. Loh, "Small-signal and signal-flow-graph modeling of switched Z-source impedance network," *IEEE Power Electr. Lett.*, vol. 3, no. 3, pp. 111–116, Sept. 2005.

99 P. C. Loh, D. M. Vilathgamuwa, C. J. Gajanayake, Y. R. Lim, and C. W. Teo, "Transient modeling and analysis of pulse-width modulated Z-source inverter," *IEEE Trans. Power Electron.*, vol. 22, no. 2, pp. 498–507, Mar. 2007.

100 Y. Li, S. Jiang, J. G. Cintron-Rivera, and F. Z. Peng, "Modeling and control of quasi-Z-source inverter for distributed generation applications," *IEEE Trans. Ind. Electron.*, vol. 60, no. 4, pp. 1532–1541, Apr. 2013.

101 A. H. Rajaei, S. Kaboli, and A. Emadi, "Sliding-mode control of z-source inverter," in *Proc. 34th Annu. Conf. IEEE Ind. Electron.*, Orlando, FL, USA, 2008, pp. 947–952.

102 J. Liu, S. Jiang, D. Cao, and F. Z. Peng, "A digital current control of quasi-Z-source inverter with battery," *IEEE Trans. Ind. Informat.*, vol. 9, no. 2, pp. 928–937, May 2013.

103 M. Mosa, R. S. Balog, and H. Abu-Rub, "High-performance predictive control of quasi-impedance source inverter," *IEEE Trans. Power Electron.*, vol. 32, no. 4, pp. 3251–3262, Apr. 2017.

104 Y. Liu, H. Abu-Rub, Y. Xue, and F. Tao, "A discrete-time average model based predictive control for quasi-Z-source inverter," *IEEE Trans. Ind. Electron.*, vol. 65, no. 8, pp. 6044–6054, Aug. 2018.

105 D. Sun *et al.*, "Modeling, impedance design, and efficiency analysis of quasi- Z source module in cascaded multilevel photovoltaic power system," *IEEE Trans. Ind. Electron.*, vol. 61, no. 11, pp. 6108–6117, Nov. 2014.

106 Y. Liu, B. Ge, H. Abu-Rub, and F.Z. Peng, "An effective control method for quasi-Z-source cascade multilevel inverter based grid-tie single-phase photovoltaic power system," *IEEE Trans. Industr. Inform.*, vol.10, no.1, pp.399–407, Feb. 2014.

107 B. Ge, Y. Liu, H. Abu-Rub, and F. Z. Peng, "State-of-charge balancing control for a battery-energy-stored quasi-Z-source cascaded-multilevel-inverter-based photovoltaic power system," *IEEE Trans. Ind. Electron.*, vol. 65, no. 3, pp. 2268–2279, Mar. 2018.

108 W. Liang, Y. Liu, and J. Peng, "A day and night operational quasi-Z source multilevel grid-tied PV power system to achieve active and reactive power control," *IEEE Trans. Power Electron.*, vol. 36, no. 1, pp. 474–492, Jan. 2021.

109 B. Ge, Q. Lei, W. Qian, and F. Z. Peng, "A family of Z-source matrix converters," *IEEE Trans. Ind. Electron.*, vol. 59, no. 1, pp. 35–46, Jan. 2012.

110 O. Ellabban, H. Abu-Rub, and S. Bayhan, "Z-source matrix converter: An overview," *IEEE Trans. Power Electron.*, vol. 31, no. 11, pp. 7436–7450, Nov. 2016.

111 Y. Liu, H. Abu-Rub, B. Ge, F. Blaabjerg, O. Ellabban, and P. Chiang Loh, "Z-source matrix converter" in *Impedance Source Power Electronic Converters*, Hoboken, NJ: John Wiley & Sons Ltd., IEEE, 2016, pp. 148–178.

112 X. Liu, P. C. Loh, P. Wang, and X. Han, "Improved modulation schemes for indirect Z-source matrix converter with sinusoidal input and output waveforms," *IEEE Trans. Power Electron.*, vol. 27, no. 9, pp. 4039–4050, Sept. 2012.

113 W. Song, Y. Zhong, H. Zhang, X. Sun, Q. Zhang, and W. Wang, "A study of Z-source dual-bridge matrix converter immune to abnormal input voltage disturbance and with high voltage transfer ratio," *IEEE Trans. Industr. Inform.*, vol. 9, no. 2, pp. 828–838, May 2013.

114 K. Park, K.-B. Lee, and F. Blaabjerg, "Improving output performance of a Z-source sparse matrix converter under unbalanced input-voltage conditions," *IEEE Trans. Power Electron.*, vol. 27, no. 4, pp. 2043–2054, Apr. 2012.

115 S. Sousa, S. Pinto, F. Silva, and J. Maia, "Extended voltage range AC drive using a Z source indirect matrix converter," in *20th Int. Conf. Electrical Machines*, Marseille, France, 2012, pp. 953–958.

116 E. Karaman, M. Farasat, and A. M. Trzynadlowski, "A 3Φ-3Φ quasi Z-source matrix converter for residential wind energy systems," in *IEEE Energy Conversion Congress and Exposition (ECCE)*, Raleigh, NC, USA, 2012, pp. 240–246.

117 S. Liu, B. Ge, X. Jiang, H. Abu-Rub, F. Z. Peng, "A novel quasi-Z-source indirect matrix converter", *Int. J. Circ. Theory Applicat.*, vol. 44, no. 4, pp. 438–454, Apr. 2015.

118 S. Liu, B. Ge, X. Jiang, H. Abu-Rub and F. Z. Peng, "Comparative evaluation of three Z-source/quasi-Z-source indirect matrix converters," *IEEE Trans. Ind. Electron.*, vol. 62, no. 2, pp. 692–701, Feb. 2015.

119 M. Guo, Y. Liu, B. Ge, S. Liu, X. Li, F. Ferreira, A. T. De Almeida, "Modeling and analysis of LC filter integrated Quasi-Z source indirect matrix converter," *Int. J. Circ. Theory Applicat.*, vol. 48, no. 4, pp. 567–586, Apr. 2020.

120 X. You, B. Ge, S. Liu, N. Nie, X. Jiang, and H. Abu-Rub, "Common mode voltage reduction of quasi-Z source indirect matriJanx converter", *Int. J. Circ. Theory Applicat.*, vol.44, no.1, pp.162–184, Jan. 2016.

121 S. Liu, B. Ge, Y. Liu, H. Abu-Rub, R. S. Balog, and H. Sun, "Modeling, analysis, and parameters design of LC-filter-integrated quasi-Z -source indirect matrix converter," *IEEE Trans. Power Electron.*, vol. 31, no. 11, pp. 7544–7555, Nov. 2016.

122 Y. Liu, W. Liang, B. Ge, H. Abu-Rub, and N. Nie, "Quasi-Z-source three-to-single-phase matrix converter and ripple power compensation based on model predictive control," *IEEE Trans. Ind. Electron.*, vol. 65, no. 6, pp. 5146–5156, Jun. 2018.

123 S. Liu, B. Ge, X. Jiang, H. Abu-Rub, F. Z. Peng, "Modeling, analysis, and motor drive application of quasi-Z-source indirect matrix converter", *Int. J. Comput. Math. Electr. Electron. Eng.*, vol. 33, no.1/2, pp.298–319, 2014.

124 M. Li, H. Abu-Rub, Y. Liu, B. Ge, and Z. Salam, "SiC power devices and applications in quasi-z-source converters/inverters," in *IEEE Conf. Energy Conversion (CENCON)*, Johor Bahru, Malaysia, 2015, pp. 331–336.

125 M. Guo, Y. Liu, B. Ge, H. Abu-Rub, "Optimum boost control of quasi-Z source indirect matrix converter," *IEEE Trans. Ind. Electron.*, vol. 65, no. 10, pp. 8393–8404, Oct. 2018.

126 M. Guo, Y. Liu, B. Ge, S. Liu, X. Li, F. J.T.E. Ferreira, A. T. de Almeida, "A quasi-Z source indirect matrix converter-fed induction motor drive," *IET Electr. Power Appl.*, vol. 14, no. 5 pp. 797–808, May 2020.

127 M. Guo, Y. Liu, B. Ge, X. Li, A. T. de Almeida, F. J. T. E. Ferreira, "Dual, three-level, quasi-Z-source, indirect matrix converter for motors with open-ended windings," *IEEE Trans. Energy Conversion*, vol. 38, no. 1, pp. 64–74, Mar. 2023.

2

Z-Source/Quasi-Z-Source Direct Matrix Converter

2.1 Introduction

Over the recent years, the matrix converter (MC) has steadily risen in significance due to its inherent ability to generate precise sinusoidal waveforms for both input and output currents. Notably, this converter offers an advantageous and adjustable input power factor, a feature elaborated upon in references [1–4]. One standout attribute of the MC is its proficiency in conducting direct AC-AC power transitions. Such capability inherently diminishes the reliance on passive energy storage mechanisms. The literature, especially references [1–18], is rich with extensive discussions on MC, including its advanced control methodologies, comprehensive theoretical underpinnings, and practical considerations for real-world applications. Moreover, the exploration into its multifaceted potential applications has been rigorous, as evident from the findings in [19–24]. Given the innovative developments surrounding reverse blocking insulated gate bipolar transistors (RB-IGBTs), the MC is positioning itself as a promising and efficient alternative to the conventional back-to-back converter systems.

As depicted in Figure 2.1, a comprehensive view of the design and layout of a 3-phase voltage-source matrix converter, abbreviated as VS-MC, unfolds before the reader. This intricately crafted converter integrates the prowess of its nine distinct AC switches. When harmoniously coupled with a meticulously devised modulation strategy, the VS-MC exhibits the capability to synthesize a particular AC output voltage. Beyond generating voltages, this converter has an added flair: it refines the morphology of the input current waveform, all while safeguarding a dependable and consistent power factor, an aspect highlighted in [1]. It's worth noting that the VS-MC doesn't operate in isolation. Drawing its lifeblood from an AC voltage source, it demands an association with an inductive-type load. Its operating principle doesn't exist in a vacuum; it draws parallels with the functional dynamics of the VS inverter. One of its operational highlights is its finesse in managing voltage delivery. It can channel one specific phase from the available three-phase input voltages, directing it with precision to a designated output conduit. When the requirement is a zero-voltage output, this converter ensures that each load terminal forms a unified bond with a singular input source terminal. Ensuring the integrity of the system, the VS-MC firmly sidesteps shoot-through switching at its gateway, and going further, refrains from permitting any open circuits on the receiving end. Reflecting the operational nuances emblematic of the VS inverter, the VS-MC, by design, commits to a buck mode of functioning, with its voltage efficiency ratio capping off at an impressive 0.866, a detail expounded upon in [1].

In the schematic laid out in Figure 2.2, one can observe the intricate design of the current-source matrix converter, often denoted as CS-MC. From a design philosophy standpoint, one might argue

Impedance Source Matrix Converters and Control, First Edition.
Yushan Liu, Xiao Li, Baoming Ge, Haitham Abu-Rub, and Frede Blaabjerg.

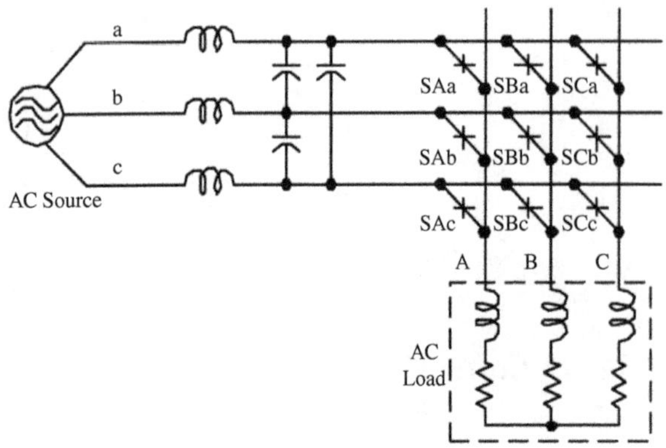

Figure 2.1 Voltage-source matrix converter.

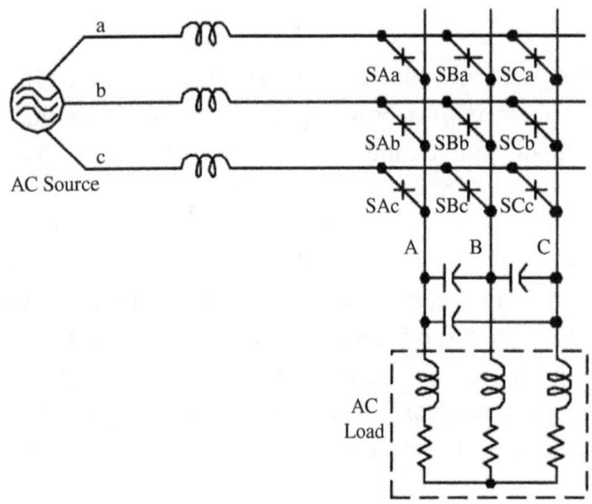

Figure 2.2 Current-source matrix converter.

that the CS-MC is conceptually analogous to its peer, the VS-MC, yet operating on an inverse logic. Unlike the VS-MC that depends on a voltage source, the CS-MC harnesses power from a current source, making a capacitive load not just beneficial, but integral for its peak performance. As we delve into the nuanced dynamics shared between the CS-MC and the VS-MC, it becomes evident that certain operational boundaries come into play: the system robustly resists open circuits at its entry point and vehemently avoids any short circuits at its exit terminal. An aspect that stands out with the CS-MC is its inherent capability: the voltage it manifests at its tail-end consistently exceeds what is introduced at the onset.

Harnessing the capabilities of both the VS-MC and CS-MC systems leads to the conceptualization of a matrix converter adept in both buck and boost functions. A closer inspection of Figure 2.3 reveals the operational nuances: by positioning the VS-MC proximal to the source side, the system achieves the desired buck functionality. On the contrary, anchoring the CS-MC closer to the load end facilitates the boost mechanism. Exploring configurations further, an inverse arrangement with the CS-MC connected upstream and the VS-MC downstream is also viable. Such a dynamic matrix converter setup amplifies versatility, rendering it particularly apt for environments where extensive voltage fluctuations either at the input or the output are commonplace. However, incorporating such a two-tiered converter comes with its share of complexities. The need for as many as 18 AC switches not only pushes up the investment but also complicates control processes, potentially compromises efficiency, and stirs up concerns regarding system robustness. Moreover, the presence of open or short circuits within these configurations isn't merely a fault; it's a genuine safety hazard.

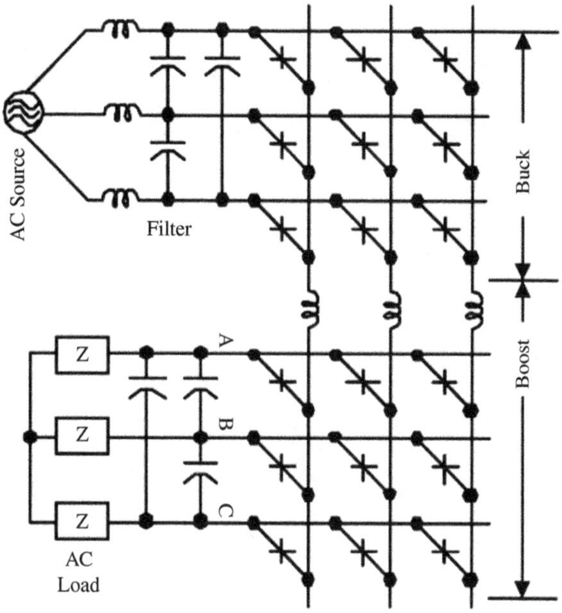

Figure 2.3 Buck–boost matrix converter using the traditional configurations [29].

It underscores the paramount importance of embedding a fail-safe commutation method, ensuring transitions occur without jeopardizing safety.

Emerging from recent innovations, the Z-source inverter, as cited in [25], has positioned itself as a cornerstone for pioneering progress across a range of power conversion circuits, a phenomenon extensively discussed in [26–30]. Scholars have exhibited significant interest in understanding the intricacies of Z-source DC/DC transformers, diving deep into the modulations in DC/AC units, and scrutinizing the complexities of AC/AC transformation modules. It's imperative to highlight the uniformity in frequency modulation, as observed across Z-source AC/AC conversion techniques, particularly as documented in [28, 29].

Within this manuscript's context, we embark on a comprehensive examination of three-phase Z-source matrix converters, often abbreviated as ZS-MCs. The aim is to shed light on their operational nuances, unraveling the mechanics that drive them, and journeying through their inherent characteristics. Spotlighting a singular ZS-MC variant, we delve into its schematic representation, elucidating methodologies for its management, and offering insights into its real-world applications. The authenticity of this ZS-MC construct and its tethered control paradigms is accentuated by both simulated results and empirical experiments.

2.2 Topology and Operating Principle

2.2.1 Topologies

As delineated in Figure 2.4, the architecture of the ZS-MCs can be parsed into three primary subdivisions: the initial MC connected to the source, the integral Z-source assembly, and the terminal MC interfacing with the load. Drawing parallels with their Z-source inverter relatives, two typologies of ZS-MCs emerge, defined by voltage and current-fed categorizations. Figure 2.4a exemplifies the voltage-centric ZS-MC design, in contrast to Figure 2.4b, which spotlights the current-infused variant. A pivotal feature in the anatomy of ZS-MCs is their capacity to modulate

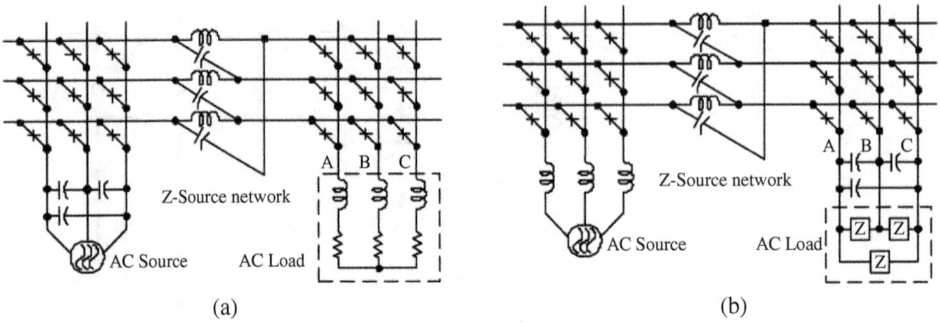

Figure 2.4 ZS matrix converters. (a) Voltage-fed ZS-MC. (b) Current-fed ZS-MC.

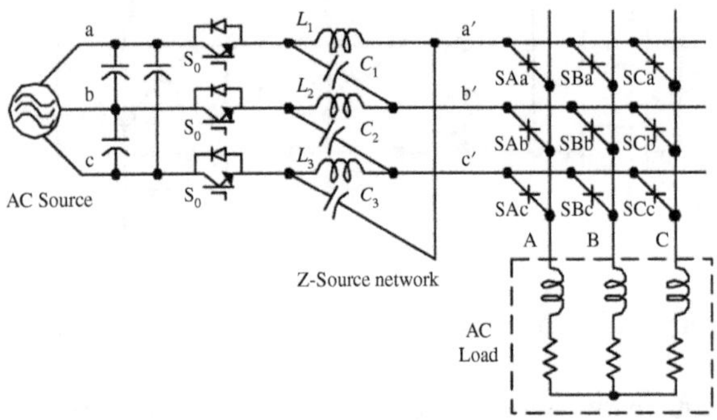

Figure 2.5 Simplified voltage-fed ZS matrix converter [29].

the source voltage, permitting both diminishing and enhancing maneuvers. While the incorporation of 18 AC switches in each ZS-MC construct is ambitious, it inevitably complicates its real-world implementation. Nonetheless, this intricate layout sets the stage for deriving refined ZS-MC models. As illustrated in Figure 2.5, an evolved rendition of the voltage-fueled ZS-MC is presented, undergoing a transformative replacement of the source-aligned MC from Figure 2.4a with a tri-phasic AC switch, labeled as S_0. Activating S_0 propels this revised ZS-MC to emulate the operational characteristics of the traditional VS-MC depicted in Figure 2.1, bestowing both primary and neutral voltage outputs to the load during its diminishing operational state. To facilitate voltage augmentation, it becomes imperative to interweave shoot-through states amid the pulse-width modulation (PWM) fluctuations of the MC, paired with rendering S_0 dormant during these epochs. The integration of shoot-through events during zero-voltage transitional phases amplifies voltage levels without compromising the integrity of the resultant waveform. The temporal density of these shoot-through episodes exerts a determinative effect on the amplification magnitude. A salient advantage of this evolved design is the curtailed switch requirement, a marked improvement from the architecture exhibited in Figure 2.3.

In our assessment of Figure 2.6, an evolution of the current-driven ZS-MC comes to light, characterized by a substitution of the source-aligned MC in Figure 2.4b with an advanced tri-phase switching apparatus. What captivates our attention is the manner in which this revamped current-driven ZS-MC seems to emulate the nuances of its voltage-oriented kin in circuit design. It deserves mention that this configuration is versatile, showcasing an aptitude for both voltage magnification and reduction.

Turning the spotlight on Figure 2.7, it unravels the intricacies of both voltage and current-centric quasi-Z-source matrix converters (qZS-MCs), bearing an uncanny semblance to the qZS inverter

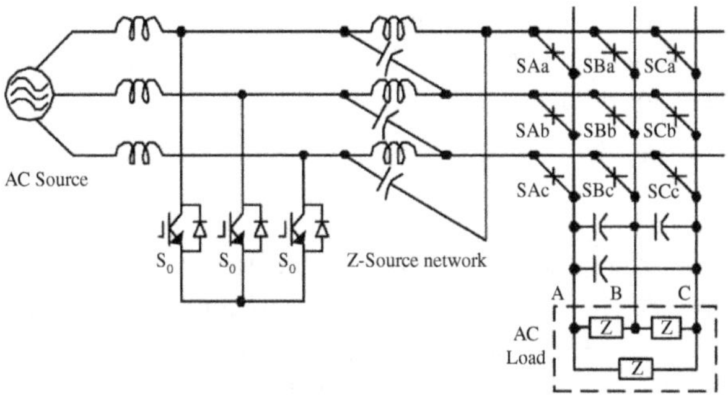

Figure 2.6 Simplified current-fed ZS matrix converter.

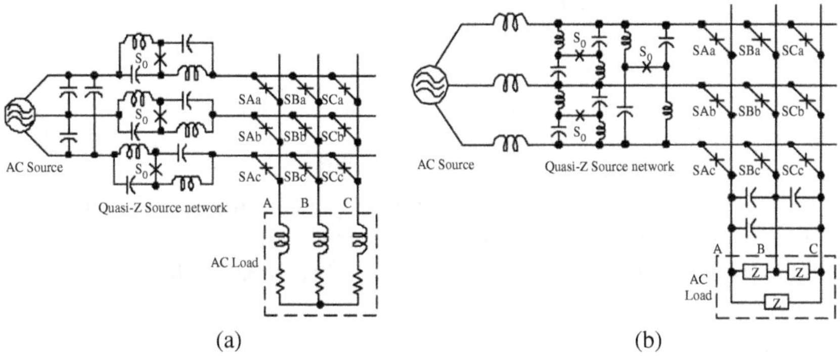

Figure 2.7 qZS-MCs. (a) Voltage-fed qZS-MC. (b) Current-fed qZS-MC.

architectures explored in reference [29]. On a functional plane, the dynamics of these qZS-MCs align with those of the earlier highlighted lean ZS-MCs. The standout facet of qZS-MCs, however, lies in their unrivaled potential for boundless voltage elevation and the refined interplay of the Z-source systems vis-a-vis voltage and current dimensions.

Peering into Figure 2.8, we're introduced to a duo of trans-ZS matrix converter blueprints. When juxtaposed with the trans-ZS inverter paradigms broached in [29], it's evident that these converters

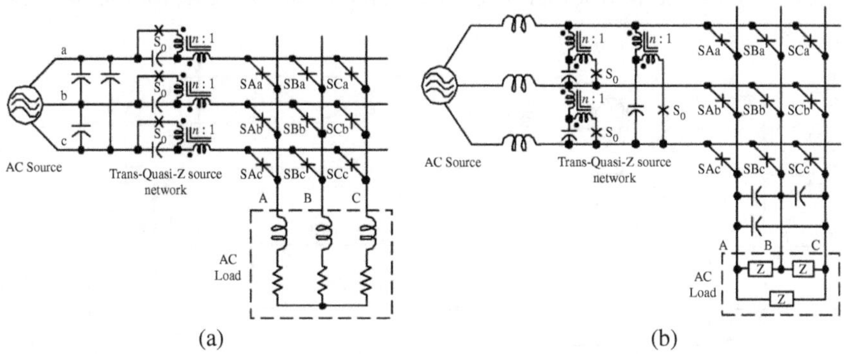

Figure 2.8 Variants of Trans-ZS-MCs. (a) Depicts the variant powered by voltage, known as the voltage-fed trans-ZS-MC. (b) Illustrates the type driven by current, termed the current-fed trans-ZS-MC.

excel in orchestrating optimal voltage amplification, all the while mitigating voltage tensions. Their governance techniques resonate with those employed in the condensed ZS-MCs, marking them as ideal candidates for contexts requiring pronounced voltage amplification.

2.2.2 Operation and Modeling

To dissect the intricacies of ZS-MCs' operation, our focus will be directed toward the refined voltage-driven ZS-MC as depicted in Figure 2.5, where we will accentuate its dual capability of bucking and boosting in the subsequent analyses. For a holistic grasp of its working principle, it becomes imperative to examine the ZS-MCs against the backdrop of two crucial switching states: namely, the shoot-through state and its opposing phase, the non-shoot-through state. Corresponding circuit illustrations during these particular states for the refined voltage-fueled ZS-MC are comprehensively captured in Figure 2.9.

2.2.2.1 Basic Model

Within the span of a designated switching cycle, which we'll denote as T_c, consider the time fraction designated for the shoot-through state as T_0. Meanwhile, the combined time intervals allocated for non-shoot-through states is expressed as T_1. This leads us to establish the relationship: T_c equals the sum of T_0 and T_1. Introducing a shoot-through duty ratio symbolized by D and calculated as the ratio T_0/T_c and referring to Figure 2.9a, it becomes apparent that during the T_1 phase when the apparatus operates in its non-shoot-through mode, specific voltage relationships emerge and can be explicated as follows:

$$\begin{bmatrix} v_{ab} \\ v_{bc} \\ v_{ca} \end{bmatrix} = \begin{bmatrix} v_{C1} \\ v_{C2} \\ v_{C3} \end{bmatrix} - \begin{bmatrix} v_{L2} \\ v_{L3} \\ v_{L1} \end{bmatrix} \tag{2.1}$$

$$\begin{bmatrix} v_{a'b'} \\ v_{b'c'} \\ v_{c'a'} \end{bmatrix} = \begin{bmatrix} v_{C1} \\ v_{C2} \\ v_{C3} \end{bmatrix} - \begin{bmatrix} v_{L1} \\ v_{L2} \\ v_{L3} \end{bmatrix} \tag{2.2}$$

Upon examining the circuitry components, we can attribute specific voltage representations for the inductors. For instance, inductors L_1, L_2, and L_3 have their voltages denoted by the symbols v_{L1}, v_{L2}, and v_{L3}. Moving on to the capacitive elements, the voltages across capacitors C_1, C_2, and C_3 are represented by symbols that mirror their respective component labels. As for the source's

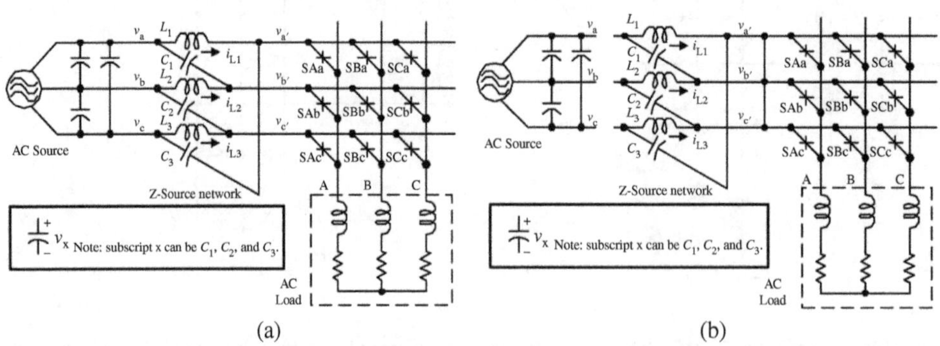

Figure 2.9 Depiction of the circuit equivalent to the streamlined voltage-fed ZS matrix converter. (a) Represents the state without shoot-through state. (b) Illustrates the state during shoot-through state.

line-to-line voltage values, they are represented by v_{ab}, v_{bc}, and v_{ca}. On the other hand, the MC bridge's line-to-line voltage values are designated by $v_{a'b'}$, $v_{b'c'}$, and $v_{c'a'}$. Drawing insights from Figure 2.9b, it's evident that

$$\begin{bmatrix} v_{C1} \\ v_{C2} \\ v_{C3} \end{bmatrix} = \begin{bmatrix} v_{L1} \\ v_{L2} \\ v_{L3} \end{bmatrix} \tag{2.3}$$

During periods where the system remains under constant conditions, it's noted that over an entire switching cycle, the average inductor voltage tends toward zero. This is especially the case when there's a considerable disparity between the switching frequency and the fundamental base frequency. Disregarding the initial voltage drop and correlating equations (2.1)–(2.3), we can extrapolate the ensuing relationship.

$$\begin{aligned} &Dv_{C1} + (1-D)(v_{C3} - v_{ca}) = 0 \\ &Dv_{C2} + (1-D)(v_{C1} - v_{ab}) = 0 \\ &Dv_{C3} + (1-D)(v_{C2} - v_{bc}) = 0 \end{aligned} \tag{2.4}$$

From (2.2) and (2.3), one has

$$\begin{aligned} v_{C1} &= (1-D)v_{a'b'} \\ v_{C2} &= (1-D)v_{b'c'} \\ v_{C3} &= (1-D)v_{c'a'} \end{aligned} \tag{2.5}$$

Given the parameters in context, it is postulated that the source functions as a balanced tri-phase system.

$$\begin{bmatrix} v_{ab} \\ v_{bc} \\ v_{ca} \end{bmatrix} = V_i \begin{bmatrix} \sin \omega t \\ \sin(\omega t - 120^\circ) \\ \sin(\omega t + 120^\circ) \end{bmatrix} \tag{2.6}$$

Given the circumstances, let V_i denote the voltage magnitude of the source, while ω signifies its angular frequency. It's crucial to note that the capacitor voltages are phase-shifted by an angle ϕ_C in relation to the source.

$$\begin{bmatrix} v_{C1} \\ v_{C2} \\ v_{C3} \end{bmatrix} = V_C \begin{bmatrix} \sin(\omega t + \phi_C) \\ \sin(\omega t + \phi_C - 120^\circ) \\ \sin(\omega t + \phi_C + 120^\circ) \end{bmatrix} \tag{2.7}$$

Taking into consideration that V_C represents the magnitude of the capacitor voltages, one can further postulate. If we infer that the voltages passing through the MC bridge exhibit a phase shift of ϕ_m relative to the source, then the following relationship can be stated:

$$\begin{bmatrix} v_{a'b'} \\ v_{b'c'} \\ v_{c'a'} \end{bmatrix} = V_m \begin{bmatrix} \sin(\omega t + \phi_m) \\ \sin(\omega t + \phi_m - 120^\circ) \\ \sin(\omega t + \phi_m + 120^\circ) \end{bmatrix} \tag{2.8}$$

Given that V_m denotes the Z-source network's output voltage amplitude, one can infer from equations (2.4), (2.6), and (2.7):

$$V_C\begin{bmatrix} D\sin(\omega t+\phi_C)+(1-D)\sin(\omega t+\phi_C+120^\circ) \\ D\sin(\omega t+\phi_C-120^\circ)+(1-D)\sin(\omega t+\phi_C) \\ D\sin(\omega t+\phi_C+120^\circ)+(1-D)\sin(\omega t+\phi_C-120^\circ) \end{bmatrix}$$

$$=(1-D)V_i\begin{bmatrix} \sin(\omega t+120^\circ) \\ \sin\omega t \\ \sin(\omega t-120^\circ) \end{bmatrix} \tag{2.9}$$

By leveraging trigonometric identities, notably the relationship $\sin^2\omega t+\sin^2(\omega t-120^\circ)+\sin^2(\omega t+120^\circ)=1.5$, and integrating this with equation (2.9), one can infer the correlation between the amplitude of the capacitor voltage and that of the source voltage.

$$\frac{V_C}{V_i}=\frac{1-D}{\sqrt{3D^2-3D+1}} \tag{2.10}$$

Drawing insights from equations (2.5), (2.7), and (2.8), one can establish a mathematical representation delineating the correlation between V_C and V_m.

$$V_C=(1-D)V_m \tag{2.11}$$

Based on the insights derived from equations (2.10) and (2.11), one can ascertain the amplitude pertaining to the line–line voltage at the MC's input interface.

$$V_m=\frac{1}{\sqrt{3D^2-3D+1}}V_i=BV_i \tag{2.12}$$

Upon evaluating the parameters, it is evident that when D reaches 0.5, the boost factor, denoted as B, achieves its maximum value, which stands at two. This signifies a voltage strength that's double the magnitude of the initial source voltage.

2.2.2.2 Buck/Boost Conversion Mode

During operations exclusively within the non-shoot-through states, illustrated in Figure 2.9a, the streamlined voltage-fed ZS-MC ensures that V_m equals V_i. In such scenarios, the ZS converter manifests characteristics akin to a traditional MC, resulting in a distinct peak for the output phase voltage.

$$v_{ln}=\frac{V_i}{\sqrt{3}}M \tag{2.13}$$

With M representing the modulation index that dictates the output voltages, one can note that the $V_i/\sqrt{3}$ consistently falls below the level of the input phase voltage. This observation accentuates the inherent buck capability of the optimized voltage-fed ZS-MC.

During the boost mode operation, the optimized voltage-fed ZS-MC operates by incorporating both the non-shoot-through and shoot-through states in every switching interval. This leads to an augmentation in the line–line voltage amplitude of the MC by a factor of B. The scale of this augmentation depends largely on the shoot-through duty ratio, denoted as D, detailed in equation (2.12). It's essential to highlight that shoot-through actions are limited only to zero switching intervals, thereby maintaining the output waveform's consistency without any deviations during the active switching intervals. The role of the shoot-through duty ratio, D, varies based on different shoot-through strategies. For instance, using basic boot control, a stable duty ratio D exists, bounded by $D \leq (2 - 4M)/3$. In contrast, with the advanced boot control approach, D varies through alternating switching intervals, aligning with the voltage outline and averaging at 1-M. A thorough exploration of this is covered in section 2.3.3. Because of the shoot-through intervals, there's an evident rise in the peak AC voltage at the MC's output end, namely

$$v_{ln} = \frac{V_m}{\sqrt{3}}M = \frac{V_i}{\sqrt{3}}BM \tag{2.14}$$

2.3 Modulation Methods

Control methodologies related to boosting, which were traditionally employed in the standard Z-source inverter, encompass techniques such as the simple boost, maximum boost, and the particularly distinguished maximum constant boost as cited in references [25, 31, 32]. These techniques show potential for application within the context of the refined voltage-fed ZS-MC. Implementing these in the new context would, however, call for subtle modifications in the configuration of the carrier envelope.

2.3.1 PWM Method for Traditional MC

In Figure 2.10, the modulation strategy for the MC, leveraging a carrier-based PWM approach, is illustrated. Here, the carrier waveform's boundary aligns with the three-phase source voltages, denoted as v_a, v_b, and v_c. The upper boundary of the waveform mirrors the highest voltage among these tri-phase inputs, while the lower one mirrors the smallest. For each switch cycle, comparisons are made between this carrier and the target output voltages, labeled v_A, v_B, and v_C. As a case in point, examine the reference voltage for phase A, represented as v_A in the given figure. When this is overlaid with the triangular carrier, it delineates the specific PWM signal for phase A. This, in turn, leads to the manifestation of a unique PWM voltage profile on the MC's output side.

2.3.2 PWM Method for the Simplified Voltage-Fed ZS-MC

For the streamlined voltage-fed ZS-MC variant, the earlier described PWM modulation strategy is still pertinent, even when it encompasses shoot-through states, as illustrated in Figure 2.10. To elucidate the concept of employing a consistent shoot-through duty ratio, Figure 2.11 presents a depiction of the adapted PWM technique suited to this ZS-MC configuration.

Within Figure 2.11, the principles of direct boost control are elucidated through two distinct shoot-through references. It is noteworthy that both these references exhibit intricate relationships with both envelopes, originating from

$$v_{st,r} = \frac{(y_{\max} - y_{\min})n + y_{\max} + y_{\min}}{2} \tag{2.15}$$

In the context presented, the coefficient "n" is instrumental in determining the shoot-through duty ratio. Within defined parameters, the minimum value resulting from the upper shoot-through reference should be confined between 0.5 p.u. and M. This implies that the variable "n" for the upper shoot-through reference is established within the confines of $(1+4M)/3$ to 1. Its inverse, represented by $(-n)$, pertains to the lower shoot-through reference. Under the outlined conditions, the modulation index is capped at a value less than 0.5. This premise is grounded on the notion that output references, denoted as v_A, v_B, and v_C, span diverse frequencies and phase shifts while

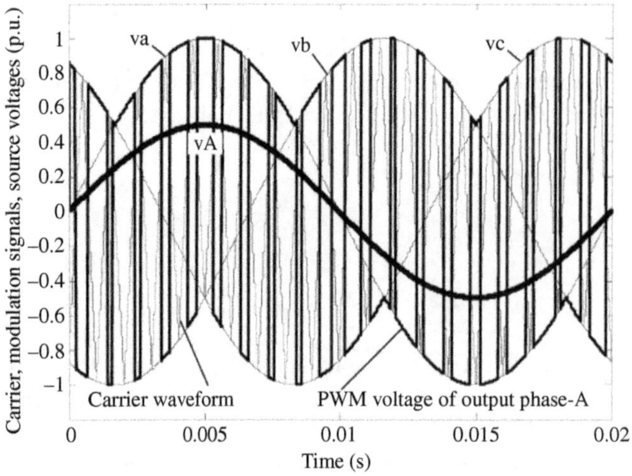

Figure 2.10 PWM method for the matrix converter.

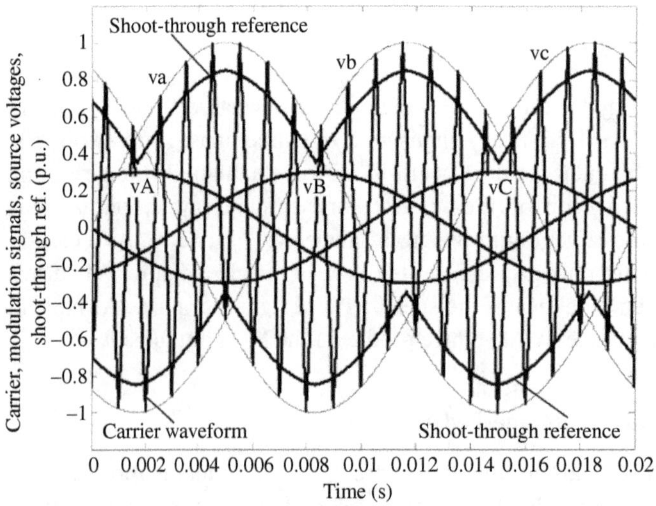

Figure 2.11 PWM control for the simplified voltage-fed ZS matrix converter.

maintaining harmony. Concluding, y_{max} and y_{min} terminologies are representative of the apex and nadir envelopes, respectively, of the source voltages.

Employing the simple boost control approach, the duration of the shoot-through phase, as referenced from the top, can be determined by the following relation.

$$T_0 = \frac{1-n}{2}T_c \tag{2.16}$$

Regarding its half-carrier cycle, the duty ratio associated with the shoot-through state can be expressed as follows:

$$D_h = \frac{1-n}{2} \tag{2.17}$$

In the representation of Figure 2.11, the voltage output references, specifically v_A, v_B, and v_C, are analyzed alongside the carrier waveform. This process generates corresponding PWM voltages,

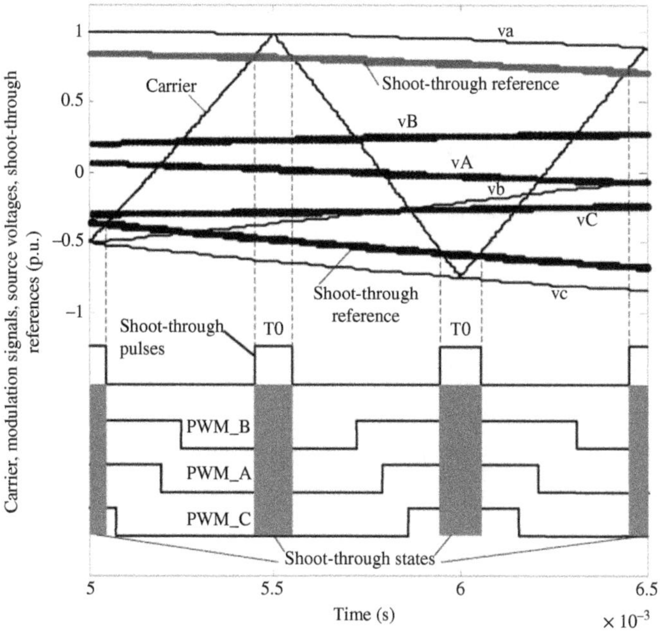

Figure 2.12 Simple boost method.

drawing parallels with the methodology described in Figure 2.10. The integration of shoot-through references ensures the inclusion of the shoot-through duty ratio in the derived PWM outputs. A comprehensive understanding of this technique can be grasped from Figure 2.12, showcasing the consequent shoot-through pulse sequences. A defined period, T_0, regulates these sequences, emerging from the juxtaposition of the shoot-through references and the carrier waveform. Within this same illustration, PWM_A, PWM_B, and PWM_C are manifested by contrasting the designated output voltage references with the aforementioned carrier waveform.

Upon meticulous examination of Figure 2.12, one discerns the distinct manifestation of shoot-through events during the junctures where the output three-phase voltages coalesce to a singular value. Such phenomena are particularly evident when v_a takes precedence on the upper envelope, while v_c commands the lower realm. Under these circumstances, the terminals of the MC's output exhibit a marked alignment with the source's phase-a or phase-c, respectively. This alignment results in the attenuation of the line-to-line output voltage, rendering it effectively nullified. It is of paramount significance to note that, despite these shoot-through intervals, the inherent contour and character of the output voltage waveform persist without deviation.

As delineated in reference [29], it is feasible to repurpose the maximum boost control approach of the conventional Z-source inverter for the optimized voltage-fed ZS-MC, which is illustrated in Figure 2.13. By implementing this specific control methodology, a consistent stability in the prevalent active states is maintained, concurrently modifying the null states into shoot-through conditions. Consequently, when adhering to a given modulation index, the peak values for T_0 and B are seamlessly secured, ensuring the output waveform remains unaffected and undistorted.

2.3.3 Voltage Gain of the Simplified Voltage-Fed ZS-MC

As depicted in Figure 2.13, the system transitions into a shoot-through state whenever the triangular carrier either exceeds the apex of the reference curves (v_A, v_B, v_C) or descends beneath

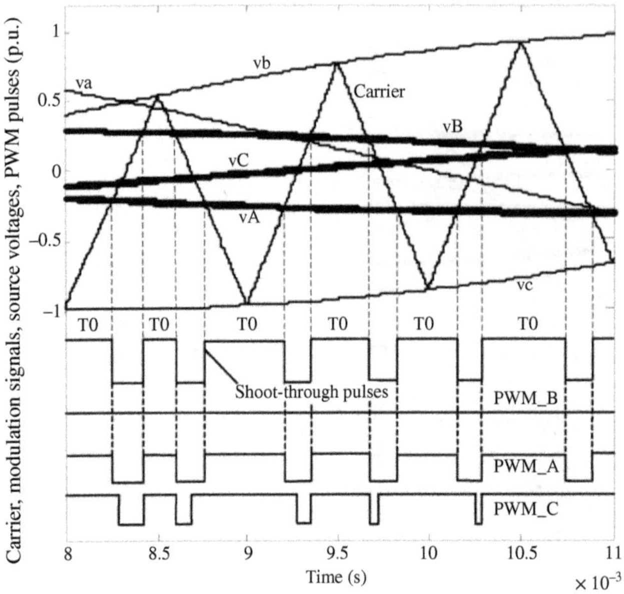

Figure 2.13 Maximum boost method.

their lowermost boundaries. It's important to emphasize that within every cycle, fluctuations are observed in the shoot-through duty ratio. The factor determining voltage amplification is inherently the average value of this duty ratio. A comprehensive derivation is essential for the formulation of its overarching equation, as emphasized in [29]. However, due to space constraints in this manuscript, a particular example will be showcased to illuminate the concept of voltage gain.

To elucidate the scenario, assume that the modulation signals align, both in terms of frequency and phase angle, with the fundamental source voltages. This assumption paves the way for a cyclical manifestation of the shoot-through state with a recurrence interval of π/3. Given the substantial disparity between the toggling and delivery frequencies, the shoot-through duty ratio within a single switch cycle—particularly from π/6 to π/2—can be represented as:

$$\frac{T_0}{T_c} = \frac{y_{max} - y_{min} - M[\sin \omega t - \sin(\omega t - 120°)]}{y_{max} - y_{min}} \tag{2.18}$$

By computing the integral of equation (2.18) over a period of π/3, one can determine the mean duty ratio associated with shoot-through state. This integration provides:

$$\frac{\overline{T_0}}{T_c} = 1 - M \tag{2.19}$$

Utilizing equation (2.12), we can deduce the associated boost factor, as illustrated by the following relation:

$$B = \frac{1}{\sqrt{3M^2 - 3M + 1}} \tag{2.20}$$

For the scenario in consideration, the optimal boost control mechanism yields a voltage amplification, as demonstrated by the subsequent relation:

$$G = BM = \frac{M}{\sqrt{3M^2 - 3M + 1}} \tag{2.21}$$

Utilizing the elementary boost control, one refers to equation (2.17) combined with the condition $n \geq (1+4M)/3$. This approach ascertains the peak shoot-through duty ratio within every carrier cycle, derived as follows:

$$D \leq \frac{2-4M}{3} \tag{2.22}$$

Taking into consideration both equations (2.12 and 2.22), we determine the voltage amplification associated with the elementary boost control strategy. This amplification is computed as:

$$G = BM = \frac{\sqrt{3}M}{\sqrt{16M^2 - 4M + 1}} \tag{2.23}$$

When analyzing various ZS-MCs, one can discern patterns that help understand their individual voltage gains. Among them, the voltage-fed qZS-MC stands out as a particularly illustrative example. This examination provides pivotal insights into the intricacies of its voltage enhancement capabilities. Furthermore, juxtaposing its performance against the simplified voltage-fed ZS-MC reveals stark contrasts and similarities. Upon closely observing Figure 2.7a, especially during the specific time frames labeled as shoot-through intervals (T_0), one can discern a clear and equivalent circuit pattern for the voltage-fed qZS-MC. This pattern is vividly depicted in Figure 2.14a, offering a detailed representation of the system's functioning.

$$\begin{bmatrix} v_{ab} \\ v_{bc} \\ v_{ca} \end{bmatrix} = \begin{bmatrix} v_{Ca1} \\ v_{Cb1} \\ v_{Cc1} \end{bmatrix} + \begin{bmatrix} v_{La2} \\ v_{Lb2} \\ v_{Lc2} \end{bmatrix} - \begin{bmatrix} v_{Lb1} \\ v_{Lc1} \\ v_{La1} \end{bmatrix} - \begin{bmatrix} v_{Cb2} \\ v_{Cc2} \\ v_{Ca2} \end{bmatrix} \tag{2.24}$$

$$\begin{bmatrix} v_{Ca1} \\ v_{Cb1} \\ v_{Cc1} \end{bmatrix} + \begin{bmatrix} v_{La2} \\ v_{Lb2} \\ v_{Lc2} \end{bmatrix} = \begin{bmatrix} v_{La1} \\ v_{Lb1} \\ v_{Lc1} \end{bmatrix} + \begin{bmatrix} v_{Ca2} \\ v_{Cb2} \\ v_{Cc2} \end{bmatrix} \tag{2.25}$$

In this context, "v" represents voltage. When referring to the capacitors within phase-x, subscript notations C_{x1} and C_{x2} designate the two capacitors, respectively. Similarly, for inductors within the same phase, L_{x1} and L_{x2} correspond to the two inductors. The phase x can be any among a, b, or c.

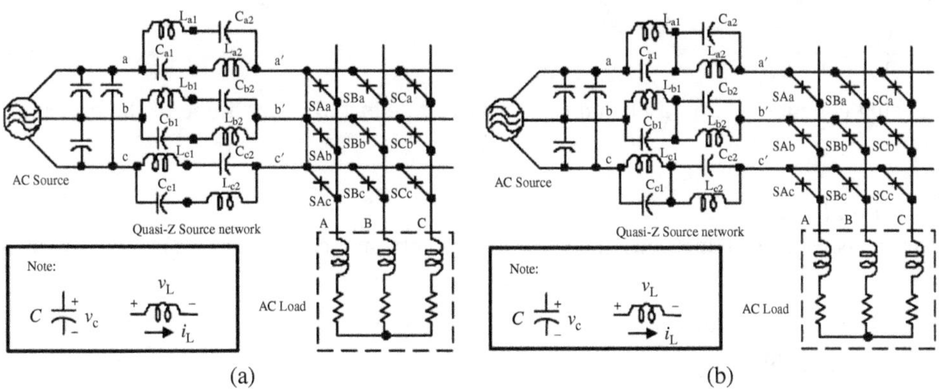

Figure 2.14 Representation of the qZS Matrix Converter under Voltage-fed Conditions. Part (a) demonstrates the state during shoot through, while part (b) highlights the scenario when not in the shoot-through state.

In the timeframe characterized by non-shoot-through states, labeled T_1, the corresponding equivalent circuit can be visualized in Figure 2.14b. From this depiction, we can infer:

$$\begin{bmatrix} v_{ab} \\ v_{bc} \\ v_{ca} \end{bmatrix} = \begin{bmatrix} v_{Ca1} \\ v_{Cb1} \\ v_{Cc1} \end{bmatrix} + \begin{bmatrix} v_{Ca2} \\ v_{Cb2} \\ v_{Cc2} \end{bmatrix} + \begin{bmatrix} v_{a'b'} \\ v_{b'c'} \\ v_{c'a'} \end{bmatrix} - \begin{bmatrix} v_{Cb1} \\ v_{Cc1} \\ v_{Ca1} \end{bmatrix} - \begin{bmatrix} v_{Cb2} \\ v_{Cc2} \\ v_{Ca2} \end{bmatrix} \tag{2.26}$$

$$\begin{bmatrix} v_{Ca1} \\ v_{Cb1} \\ v_{Cc1} \end{bmatrix} = \begin{bmatrix} v_{La1} \\ v_{Lb1} \\ v_{Lc1} \end{bmatrix}, \begin{bmatrix} v_{Ca2} \\ v_{Cb2} \\ v_{Cc2} \end{bmatrix} = \begin{bmatrix} v_{La2} \\ v_{Lb2} \\ v_{Lc2} \end{bmatrix} \tag{2.27}$$

Given the inherent symmetry present in the quasi-Z network structure, it is evident that:

$$\begin{bmatrix} v_{La2} \\ v_{Lb2} \\ v_{Lc2} \end{bmatrix} = \begin{bmatrix} v_{La1} \\ v_{Lb1} \\ v_{Lc1} \end{bmatrix}, \begin{bmatrix} v_{Ca1} \\ v_{Cb1} \\ v_{Cc1} \end{bmatrix} = \begin{bmatrix} v_{Ca2} \\ v_{Cb2} \\ v_{Cc2} \end{bmatrix} \tag{2.28}$$

Under a steady-state condition, over a complete switching cycle, the mean voltage across the inductors should register as zero. By combining insights from (2.24), (2.27), and (2.28), one can deduce (2.29).

$$\begin{bmatrix} v_{Cb1} \\ v_{Cc1} \\ v_{Ca1} \end{bmatrix} - \begin{bmatrix} v_{Ca1} \\ v_{Cb1} \\ v_{Cc1} \end{bmatrix} = \frac{D}{1-2D} \begin{bmatrix} v_{ab} \\ v_{bc} \\ v_{ca} \end{bmatrix} \tag{2.29}$$

Considering the balanced voltage conditions across the capacitors in a three-phase system, equation (2.29) can be reformulated as

$$\begin{bmatrix} v_{Ca1} \\ v_{Cb1} \\ v_{Cc1} \end{bmatrix} = \frac{D}{6D-3} \begin{bmatrix} v_{ab} - v_{ca} \\ v_{bc} - v_{ab} \\ v_{ca} - v_{bc} \end{bmatrix} \tag{2.30}$$

Integrating the expressions from (2.26) and (2.28), we derive equation (2.30), which reveals

$$\begin{bmatrix} v_{a'b'} \\ v_{b'c'} \\ v_{c'a'} \end{bmatrix} = \frac{1}{1-2D} \begin{bmatrix} v_{ab} \\ v_{bc} \\ v_{ca} \end{bmatrix} \tag{2.31}$$

Utilizing equations (2.6), (2.8), and (2.31) and drawing upon the methodology for the simplified voltage-fed ZS-MC, we delineate the boost factor pertaining to the voltage-fed qZS-MC.

$$B = \frac{V_m}{V_i} = \frac{1}{1-2D} \tag{2.32}$$

By analyzing equations (2.6), (2.7), and (2.30), we can deduce the amplitude of the capacitor voltage as

$$V_C = \frac{D}{1-2D} V_i \tag{2.33}$$

With varying boost control methods, the voltage gains corresponding to the voltage-fed qZS-MC vary:

1) When utilizing the maximum boost control strategy, one can refer to equation (2.19) to determine the voltage gain.

$$G = BM = \frac{M}{2M-1} \tag{2.34}$$

2) When adopting a straightforward boost control method, one can derive the voltage gain from equation (2.22).

$$G = BM = \frac{3M}{8M - 1} \tag{2.35}$$

Figure 2.15 serves as a comprehensive graphical representation, with traces 1–4 elucidating the intricate linkage between voltage gain and modulation index. This figure accentuates the nuanced differences between the streamlined voltage-fed ZS-MC and the qZS-MC. Each trace captures the behavior of these systems when interacting with distinct control strategies, namely maximum and direct boost controls. On the other side of the spectrum, trace 5 delves deeper into the gain characteristics of the conventional MC, providing a point of reference. A standout observation is the superior performance of both ZS-MC variations when compared to the conventional MC, especially in terms of gain. This superiority becomes even more pronounced when the ZS-MCs operate under the maximum boost control. This specific control condition induces a consistent shoot-through during all zero-state durations, leading to an optimized voltage performance [29]. Focusing on the streamlined design, there's a notable voltage surge that peaks at 1.155 when the modulation index, represented as M, stands at 0.667. Also, an interesting phenomenon occurs when M reaches 0.5 – the gain of this streamlined design converges to unity. This is further substantiated by an empirical finding, symbolized by a circle in Figure 2.15, which resonates seamlessly with the theoretical projection presented by curve 4 during heightened boost scenarios. Yet, when transitioning to a more rudimentary boost method, the streamlined design's gain exhibits a decline, becoming particularly evident when the modulation index is confined to 0.5 or lower. A specific empirical data point, represented as $G = 0.43$ and highlighted as a star, offers a contrasting view when placed beside the streamlined design under the same modulation condition. Lastly, a recurring theme in the data is the qZS-MC's consistently superior voltage gains over its streamlined counterpart, a testament to its design efficacy. The introduction of third harmonic waves seems to be a promising avenue, potentially widening the modulation index spectrum and further augmenting voltage gains, regardless of the control methodology in place.

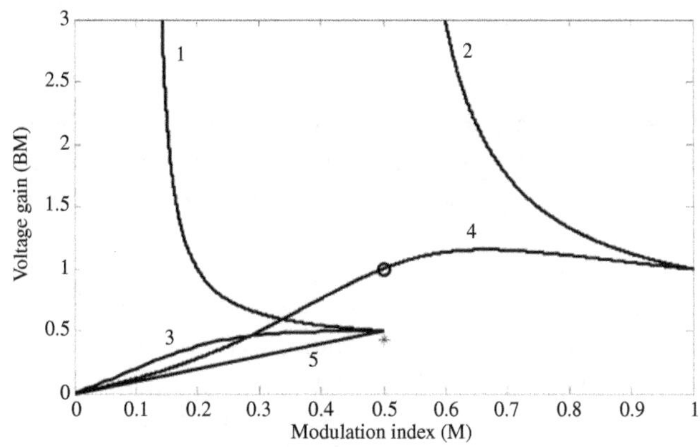

Figure 2.15 Voltage Gain Analysis: Tracks 1 and 2 depict the voltage-fed qZS-MC, whereas tracks 3 and 4 represent the simplified voltage-fed ZS-MC. Track 5 illustrates the conventional MC. Specifically, tracks 1 and 3 derive from the simple boost control methodology, while tracks 2 and 4 emanate from the maximum boost control approach [29].

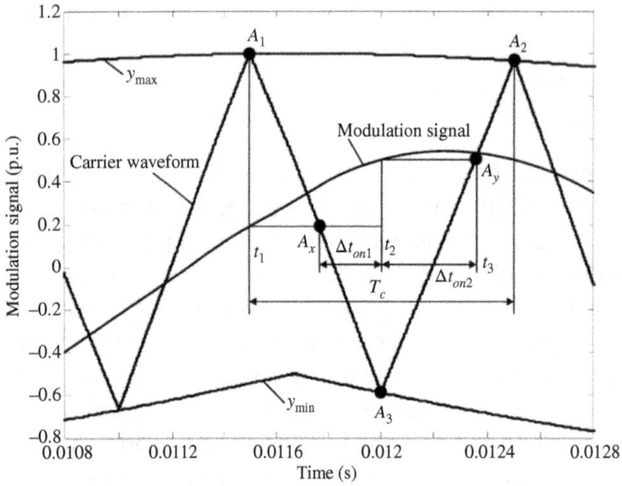

Figure 2.16 Calculation of duty cycle.

2.3.4 Implementation of Control Method

The PWM control technique tailored for the streamlined voltage-fed ZS-MC becomes more intuitive when implemented using advanced tools, including a digital signal processor (DSP) alongside a complex programmable logic device (CPLD). For a clearer understanding of how the PWM duty cycle is determined, one might consult Figure 2.16, serving as an illustrative example. In this representation, the variable T_c is indicative of the carrier wave's duration. Sequential time intervals are defined by t_1 and t_2, which capture the carrier wave's apex and nadir, respectively. Building upon this configuration, the duty cycle, denoted by "t_{on}," can be ascertained through a specific equation:

$$\Delta t_{on1} = \frac{M \sin \omega t_1 - MN}{MX_1 - MN} \cdot \frac{T_c}{2} \tag{2.36}$$

$$\Delta t_{on2} = \frac{M \sin \omega t_2 - MN}{MX_2 - MN} \cdot \frac{T_c}{2} \tag{2.37}$$

$$\Delta t_{on} = \Delta t_{on1} + \Delta t_{on2} \tag{2.38}$$

Within this framework, the peak voltage value, as captured at the instance denoted by t_1, bears the label MX_1. On the other hand, the crest's voltage value at the timestamp t_2 is symbolized by MX_2. Complementing this, the voltage value at the lowest point of the envelope during the t_2 interval is articulated as MN.

For generating the PWM sequences, specifically S_A, S_B, and S_C, one might resort to Equations (2.36) through (2.38). Once crafted, these sequences become inputs for the 9 AC switches, ensuring the creation of the desired PWM pulses. To streamline this process, an array of six auxiliary logic signals have been integrated into Figure 2.17. Among these, S_{z1}, S_{x1}, and S_{y1} act as representations for the upper voltage envelope's indicators for phases c, a, and b, in that order. For instance, should phase-c's voltage emerge as the leading one within the group of three, S_{z1} will be activated to 1. In a contrasting setup, S_{y2}, S_{z2}, and S_{x2} embody the indicators for the bottom voltage envelopes of phases b, c, and a correspondingly. As a case in point, should phase-b's voltage register as the minimal among the trio, S_{y2} is activated to 1. By merging these six indicators with

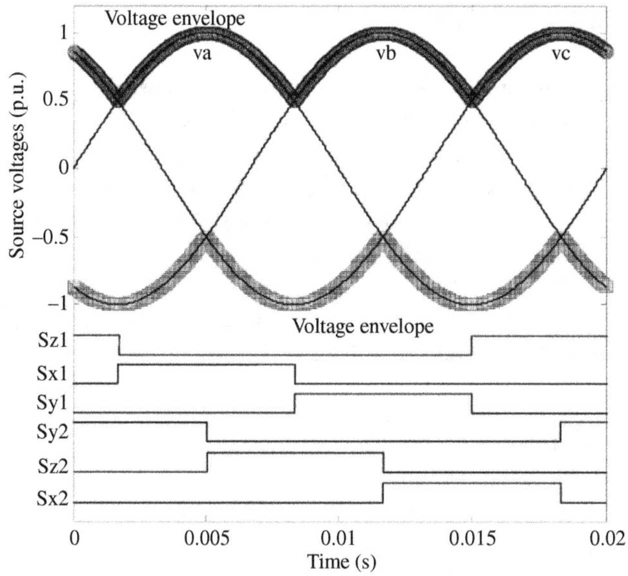

Figure 2.17 Indicators of the voltage envelopes.

the three PWM sequences, it's feasible to construct 9 unique switch signals, which then follow an ensuing logical pattern:

$$S'_{Aa} = S_{x1}S_A + S_{x2}\overline{S}_A,\quad S'_{Ab} = S_{y1}S_A + S_{y2}\overline{S}_A,$$
$$S'_{Ac} = S_{z1}S_A + S_{z2}\overline{S}_A,\quad S'_{Ba} = S_{x1}S_B + S_{x2}\overline{S}_B,$$
$$S'_{Bb} = S_{y1}S_B + S_{y2}\overline{S}_B,\quad S'_{Bc} = S_{z1}S_B + S_{z2}\overline{S}_B,$$
$$S'_{Ca} = S_{x1}S_C + S_{x2}\overline{S}_C,\quad S'_{Cb} = S_{y1}S_C + S_{y2}\overline{S}_C,$$
$$S'_{Cc} = S_{z1}S_C + S_{z2}\overline{S}_C,$$

where the symbol "+" denotes the logic OR.

While the logical procedures detailed previously are apt for steering the standard voltage-fed MC, the simplified voltage-fed ZS-MC requires a nuanced approach. More pointedly, it's necessary to integrate shoot-through states into the process before progressing to the driving operations. When leveraging the maximum boost control methodology, the resultant gate drive signals can be described in the following manner:

$$S_{st\_1} = S'_{Aa}S'_{Ba}S'_{Ca},\ S_{st\_2} = S'_{Ab}S'_{Bb}S'_{Cb},\ S_{st\_3} = S'_{Ac}S'_{Bc}S'_{Cc}$$
$$S_{st} = S_{st\_1} + S_{st\_2} + S_{st\_3}$$
$$S_{Aa} = S'_{Aa}S_{st},\ S_{Ab} = S'_{Ab}S_{st},\ S_{Ac} = S'_{Ac}S_{st}$$
$$S_{Ba} = S'_{Ba}S_{st},\ S_{Bb} = S'_{Bb}S_{st},\ S_{Bc} = S'_{Bc}S_{st}$$
$$S_{Ca} = S'_{Ca}S_{st},\ S_{Cb} = S'_{Cb}S_{st},\ S_{Cc} = S'_{Cc}S_{st}$$

Here, the operator "+" signifies a logical OR operation. The execution of all these logical functionalities will be orchestrated within the confines of the CPLD.

2.4 Simulation and Experimental Results

In the lab, a prototype of the streamlined voltage-fed ZS-MC was developed. To corroborate the conceptual design and theoretical interpretations, both simulations and hands-on experiments were undertaken. During these tests, the grid supplied power to the streamlined voltage-fed ZS-MC, while an R-L load was linked to the ZS-MC's output. Shared parameters for both virtual and physical experiments encompass: an L_z value of 1 mH and a C_z value of 330 μF for the Z-source network; for the R-L load, R was set at 20 Ω and L at 6.3 mH; and the power source from the grid was at a frequency of 60 Hz. To illustrate the ZS-MC's performance and validate the proposed control strategy, we delved into three specific scenarios: (i) operation in buck mode excluding shoot-through instances; (ii) operation in the boost mode; and (iii) a shift from buck operation to its boost counterpart.

1) Case 1: buck mode without shoot-through state
   When operating without the integration of shoot-through states, the optimized voltage-fed ZS-MC adopts a buck conversion mode, drawing parallels with the behavioral characteristics of a conventional MC. Figure 2.18a,b presents a visual representation of this mode, capturing both the simulation and experimental results. These results were obtained under conditions defined by a specific shoot-through duty ratio, with the boost factor B meticulously adjusted to 1, and another parameter. The observed voltage gain, hovering around 0.43, offers a satisfactory correlation with the projected theoretical figure of 0.5. This alignment becomes even more pronounced when acknowledging the inevitable voltage reductions arising from factors like line impedance and the inherent transitions witnessed within the MC framework. Moving on to Figure 2.19, it offers insights into the PWM duty cycles, which are inferred from the mathematical formulations (2.36)–(2.38). Furthermore, the figure also presents the resulting three-phase current dynamics and the consequent phase voltage variations. A salient feature of the tri-phase currents is their pristine sinusoidal form, which not only attests to the high level of accuracy of the aforementioned mathematical formulations but also underscores the consistent reliability of the logic operations tied to them.
2) Case 2: boost mode with shoot-through state
   Figure 2.20a,b presents the outcomes from both simulation and hands-on experimentation conducted under conditions where M is defined at 0.5, employing the maximum boost control as a voltage amplification strategy. With reference to the mathematical formulation of equation (2.21) and analytical data from Figure 2.15, there's an expectation for the boost factor B to settle around the 2.0 mark. This, in turn, suggests a voltage gain, denoted as G, equivalent to MB or more precisely, 1.0. It's noteworthy that practical experiments reported a voltage gain value hovering near 0.992, resonating well with the theoretical estimations. The simulation stage was meticulously designed, incorporating elements like inherent source harmonics and line impedance, ensuring a faithful representation of power derived from a real-world grid source. There was a notable symmetry between the calculated and empirical results.

   Upon examining Figure 2.20, one can observe noticeable ripples manifesting in the output current. These ripples stem from the periodic shifts in the shoot-through duty ratio, symbolized as D, especially when maximizing the boost control. In essence, such perturbations in D bring forth a line–line voltage to the matrix converter that deviates from the ideal sinusoidal nature, leading to the manifestation of the aforementioned current ripples. However, potential improvements in the output current's quality are within reach. A pathway to a more harmonious sinusoidal

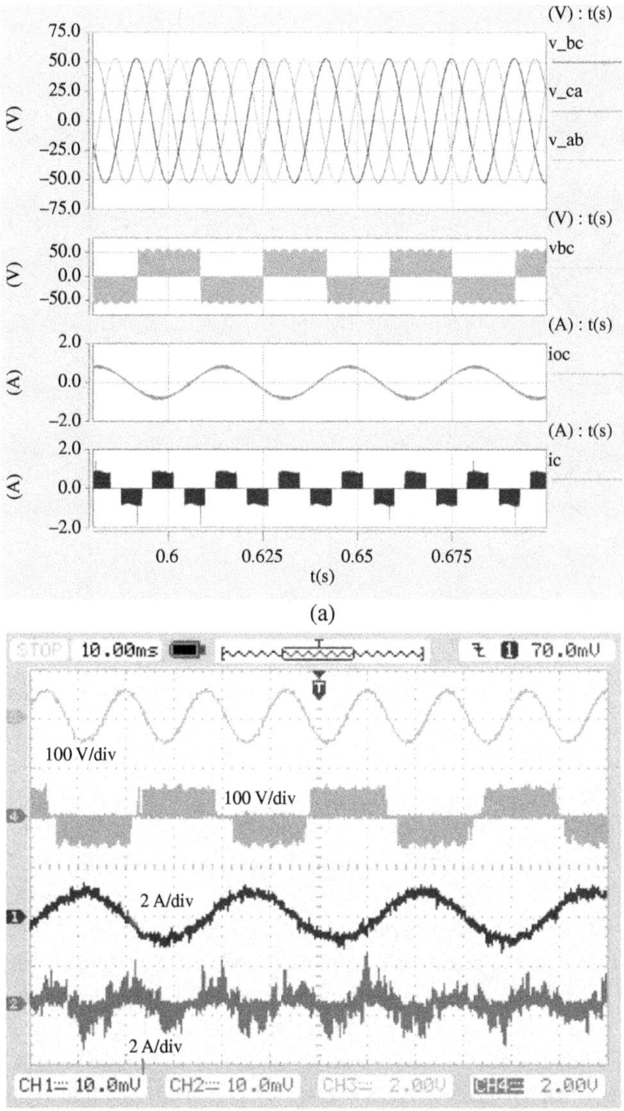

(a)

(b)

Figure 2.18 Depicting the buck mode within the streamlined voltage-fed ZS matrix converter framework: Beginning at the uppermost section, the diagram portrays the inter-line voltage of the source, followed by the line–line voltage and output current specifics of the MC, and concluding with the MC's intake phase current. Subfigure (a) encapsulates the simulated data points, whereas subfigure (b) delineates the empirical observations gathered from hands-on trials.

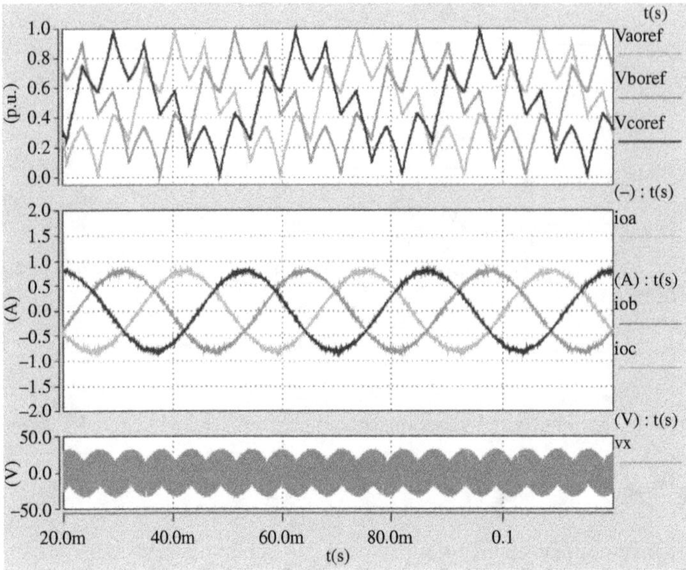

Figure 2.19 PWM duty cycles, output currents and phase voltage of the MC.

profile lies in exploiting the boost control technique illustrated in Figure 2.11, known for its unwavering shoot-through duty ratio *D*.

3) Case 3: transition from buck mode to boost mode
Within this context, the changeover from buck-mode operation to a boosted one is evident. At the outset, the streamlined voltage-fed ZS-MC predominantly operates within its buck regime. But a transformation is triggered by the sudden infusion of shoot-through states into the PWM sequences, a transformation captured lucidly in Figure 21. Marrying the max boost control with this shoot-through state brings about a pronounced escalation in both the resulting voltage and the current coursing through the load. To put it in numbers, there's an elevation in the voltage gain, catapulting it from an initial 0.43 to a subsequent 0.992. The harmony witnessed between the simulated and actual empirical findings accentuates the proficient operation of our streamlined voltage-fed ZS-MC model.
However, embracing the max boost control isn't devoid of its associated ramifications. It doesn't just act as a catalyst for harmonics in the exit voltage and current but also throws the grid current harmonics into a state of turmoil, a phenomenon elucidated in Figure 2.21a. As the grid impedance escalates, the quality of line-to-line voltage provisioned to the Z-source circuitry wanes. From an advisory standpoint, resorting to max boost control ought to be a strategy for situations demanding paramount boost levels where the side effect of harmonic intrusions can be momentarily sidestepped. Should the preservation of harmonic fidelity be of essence, pivoting toward the elementary boost control emerges as a sagacious choice, promising unblemished voltage and current profiles. Moreover, under scenarios with lighter load demands, the dominant grid current, ushered in by ZS network's capacitive nature, becomes starkly apparent.

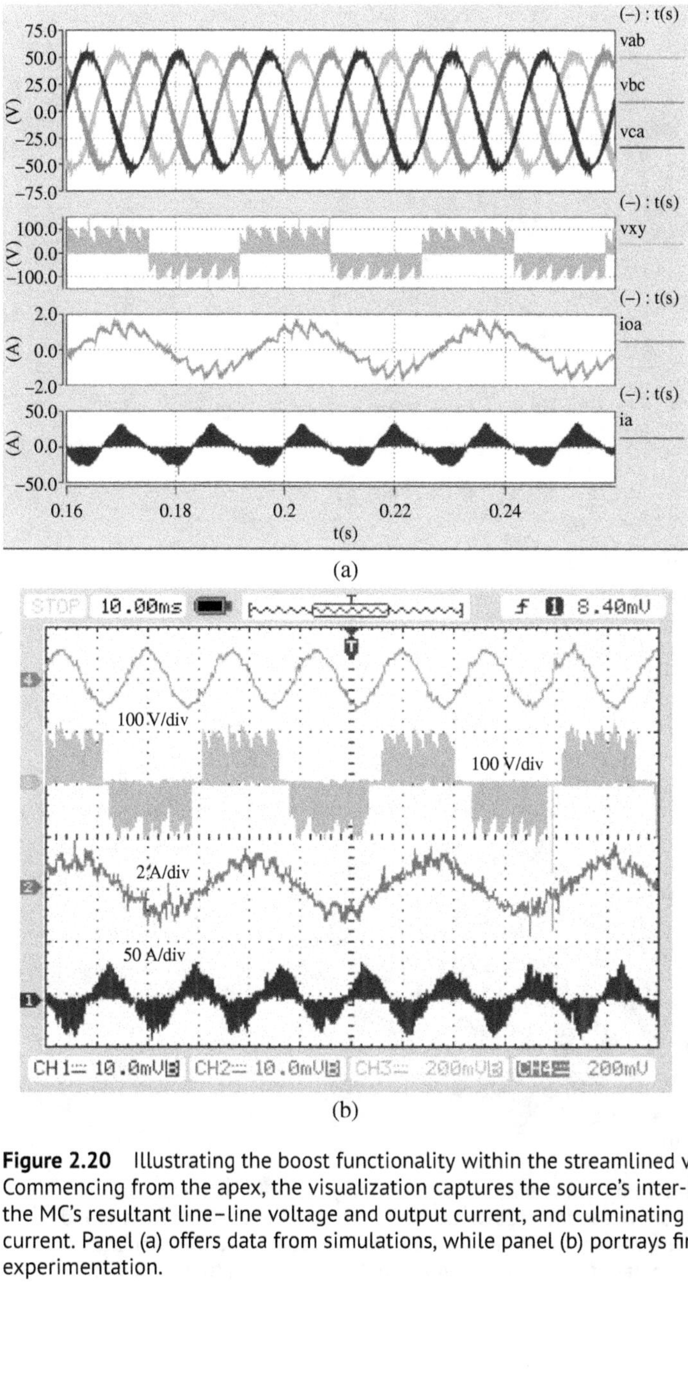

Figure 2.20 Illustrating the boost functionality within the streamlined voltage-fed ZS matrix converter: Commencing from the apex, the visualization captures the source's inter-line voltage, succeeded by the MC's resultant line–line voltage and output current, and culminating with the MC's intake phase current. Panel (a) offers data from simulations, while panel (b) portrays findings from hands-on experimentation.

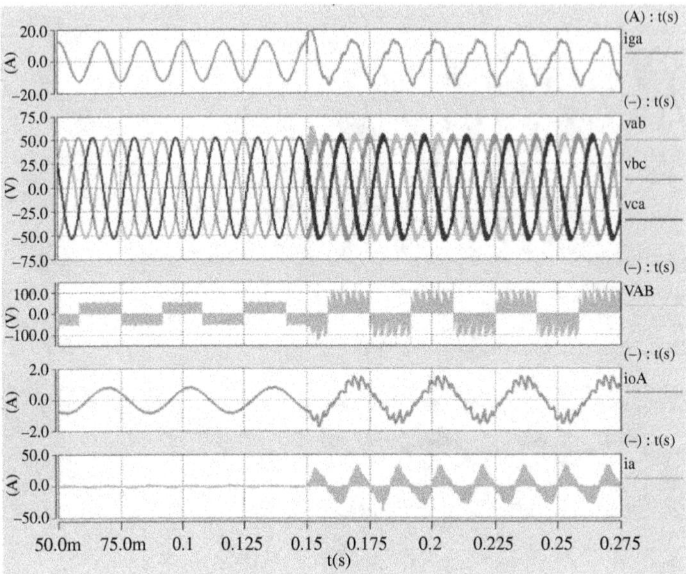

(a)

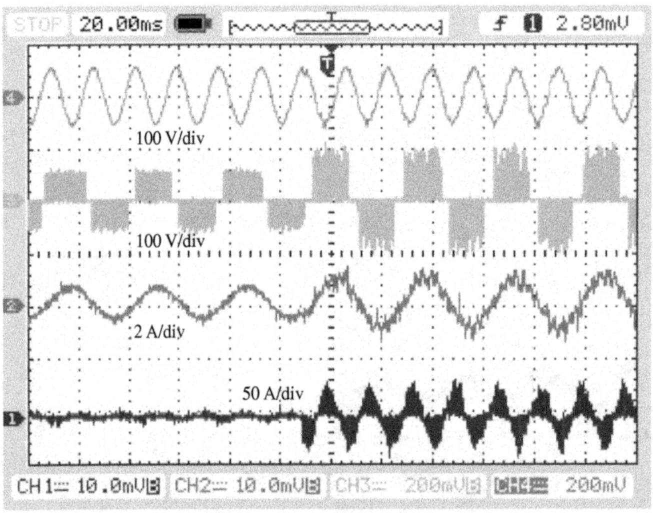

(b)

Figure 2.21 Depicting the evolution from buck operation to the employment of max boost control: ascending from the base, the visuals showcase the MC's input phase current, the resultant current and inter-line voltage from the MC, the originating line–line voltage from the source, and the simulated grid current flow. Panel (a) presents the simulation-derived outcomes, while panel (b) captures empirical observations.

2.5 Conclusion

With the advent of RB-IGBTs, the allure of MCs has seen a surge. While conventional voltage-fed MCs generate an AC voltage restrained below input source levels, the current-fed counterparts surpass the source voltage in their output. This discourse introduces a suite of ZS-MCs, equipped to deliver desired output voltage levels. Delving into their foundational topologies and operational dynamics, the chapter uses a streamlined voltage-fed ZS-MC to elucidate its intricate functioning, analytical aspects, PWM techniques, and boost controls. To empirically substantiate the ZS-MC's mechanisms and control strategies, a prototype leveraging RB-IGBTs took shape in the laboratory. Both simulated and hands-on test results underscore its prowess in voltage regulation, both in buck and boost modalities. The ZS-MCs, with a mere addition of three switches, encapsulate buck–boost capabilities, offering a cost-effective and reliable solution, especially beneficial for scenarios demanding expansive input–output voltage bandwidths.

References

1 K. Siyoung, S. Seun-Ki, and T. A. Lipo, "AC/AC power conversion based on matrix converter topology with unidirectional switches," *IEEE Trans. Ind. Appl.*, vol. 36, no. 1, pp. 139–145, Jan./Feb. 2000.

2 X. Dan and M. F. Rahman, "Sensorless direct torque and flux controlled IPM synchronous machine fed by matrix converter over a wide speed range," *IEEE Trans. Ind. Informat.*, vol. 9, no. 4, pp. 1855–1867, Nov. 2013.

3 M. Rivera *et al.*, "A comparative assessment of model predictive current control and space vector modulation in a direct matrix converter," *IEEE Trans. Ind. Electron.*, vol. 60, no. 2, pp. 578–588, Feb. 2013.

4 B. Metidji, N. Taib, L. Baghli, T. Rekioua, and S. Bacha, "Phase current reconstruction using a single current sensor of three-phase AC motors fed by SVM-controlled direct matrix converters," *IEEE Trans. Ind. Electron.*, vol. 60, no. 12, pp. 5497–5505, Dec. 2013.

5 P. G. Potamianos, E. D. Mitronikas, and A. N. Safacas, "Open-circuit fault diagnosis for matrix converter drives and remedial operation using carrier-based modulation methods," *IEEE Trans. Ind. Electron.*, vol. 61, no. 1, pp. 531–545, Jan. 2014.

6 K. Kobravi, R. Iravani, and H. A. Kojori, "Three-leg/four-leg matrix converter generalized modulation strategy—Part I: A new formulation," *IEEE Trans. Ind. Electron.*, vol. 60, no. 3, pp. 848–859, Mar. 2013.

7 K. Kobravi, R. Iravani, H. A. Kojori, "Three-leg/four-leg matrix converter generalized modulation strategy—Part II: Implementation and verification," *IEEE Trans. Ind. Electron.*, vol. 60, no. 3, pp. 860–872, Mar. 2013.

8 F. Villarroel, J. R. Espinoza, C. A. Rojas, J. Rodriguez, M. Rivera, and D. Sbarbaro, "Multiobjective switching state selector for finite-states model predictive control based on fuzzy decision making in a matrix converter," *IEEE Trans. Ind. Electron.*, vol. 60, no. 2, pp. 589–599, Feb. 2013.

9 J. Wang, B. Wu, D. Xu, and N. R. Zargari, "Indirect space-vector-based modulation techniques for high-power multimodular matrix converters," *IEEE Trans. Ind. Electron.*, vol. 60, no. 8, pp. 3060–3071, Aug. 2013.

10 L. Zarri, M. Mengoni, A. Tani, and J. O. Ojo, "Range of the linear modulation in matrix converters," *IEEE Trans. Power Electron.*, vol. 29, no. 6, pp. 3166–3178, Jun. 2014.

11 F. Gruson, P. Le Moigne, P. Delarue, A. Videt, X. Cimetiere, and M. Arpilliere, "A simple carrier-based modulation for the SVM of the matrix converter," *IEEE Trans. Ind. Informat.*, vol. 9, no. 2, pp. 947–956, May 2013.

12 T. D. Nguyen and H. -H. Lee, "A new SVM method for an indirect matrix converter with common-mode voltage reduction," *IEEE Trans. Ind. Informat.*, vol. 10, no. 1, pp. 61–72, Feb. 2014.

13 C. Xia, J. Zhao, Y. Yan, and T. Shi, "A novel direct torque control of matrix converter-fed PMSM drives using duty cycle control for torque ripple reduction," *IEEE Trans. Ind. Electron.*, vol. 61, no. 6, pp. 2700–2713, Jun. 2014.

14 L. Empringham, J. W. Kolar, J. Rodriguez, P. W. Wheeler, and J. C. Clare, "Technological issues and industrial application of matrix converters: a review," *IEEE Trans. Ind. Electron.*, vol. 60, no. 10, pp. 4260–4271, Oct. 2013.

15 J. Rodrguez, M. Rivera, J. W. Kolar, and P. W. Wheeler, "A review of control and modulation methods for matrix converters," *IEEE Trans. Ind. Electron.*, vol. 59, no. 1, pp. 58–70, Jan. 2012.

16 K. Basu and N. Mohan, "A single-sage power electronic transformer for a three-phase PWM AC/AC drive with source-based commutation of leakage energy and common-mode voltage suppression," *IEEE Trans. Ind. Electron.*, vol. 61, no. 11, pp. 5881–5893, Nov. 2014.

17 J. Espina, C. Ortega, L. de Lillo, L. Empringham, J. Balcells, and A. Arias, "Reduction of output common mode voltage using a novel SVM implementation in matrix converters for improved motor lifetime," *IEEE Trans. Ind. Electron.*, vol. 61, no. 11, pp. 5903–5911, Nov. 2014.

18 L. Wei and T. A. Lipo, "A novel matrix converter topology with simple commutation," in *36th IAS Annual Meeting*., vol. 3, pp. 1749–1754, Sept. 30 2001–Oct. 4 2001.

19 E. Karaman, M. Farasat, and A. Trzynadlowski, "A comparative study of series and cascaded Z-source matrix converters," *IEEE Trans. Ind. Electron.*, vol. 61, no. 10, pp. 5164–5173, Oct. 2014.

20 C. Klumpner, F. Blaabjerg, I. Boldea, and P. Nielsen, "New modulation method for matrix converters," *IEEE Trans. Ind. Appl.*, vol. 42, no. 3, pp. 797–806, May 2006.

21 F. Blaabjerg, D. Casadei, C. Klumpner, and M. Matteini, "Comparison of two current modulation strategies for matrix converters under unbalanced input voltage conditions," *IEEE Trans. Ind. Electron.*, vol. 49, no. 2, pp. 289–296, Apr. 2002.

22 X. Wang, H. Lin, H. She, B. Feng, "A research on space vector modulation strategy for matrix converter under abnormal input-voltage conditions," *IEEE Trans. Ind. Electron.*, vol. 59, no. 1, pp. 93–104, Jan. 2012.

23 M. Rivera, J. Rodriguez, J. R. Espinoza, and H. Abu-Rub, "Instantaneous reactive power minimization and current control for an indirect matrix converter under a distorted ac supply," *IEEE Trans. Ind. Informat.*, vol. 8, no. 3, pp. 482–490, Aug. 2012.

24 Y. D. Yoon and S. K. Sul, "Carrier-based modulation technique for matrix converter," *IEEE Trans. Power Electron.*, vol. 21, no. 6, pp. 1691–1703, Nov. 2006.

25 G. T. Chiang and J. Itoh, "Comparison of two overmodulation strategies in an indirect matrix converter," *IEEE Trans. Ind. Electron.*, vol. 60, no. 1, pp. 43–53, Jan. 2013.

26 J. Kaniewski, Z. Fedyczak, and G. Benysek, "AC voltage sag/swell compensator based on three-phase hybrid transformer with buck-boost matrix-reactance chopper," *IEEE Trans. Ind. Electron.*, vol. 61. no. 8, pp. 3835–3846, Aug. 2014.

27 W. Song, Y. Zhong, H. Zhang, X. Sun, Q. Zhang, and W. Wang, "A study of Z-source dual-bridge matrix converter immune to abnormal input voltage disturbance and with high voltage transfer ratio," *IEEE Trans. Ind. Informat.*, vol. 9, no. 2, pp. 828–838, May 2013.

28 X. Liu, P. C. Loh, P. Wang, and X. Han, "Improved modulation schemes for indirect Z-source matrix converter with sinusoidal input and output waveforms," *IEEE Trans. Power Electron.*, vol. 27, no. 9, pp. 4039–4050, Sept. 2012.

29 B. Ge, Q. Lei, W. Qian, and F. Z. Peng, "A family of Z-source matrix converters," *IEEE Trans. Ind. Electron.*, vol. 59, no. 1, pp. 35–46, Jan. 2012.

30 S. Liu, B. Ge, H. Abu-Rub, F. Z. Peng, and Y. Liu, "Quasi-Z-source matrix converter based induction motor drives," in *IECON 2012-38th Annu. Conf. IEEE Industrial Electronics Society*, pp. 5303–5307, 25–28 Oct. 2012.

31 L. Huber and D. Borojevic, "Space vector modulated three-phase to three-phase matrix converter with input power factor correction," *IEEE Trans. Ind. Appl.*, vol. 31, no. 6, pp. 1234–1246, Nov./Dec. 1995.

32 M. Shen, A. Josph, J. Wang, F. Z. Peng, and D. J. Adams, "Comparison of traditional inverters and Z-source inverter for fuel cell vehicles," *IEEE Trans. Power Electron.*, vol. 22, no. 4, pp. 1453–1463, Jul. 2007.

3

Z-Source/Quasi-Z-Source Indirect Matrix Converter (Non-All SiC Solution)

3.1 Introduction

Matrix Converters (MCs) introduce a notably refined technology in providing a straightforward and continuous connection between an AC source and the subsequent load, proficiently abolishing the requirement of a DC-link capacitor. The architectural brilliance of MCs is encapsulated in their elegantly simple and space-efficient power circuit, achieved through the deliberate avoidance of a DC-link capacitor. These converters not only assure the derivation of output voltages that can adeptly adjust in amplitude and frequency but also safeguard the preservation of sinusoidal currents, maintaining their integrity at both input and output domains. The schematic functionality of MCs can be adeptly partitioned into two predominant classifications, namely Direct Matrix Converters (DMCs) and Indirect Matrix Converters (IMCs), each meticulously illustrated in Figure 3.1. The DMCs uniquely enable both voltage and current conversion through a singular, cohesive power conversion stage, a feature that markedly diverges from the operational path of IMCs. The latter implements a uniquely structured, bifurcated indirect power conversion method, navigating through two distinctive power conversion phases. Even though they harbor comparable functionalities, IMCs endure a conspicuously lower commutation stress than their counterparts, a result of their inherent two-stage structure. However, it's pivotal to underscore that while this approach may tout simplicity, it concurrently necessitates the integration of an escalated number of power devices, meticulously connected in series, within the high-current conduit. This particular configuration inadvertently induces an enhancement in semiconductor losses and, customarily, incurs a minor decrement in overall efficiency, especially when juxtaposed with the operational metrics of DMCs. It warrants mention that, in practical scenarios, the divergences in control performances between DMCs and IMCs rarely impact significantly, maintaining a semblance of operational parity [1].

Even with the intriguing attributes of matrix converters, their incorporation into the industrial domain has unfolded with a measured approach, attributed primarily to a host of enduring and yet-unaddressed issues. The core predicament revolves around the constrained voltage transfer ratio, expressly defined as the output-to-input voltage ratio, which has been historically restricted to a numerical value of 0.866 during the application of linear modulation techniques [1]. An expansive corpus of research has strategically veered toward employing the over-modulation method as a viable strategy to circumnavigate the conundrum emanating from an inadequate voltage transfer ratio. While it remains essential to underscore, the activation of over-modulation unequivocally demands a trade-off, specifically affecting the purity of both the output voltage and input current waveforms and thereby yielding a notable decrement in their associated quality metrics [2].

Impedance Source Matrix Converters and Control, First Edition.
Yushan Liu, Xiao Li, Baoming Ge, Haitham Abu-Rub, and Frede Blaabjerg.

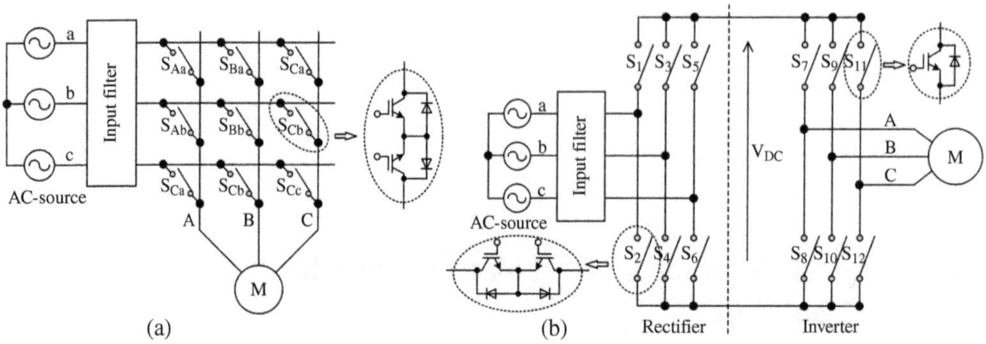

Figure 3.1 Illustration of AC/AC converter architectures: (a) Direct matrix converter topology and (b) topology related to indirect matrix converter.

This scenario propels the enhancement of the voltage transfer ratio into the spotlight, identifying it as a preeminent research avenue. A solution, ostensibly straightforward, can be realized by integrating a transformer to establish an intermediary interface between the power source and the matrix converter (MC). However, this tactic of enlisting a mains transformer is not without its distinct complications. It not only emerges as a fiscal commitment and prompts an amplification in the physical dimensions of the system but also casts a conspicuous, albeit indirect, influence on the holistic system efficiency.

In the field of power conversion, both converters and inverters, particularly those embracing Z-source (ZS) or quasi-Z-source (QZS) configurations, have become subjects of interest due to their inherent capability to achieve marked voltage amplification within a singular power conversion stage. This interest primarily emanates from their intrinsic boosting functionalities [3, 4]. A quintessential example is manifested through the methodical incorporation of the ZS/QZS network, which is astutely situated between the DC input source and the inverter. This strategic positioning not only facilitates a noticeable elevation in voltage but also obviates the requirement for multiple conversion stages, thereby streamlining the conversion process without compromising on the energy efficiency or operability of the system.

The innovative incorporation of the ZS/QZS network into several matrix converter (MC) topologies propels the development of a newly conceptualized variant of matrix converters, namely the Z-source matrix converter (ZS-MC). Such inclusion tactfully circumvents the traditional matrix converter's impediment of a constrained voltage gain. Further, it allows for a tolerable short circuit, systematically easing the commutation process intrinsic to the ZS-MC. Subsequently, the ZS-MC emerges, not merely as a concept, but a tangibly cost-effective, reliable, and efficiency-oriented structural approach for performing AC/AC conversion operations, which expansively encompasses both buck and boost procedures. It is imperative to underscore that the Z-source Indirect Matrix Converter (ZS-IMC) topology can be dissected into two fundamentally distinct configurations, each distinguished by the judicious placement of the ZS/QZS network. In the first configuration, the ZS/QZS network resides between the rectifier and inverter at the intermediate DC-link of the indirect matrix converter. In the pursuit of amplifying voltage gain through meticulous positioning, tangible trade-offs emerge, mandating a spatial augmentation due to the integration of substantially larger inductors and capacitors within the structural confines of the DC-link bus and consequently deviating from adhering to a wholly silicon solution. Conversely, adopting an alternative configuration, which strategically allocates the ZS/QZS network between the input AC source and the rectifier component of the indirect matrix converter, proficiently carves a pathway toward seamlessly achieving a wholly silicon solution, thereby circumventing some previously encountered limitations. However, this arrangement does not come without its own setbacks, as it mandates the integration of a substantial number of switches, inherently inviting an elevation in overall cost.

This chapter is dedicated to offering a thorough analysis of the ZS-IMC, with a special emphasis placed on configurations that do not exclusively employ silicon-based technologies. Whereas configurations deploying an all-silicon solution will be the subject of a meticulous discussion in the forthcoming chapter, ensuring a clear demarcation of focus and exploration within each segment.

3.2 Topologies and Operating Principle

3.2.1 Topologies

Figure 3.2b,c serves to unequivocally display two refined variations of AC-AC Z-Source converter topologies, each one carefully evolved from their originating sparse and ultra-sparse matrix converter topologies, respectively. Moving forward, as discerningly outlined in the scholarly literature [6–8], and visually elucidated in Figure 3.2d, a supplementary power switch (S_1) finds its precise placement within the DC-link, judiciously situated before the Z-source network, thus ensuring a structured pathway for bidirectional power flow across the comprehensive system. When engaging in operations under the conventional motoring mode, the DC-link current navigates adeptly through the Insulated-Gate Bipolar Transistor's (IGBT) integrated anti-parallel diode, steadfastly delivering power to the load through the conscientious operation of the Z-source inverter. Alternatively, during phases that employ power regenerative braking mode, the current demonstrates its capability to fluidly transition from the IGBT's collector to its emitter, facilitating a crucial energy feedback mechanism to the electrical grid. Each component enmeshed within these configurations executes a significant role, rigorously maintaining not only the operational integrity but also fostering the enhancement of the efficiency in the power conversion process. Engaging inductors L_1 and L_2, in harmonious conjunction with capacitors C_1 and C_2, skillfully shapes a Z-Source network, which exudes symmetrical attributes; these prove indispensable in assuaging power disturbances and delivering a robust, unswerving power supply to the load. The methodical employment of four-quadrant switches and IGBTs, each with its anti-parallel diode during the respective rectification and inversion phases, strategically enhances the reliability and versatility of these converter topologies. Notably, the topologies exemplified in Figure 3.2b,c significantly underscore streamlined design approaches, which inherently amalgamate both functional and practical aspects, ensuring the efficacy of the AC-AC Z-source converters in diverse operational scenarios. This comprehensive analysis and depiction therein affirm the viability and robustness of such converter topologies in facilitating enhanced voltage management across various applications in the power electronics domain.

The scholarly works, as particularly highlighted in references [9, 10], pave the way for the conceptualization of a meticulously designed three-phase ultra-sparse matrix converter. This innovative converter is characterized by the strategic incorporation of a series of Z-source configurations—specifically, the series Z-source, quasi Z-source, and switched inductor Z-source, which are explicitly depicted in Figure 3.3. With a deliberate embedding into the indirect matrix converter's respective rails, these divergent Z-source variations present a groundbreaking methodology to forge ahead in the realm of power conversion processes.

One of the defining features of these bespoke converters lies in their systematic design and engineering, enabling a suite of advancements that perceptibly move beyond the realms explored by the preceding cascaded Z-source matrix converter. A myriad of technical enhancements is witnessed herein, such as a discernible mitigation of the voltage navigating through the Z-source's

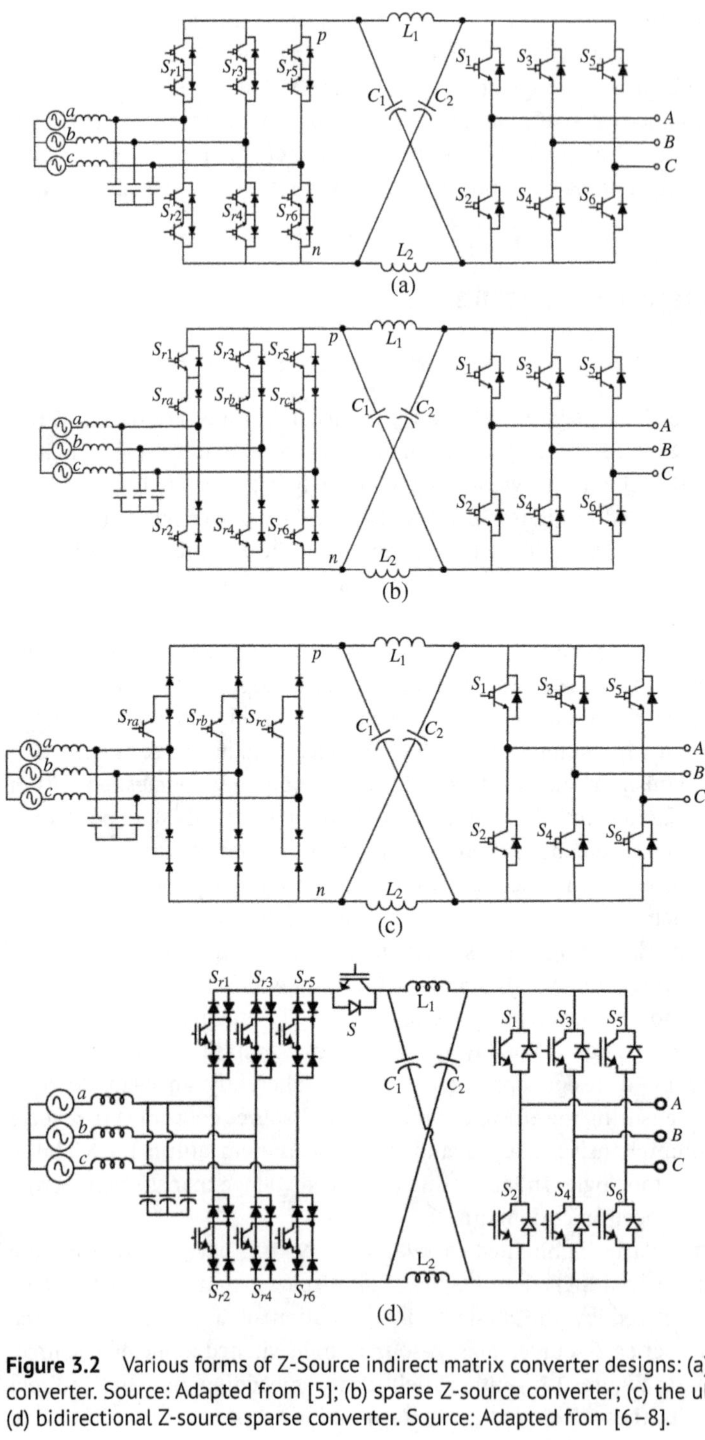

Figure 3.2 Various forms of Z-Source indirect matrix converter designs: (a) Z-source indirect matrix converter. Source: Adapted from [5]; (b) sparse Z-source converter; (c) the ultra-sparse Z-source converter; (d) bidirectional Z-source sparse converter. Source: Adapted from [6–8].

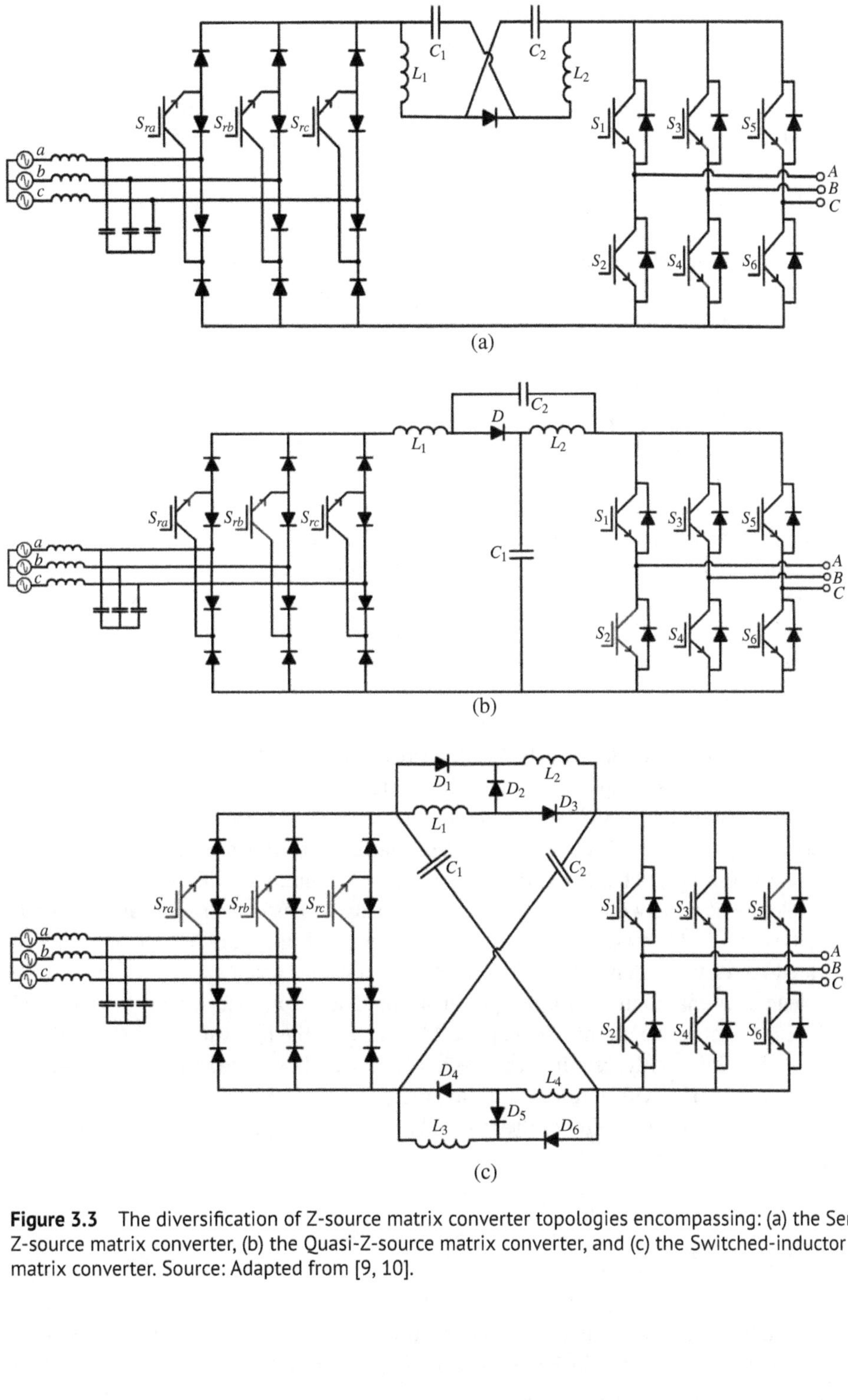

Figure 3.3 The diversification of Z-source matrix converter topologies encompassing: (a) the Series Z-source matrix converter, (b) the Quasi-Z-source matrix converter, and (c) the Switched-inductor Z-source matrix converter. Source: Adapted from [9, 10].

capacitor and a meticulously engineered limitation of the inrush current during the initial startup phase, with a distinct emphasis on the series Z-source. Moreover, a tangible amplification in the boost ratio is observable for the quasi Z-source and switched inductor Z-source matrix converters. Notably, the latter demonstrates a particularly exemplary elevation in its boost ratio, adhering to rigorous engineering standards. Further insights into the Fast Fourier Transform (FFT) analysis, detailed in references [11, 12], facilitate an understated yet insightful exploration of the input and output currents pertinent to these converters. Within this analytical context, a nuanced yet detectable enhancement in the performance trajectory of the switched inductor Z-source matrix converter emerges, especially when positioned against the quasi Z-source converter in a comparative analytical framework. In a parallel comparative scenario, a discernible yet nuanced superiority of the series Z-source matrix converter emerges, particularly when juxtaposed with its cascaded variant. Through detailed analysis, when we study the qualitative characteristics associated with the input currents, we can identify empirical data that underscores the effectiveness of the series Z-source matrix converter. Notably, this converter showcases a notable technical edge. However, it's essential to consider that its true potential is realized when specific operational parameters are taken into account.

Within the sphere of power electronics, the widely used three-level indirect matrix converter has seen advancements with the incorporation of a Z-source impedance network. This specific network finds its placement bridging the primary current-source rectifier (CSR) and the subsequent neutral point clamped (NPC) Voltage-Source Inverter (VSI), a design choice reflected in references [13, 14]. A major advantage of this modified design is its ability to reduce dependence on the customary split-DC capacitors that are central to many traditional NPC converters. The design leverages filtering capacitors already present in the primary CSR. For a visual elucidation, Figure 3.4a offers a clear portrayal of the bidirectional topology that defines the three-level Z-source indirect matrix converter. Simultaneously, Figure 3.4b illustrates the corresponding unidirectional design, together furnishing a detailed understanding of these updated configurations.

The integration and assimilation of Z-source and the switched-inductor Z-source (SIZS) circuits, when synergistically combined with the super-sparse matrix converter topology, is thoroughly explored and detailed in [15]. This in-depth exploration and examination led to the innovative design and development of new models of three-phase to single-phase matrix converters, specifically identified as the ZS-MC and SIZS-MC. Elegantly presented in Figure 3.5, these advanced and sophisticated converter designs adeptly demonstrate a notable capability to enhance voltage levels, while steadfastly maintaining a unity power factor. These converters, with their refined and practical design, emerge as prominently applicable and versatile in a variety of applications, especially those that inherently require unidirectional power flow. Elegantly amalgamating proficient energy conversion with pragmatic utility, the converters not only symbolize but also exemplify a cooperative, efficient blend of theoretical knowledge and practical application across diverse energy utilization scenarios.

3.2.2 Operating Principle

Figure 3.6 offers a detailed depiction of the various operational phases of the quasi-Z-source matrix converter, specifically navigating through Non-Shoot-Through (NST) and Shoot-Through (ST) situations. In the condition illustrated in Figure 3.6a, diode D is activated (ON), thereby enabling capacitors C_1 and C_2 to charge concurrently through the rectified input voltages. Subsequent to this, as the converter transitions, as seen in Figure 3.6b, it assumes an operational posture under

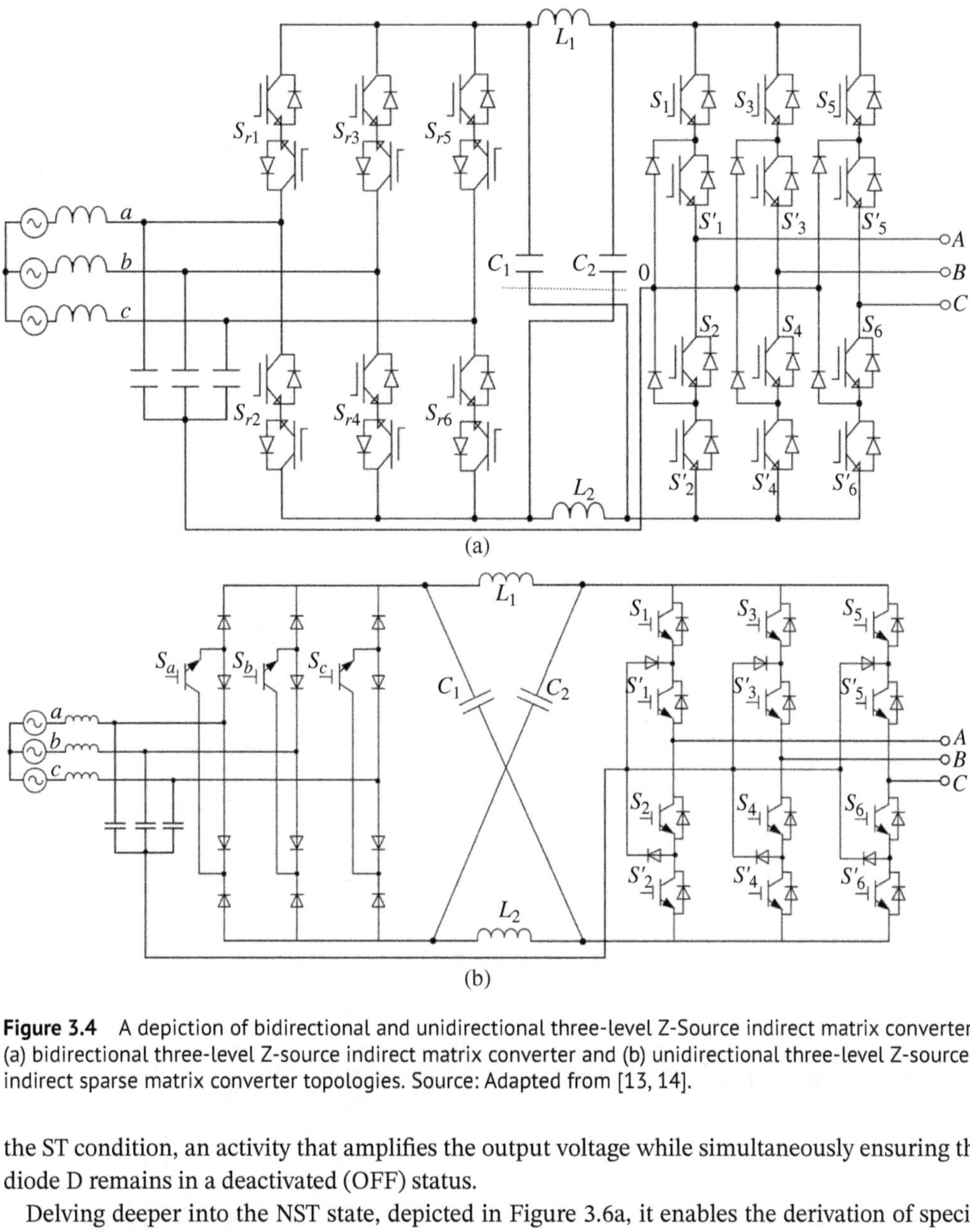

Figure 3.4 A depiction of bidirectional and unidirectional three-level Z-Source indirect matrix converters: (a) bidirectional three-level Z-source indirect matrix converter and (b) unidirectional three-level Z-source indirect sparse matrix converter topologies. Source: Adapted from [13, 14].

the ST condition, an activity that amplifies the output voltage while simultaneously ensuring that diode D remains in a deactivated (OFF) status.

Delving deeper into the NST state, depicted in Figure 3.6a, it enables the derivation of specific voltage and current equations as articulated in [16].

$$\begin{cases} C_1 \dfrac{du_{C1}}{dt} = i_{L1} - i_o, \quad C_2 \dfrac{du_{C2}}{dt} = i_{L2} - i_o \\ L_1 \dfrac{di_{L1}}{dt} = u_{in} - u_{C1}, \quad L_2 \dfrac{di_{L1}}{dt} = -u_{C2} \end{cases} \tag{3.1}$$

In this context, i_{L1}, i_{L2}, and i_o represent the currents traversing through two distinct inductors and the DC-link bus, respectively, the voltages enveloping the two capacitors are symbolized by u_{C1} and u_{C2}. Simultaneously, u_{in} signifies the voltage entering the Quasi-Z-Source (QZS) network. Crucially, u_{in} is identified as the virtual DC link voltage, which is subsequently generated by the

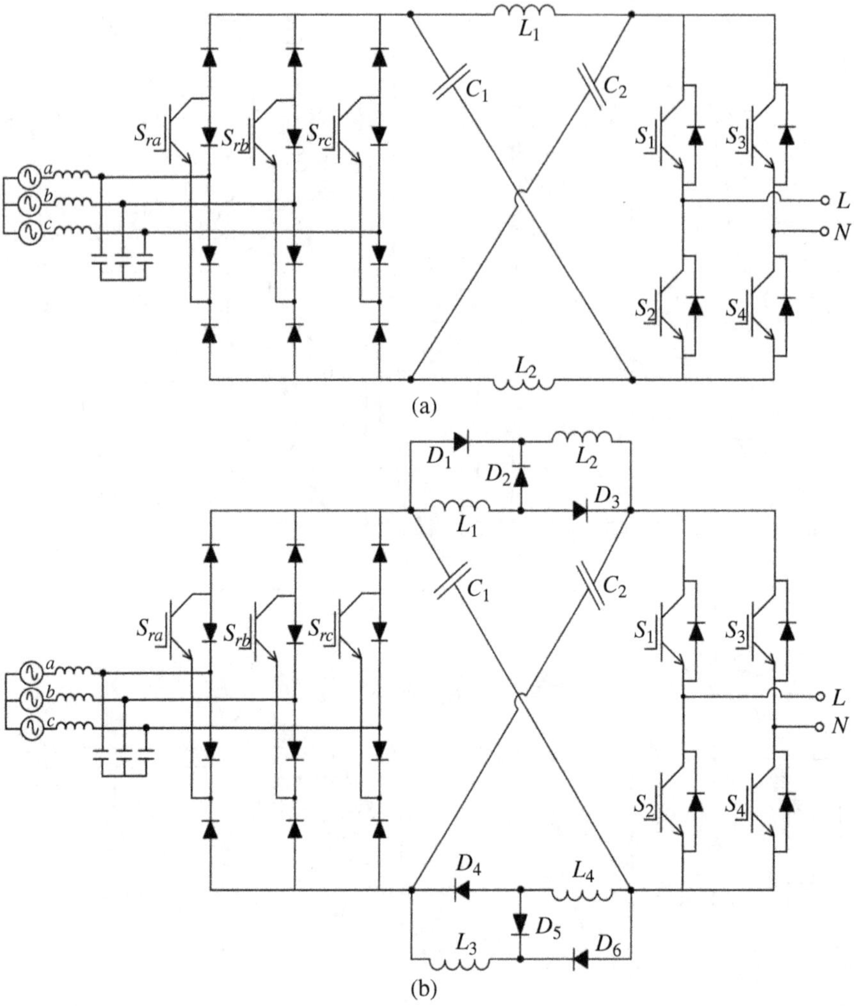

Figure 3.5 Three-phase to single-phase conversion through super-sparse topologies: (a) Z-source matrix converter (ZS-MC) and (b) switched-inductor Z-source matrix converter (SIZS-MC) designs, as delineated in [15]. Source: Karaman et.al. (2012)/with permission of IEEE.

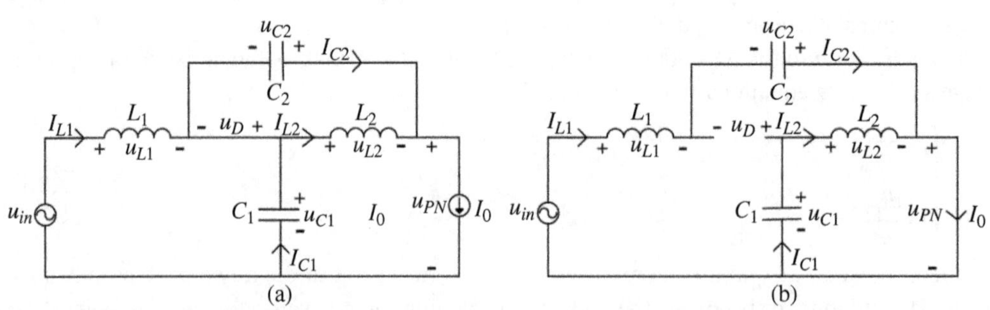

Figure 3.6 Operation modes of the quasi-Z-source matrix converter involving: (a) the non-shoot-through state and (b) the shoot-through state.

initial rectification state, serving a pivotal role in the functioning of the converter by establishing the required voltage levels.

Examining Figure 3.6b, which depicts the condition during the ST state, it enables the derivation of the subsequent equations as presented in source [16].

$$\begin{cases} C_1\dfrac{du_{C1}}{dt} = -i_{L2}, & C_2\dfrac{du_{C2}}{dt} = -i_{L1} \\ L_1\dfrac{di_{L1}}{dt} = u_{in} + u_{C2}, & L_2\dfrac{di_{L1}}{dt} = u_{C1} \end{cases} \tag{3.2}$$

Within a single switching cycle, designated as T_s, and recognizing that T_0 characterizes the duration of the ST state, the definition of the ST duty cycle, denoted as D, can be systematically established through the mathematical relationship $D = T_0/T_s$. A crucial observation here involves recognizing that during a condition of steady state, the average voltages across inductors and the average currents passing through the capacitors are expected to be zero throughout a full switching period. Thus, by employing equations (3.1) and (3.2), a window opens to determine the subsequent relationships, providing a mathematical foundation for further analyses and applications.

$$\begin{cases} u_{C1} = \dfrac{1-D}{1-2D}u_{in} \\ u_{C2} = \dfrac{D}{1-2D}u_{in} \end{cases} \tag{3.3}$$

The voltage output, emanating from the Quasi-Z-Source network, is articulated as follows:

$$u_{PN} = u_{C1} + u_{C2} = \frac{1}{1-2D}u_{in} \tag{3.4}$$

The boost factor of the voltage, denoted as B, is defined mathematically as:

$$B = \frac{u_{PN}}{u_{in}} = \frac{1}{1-2D} \tag{3.5}$$

The determination of the voltage gain, termed G, for the QZS-IMC illustrated in Figure 3.3b, is subject to the ensuing calculation:

$$G = Bm \tag{3.6}$$

where $m = m_i\, m_o$ represents the modulation index affiliated with the IMC, m_i signifies the modulation index pertaining to the preliminary rectifier, and m_o denotes the modulation index associated with the subsequent inverter.

3.2.3 Parameters Design of the QZS Network

The capacitance value within the QZS network is judiciously chosen, predicated upon the target voltage ripple and the current coursing through the capacitor. This selection adheres to the equations delineated in [16]:

$$u_L = L\frac{di_L}{dt}, \quad i_C = C\frac{du_C}{dt} \tag{3.7}$$

Deriving from equation (3.2), the subsequent formulation is obtained:

$$\Delta u_{C1} = \frac{i_{L2}DT_s}{C_1}, \quad \Delta u_{C2} = \frac{i_{L1}DT_s}{C_2}, \quad \Delta i_{L1} = \frac{(u_{in} + u_{C2})DT_s}{L_1}, \quad \Delta i_{L2} = \frac{u_{C1}DT_s}{L_2} \tag{3.8}$$

In this context, $\Delta u_C = r_C\, u_C$ and $\Delta i_L = r_L\, i_L$ signify the maximum oscillations in the voltage and current ripples, respectively. Furthermore, r_C represents the ratio of the capacitor voltage ripple, while r_L exemplifies the ratio pertinent to the inductor current ripple.

By meticulously applying equations (3.3) through (3.5), along with equation (3.8), a derived solution can be systematically obtained.

$$C_1 \geq \frac{PD(1-2D)T_s}{r_C(1-D)u_{in}^2}, \quad C_2 \geq \frac{P(1-2D)T_s}{r_C u_{in}^2}, \quad L_1 = L_2 \geq \frac{u_{in}^2(1-2D)DT_s}{P(1-D)r_L} \tag{3.9}$$

In this context, the variable P represents the power input into the system, reflecting the total energy supplied to it per unit of time.

3.3 Modulation Methods

A variety of modulation techniques for the QZS-IMC, which does not employ an all-silicon solution, have been diligently developed and studied. One notable instance involves an optimized modulation specifically tailored for a series Z-source very sparse matrix converter, thoroughly documented in [17]. Furthermore, the investigation into the space vector modulation (SVM) of Z-source ultrasparse matrix converter has been conducted, considering scenarios with and without the inclusion of a zero state, as detailed in [18]. Additionally, exploration into modulation methodologies aimed at reducing the common-mode voltage of Z-source/Quasi-Z-source ultrasparse matrix converters has been undertaken, as elaborately explored in [19–21].

In this context, the fundamental SVM technique, particularly for the QZS-IMC that does not utilize an all-silicon solution, is elucidated by adeptly integrating both the space vector rectification (SVR) and the space vector inversion (SVI) methodologies. Following the determination of the SVR duty cycles, namely $d_{\alpha 1}$, $2d_{\alpha 2}$, and $d_{\alpha 0}$, as well as the SVI duty cycles, specified as $d_{\beta 1}$, $d_{\beta 2}$, $d_{\beta 0}$, and d_{st}, the duty cycles relevant to the QZS-IMC are subsequently acquired through the method of cross-multiplying the duty cycles inherent to both the rectifier and inverter stages.

Let's conceptualize the following variables as representative of the three-phase input voltages:

$$\begin{aligned} v_a(t) &= V_{im}\cos(\theta_i) \\ v_b(t) &= V_{im}\cos\left(\theta_i - \frac{2\pi}{3}\right) \\ v_c(t) &= V_{im}\cos\left(\theta_i + \frac{2\pi}{3}\right) \end{aligned} \tag{3.10}$$

In this context, V_{im} signifies the amplitude of the input phase voltages, and θ_i denotes the angular displacement of the voltage in phase a. Figure 3.7 graphically illustrates both the normalized line-to-line voltages and the phase input voltages for enhanced visualization and comprehension.

Depicted in Figure 3.8 is an exhaustive representation of the spatial vector diagram, underlining the criticalities associated with SVR and SVI. Observations within this space vector framework highlight that vectors symbolizing the reference input current and output voltage demonstrate a rotational behavior. Maintaining an outlook that appreciates widespread relevance, there's an embedded hypothesis suggesting that both the reference output voltage and input current space vectors are deliberately aligned within the confines of sector 1, a positioning evident in Figure 3.8. In such a scenario, u_{in} finds a harmonious alignment with v_{ab}, v_{ac}, or in certain scenarios, 0 V. Concurrently, a scrutiny of the NST mode indicates that the DC current introduced into the inverter,

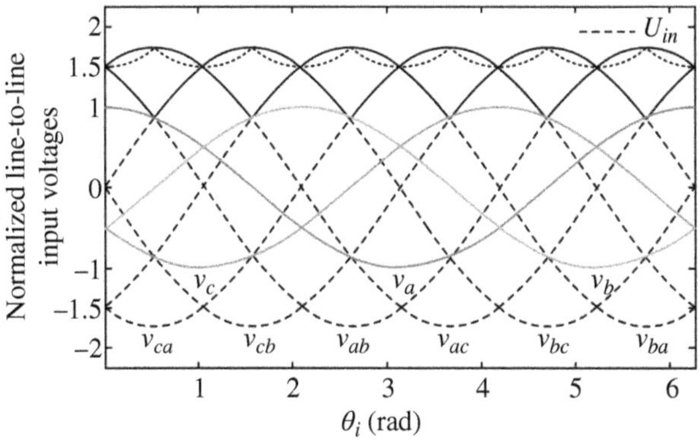

Figure 3.7 The normalization of line-to-line and phase input voltages coupled with the local average of rectifier output voltage (U_{in}) during instances when d_{st} and $d_{\alpha 0}$ assume a zero value [18]. Source: Bozorgi et.al. (2018)/with permission of IEEE.

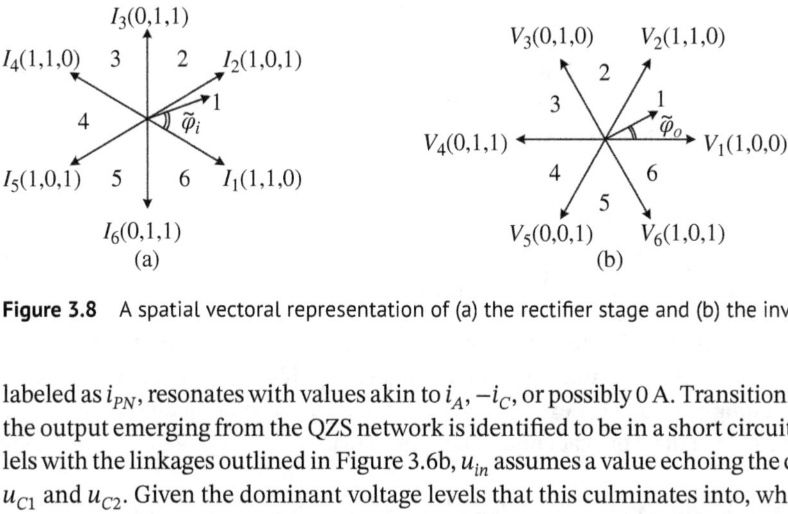

Figure 3.8 A spatial vectoral representation of (a) the rectifier stage and (b) the inverter stage.

labeled as i_{PN}, resonates with values akin to i_A, $-i_C$, or possibly 0 A. Transitioning to the ST segment, the output emerging from the QZS network is identified to be in a short circuit state. Drawing parallels with the linkages outlined in Figure 3.6b, u_{in} assumes a value echoing the combined strengths of u_{C1} and u_{C2}. Given the dominant voltage levels that this culminates into, when juxtaposed against the input voltages, the rectifier's diodes transition into a reverse bias mode. This action effectively delineates the QZS network from its primary input voltage catalyst. As a logical progression, during the active ST phase, i_{PN} harmonizes with a value twice that of i_L, which in this context signifies the current coursing through the QZS inductor.

The determination of duty cycles, pertaining to the switching states of the rectifier, can be accurately computed through a prescribed method, which involves

$$\begin{aligned} d_{\alpha 1} &= m_i \sin\left(\frac{\pi}{3} - \tilde{\varphi}_i\right) \\ d_{\alpha 2} &= m_i \sin(\tilde{\varphi}_i) \\ d_{\alpha 0} &= 1 - d_{\alpha 1} - d_{\alpha 2} \end{aligned} \tag{3.11}$$

In the equation, the parameter $0 \le m_i \le 1$ signifies the modulation index related to the rectifier stage, and this index plays a pivotal role in

$$\tilde{\varphi}_i = \varphi_i - (N_i - 1)\frac{\pi}{3} + \frac{\pi}{6} \tag{3.12}$$

In the discussed context, the symbol β_i distinctly represents the angular coordinate of the reference input current vector, whereas N_i concretely denotes the specific sector wherein the aforementioned current vector finds its position. For the explicit purpose of attaining an operation characterized by unity power factor, it is determined that β_i is aligned to be equal to θ_i.

The determination of duty cycles, pertinent to the inverter switching states, is achievable as illustrated in the subsequent methodology and mathematical representations:

$$
\begin{aligned}
d_{\beta 1} &= \frac{\sqrt{3}(1-d_{st})V_{om}}{U_{PN}} \sin\left(\frac{\pi}{3} - \tilde{\varphi}_o\right) \\
d_{\beta 2} &= \frac{\sqrt{3}(1-d_{st})V_{om}}{U_{PN}} \sin(\tilde{\varphi}_o) \\
d_{\beta 0} &= 1 - d_{\beta 1} - d_{\beta 2} - d_{st}
\end{aligned}
\tag{3.13}
$$

within this framework, $\tilde{\varphi}_o$ is discerned according to the guidelines provided below:

$$
\tilde{\varphi}_o = \varphi_o - (N_o - 1)\frac{\pi}{3}
\tag{3.14}
$$

The angle, symbolized as φ_o, related to the reference output voltage vector, and its associated sector, labeled as N_o, within which the former is situated, emerge as pivotal elements for consideration. It is imperative to underline that through the utilization of a cross-multiplication strategy, applied to the duty cycles inherent in both the rectifier and inverter stages—articulated comprehensively in expressions (3.8) and (3.13)—the duty cycles that are relevant to the QZS-IMC can be judiciously derived.

$$
\begin{cases}
d_{\alpha 1\beta 1} = d_{\alpha 1}d_{\beta 1} \\
d_{\alpha 1\beta 2} = d_{\alpha 1}d_{\beta 2} \\
d_{\alpha 1\beta 0} = d_{\alpha 1}d_{\beta 0}
\end{cases}
\quad
\begin{cases}
d_{\alpha 2\beta 1} = d_{\alpha 2}d_{\beta 1} \\
d_{\alpha 2\beta 2} = d_{\alpha 2}d_{\beta 2} \\
d_{\alpha 2\beta 0} = d_{\alpha 2}d_{\beta 0}
\end{cases}
\quad
\begin{cases}
d_{\alpha 1st} = d_{\alpha 1}d_{st} \\
d_{\alpha 2st} = d_{\alpha 2}d_{st} \\
d_{\alpha 0st} = d_{\alpha 0}d_{st}
\end{cases}
\tag{3.15}
$$

Expanding upon a specific assumption regarding the spatial orientation of the input current vector, it becomes possible to strategically extract the average voltage, signified as U_{in}, from the rectifier stage over a single switching period. This extraction is validated by the subsequent derivation, a topic which has found its place in prior academic discussions [18].

$$
U_{in}U_{in} = (d_{\alpha 1\beta 1} + d_{\alpha 1\beta 2} + d_{\alpha 1\beta 0})v_{ab} + (d_{\alpha 2\beta 1} + d_{\alpha 2\beta 2} + d_{\alpha 2\beta 0})v_{ac} + (d_{\alpha 1st} + d_{\alpha 2st} + d_{\alpha 0st})(u_{C1} + u_{C2})
\tag{3.16}
$$

Given that the sum of $d_{\beta 1}$, $d_{\beta 2}$, $d_{\beta 0}$, and d_{st} is equal to 1, it becomes feasible to recast equation (3.13) in an alternative form, elucidating the mathematical relationships in a differently structured manner.

$$
\begin{aligned}
U_{in} &= (1-d_{st})d_{\alpha 1}v_{ab} + (1-d_{st})d_{\alpha 2}v_{ac} + d_{st}(u_{C1} + u_{C2}) \\
&= (1-d_{st})\frac{3}{2}V_{im} + d_{st}(u_{C1} + u_{C2})
\end{aligned}
\tag{3.17}
$$

Leveraging the foundational principles of SVM theory, it is plausible to derive the maximal amplitude pertaining to the output voltage, articulated in the following manner:

$$
V_{om} = \frac{\sqrt{3}}{2}\frac{2U_{PN}}{3} = \frac{\sqrt{3}V_{im}(1-d_{st})}{2(1-2d_{st})}
\tag{3.18}
$$

In this context, V_{om} signifies the amplitude of the output voltage. Given that both U_{PN} and V_{om} maintain a constant nature, the utilization of this modulation technique facilitates the generation of sinusoidal output voltages. Consequently, the voltage transfer ratio, represented symbolically as G, can be extrapolated as outlined below:

$$G = \frac{V_{om}}{V_{im}} = \frac{\sqrt{3}(1 - d_{st})}{2(1 - 2d_{st})} \tag{3.19}$$

In the event that d_{st} is equal to zero within equation (3.19), the equation simplifies to $G = \sqrt{3}/2$, representing the apogee of the voltage transfer ratio pertinent to the QZS-IMC. Proceeding similarly as in previous discussions, it becomes feasible to ascertain the direct current (DC) present at the inverter stage's input as delineated in the ensuing steps and equations:

$$\begin{aligned} I_{PN} &= (d_{\alpha 1\beta 1} + d_{\alpha 2\beta 1} + d_{\alpha 0\beta 1})i_A - (d_{\alpha 1\beta 2} + d_{\alpha 2\beta 2} + d_{\alpha 0\beta 2})i_C + (d_{\alpha 1st} + d_{\alpha 2st} + d_{\alpha 0st})\cdot 2i_L \\ &= d_{\beta 1}i_A - d_{\beta 2}i_C + 2d_{st}i_L \end{aligned} \tag{3.20}$$

Utilizing the capacitor charge equilibrium, often referred to as the ampere-second balance rule, a plausible assumption surfaces: the average current coursing through the capacitor, denoted as I_C, is effectively zero. Concurrently, the assumption that i_L is equivalent to I_L is presented, wherein I_L symbolizes the average current traversing the QZS network inductors during a single switching period. In light of these assumptions, the subsequent equations can be deduced:

$$I_L = I_{in} = I_{PN} = \frac{\sqrt{3}I_o \cos(\theta_o)(1 - d_{st})}{2(1 - 2d_{st})} \tag{3.21}$$

$$I_{im} = m_i I_{in} = \frac{\sqrt{3}I_o \cos(\theta_o)(1 - d_{st})}{2(1 - 2d_{st})} \tag{3.22}$$

In a scenario where θ_o serves as a representation of the phase displacement occurring between the output current and voltage, several essential notations necessitate a thorough definition for clarity: the symbol I_{im} is utilized to signify the amplitude of the input currents, while I_{in} is employed to characterize the amplitude of the output current in the rectifier stage, and I_{PN} is utilized to represent the amplitude of the inverter input current. Upon inspecting the structural framework outlined by equation (3.22), it becomes apparent that the amplitude of the input currents maintains a steadfast constancy. An essential aspect that warrants specific emphasis involves the derivation of the equations displayed in this segment: they originate from a clearly articulated assumption which is that the modulation index, represented as m_i, in the rectifier stage is intentionally set equal to 1. This intentional designation is crafted with a strategic objective, namely, to optimize and thereby maximize the boosting capability intrinsic to the converter.

3.4 Simulation Results and Applications

3.4.1 Applications

The ZS-IMC emerges as a distinctive alternative, presenting a departure from the traditionally silicon-centric designs and finds its apt integration within a dynamic speed drive framework. This framework, leveraging the capabilities of a tri-phase induction motor (IM), is meticulously elucidated in Figure 3.9. At its core lies an innovative control mechanism, devised with dual aims: to expand the spectrum of operating voltage and to uphold a consistent unitary input power factor, factors that are imperative for the seamless operation of a motor drive grounded on the principles of the indirect matrix converter. Delving deeper into the control architecture, it unveils a structure that, while appearing intricate, systematically segments into two primary facets. Central to both

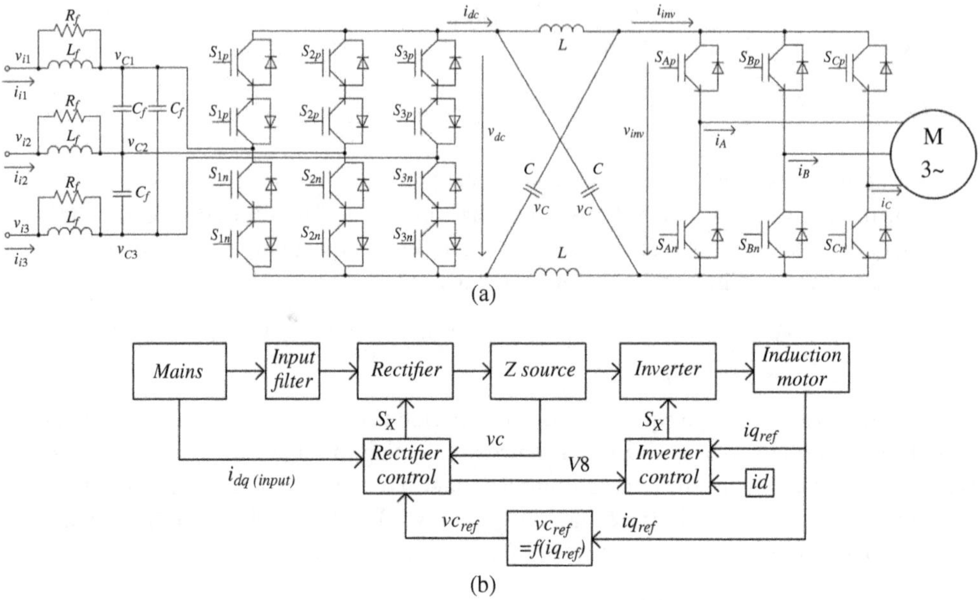

Figure 3.9 Z-source indirect matrix converter underpinning the IM Drive system: (a) exposition of system topology and (b) overview of control algorithm methodology [22]. Source: Sousa et.al. (2012) / with permission of IEEE.

segments is the pivotal role played by the DC link Z-source capacitor voltage. The preliminary rectification phase witnesses a detailed modulation of the current vectors, achieving a balance in phase and amplitude adjustments, and thus, gravitating toward an ideal unity power factor. Contrarily, as the system transitions to the inverter phase, aligned with the principles of closed-loop control for the AC currents emanating from the tri-phase induction motor, specific preordained operational methodologies are activated. Such activations are instrumental in crafting the essential stator voltage, ensuring a robust energy relay pathway from the foundational power supply to the intended load endpoint. The depth and effectiveness of this multi-faceted approach, along with its tangible outcomes, are elaborately chronicled in [22].

In the explicit realm of technological application, particularly within the framework of induction motor drive systems, the QZS-IMC emerges as a pivotal tool, facilitated by a method succinctly delineated as indirect field-oriented control, a detailed visual of which is depicted in Figure 3.10 [16]. Venturing into a distinctly different technological domain, particularly when delving into the intricacies of gearless wind energy systems, one observes a unique application that leans heavily on an indirect sparse-matrix topology, robustly complemented with inductive-capacitive-diode (LCD) networks. This application is insightfully illustrated in Figure 3.11 [23]. Initiation of the process encompasses the wind energy being adroitly harnessed by the turbine, subsequently being converted by the permanent magnet synchronous generator (PMSG), and eventually, being transmitted to the grid by means of a systematically structured matrix converter. A crucial aspect to highlight involves the rectification stage of this converter, which is tasked with preserving a meticulously regulated power factor by extracting sinusoidal currents from the PMSG through a well-orchestrated, controlled methodology. The tactical employment of the LCD network, in unison with the inversion stage, propels a deliberate elevation in the grid-side voltage. Additionally, it's vital to highlight the generator segment's commitment to a meticulously designed control approach, rooted deeply in the zero d-axis current control ethos. Concurrently, in the rectification phase, there's a focus on

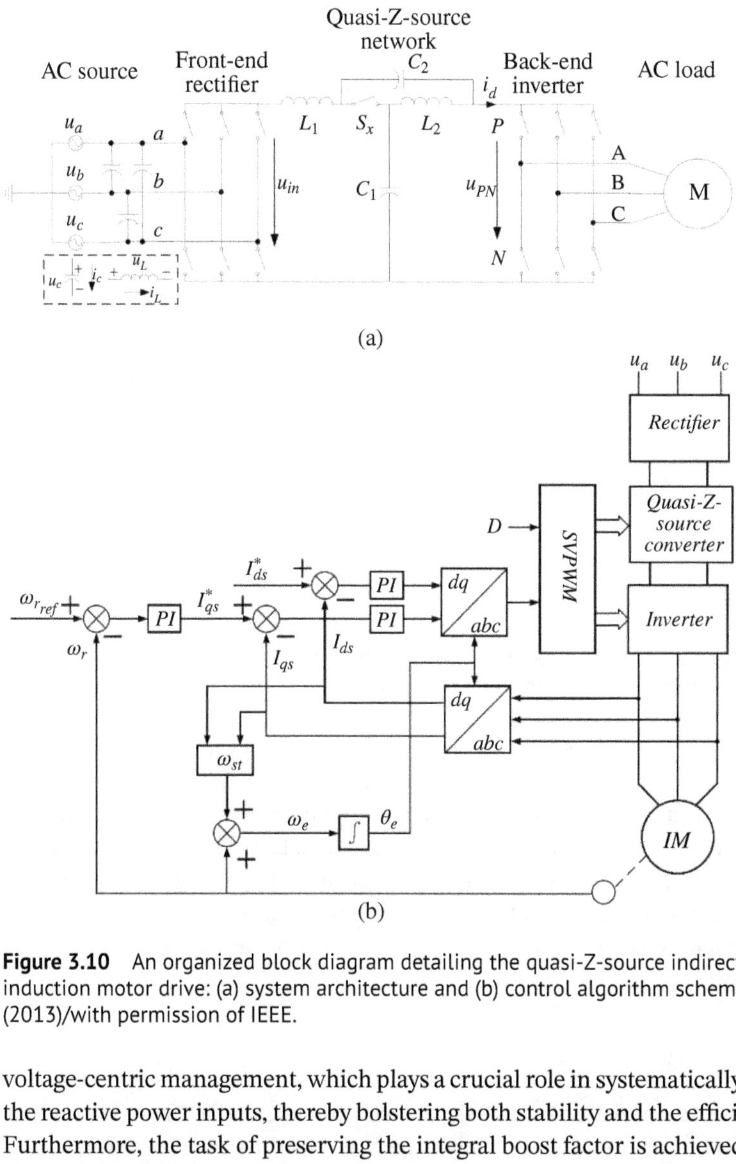

Figure 3.10 An organized block diagram detailing the quasi-Z-source indirect matrix converter-based induction motor drive: (a) system architecture and (b) control algorithm schematics [16]. Source: Liu et.al. (2013)/with permission of IEEE.

voltage-centric management, which plays a crucial role in systematically separating the active from the reactive power inputs, thereby bolstering both stability and the efficiency of power conveyance. Furthermore, the task of preserving the integral boost factor is achieved by carefully adjusting the shoot-through duty cycles present in the inverter segment, guaranteeing the system's performance at its peak potential.

In applications related to electrical motor systems, Particle Swarm Optimization (PSO) prominently stands out as an essential tool, fundamentally employed with the primary aim of elevating the quality of motor stator currents, specifically in instances where discernible distortions are witnessed in the input voltages [24]. An illustrative, graphical depiction of the ultra-sparse Z-source matrix converter (USZS-MC) engaged to actuate an induction motor can be meticulously examined in Figure 3.12a. Throughout the entirety of its operational phase, it is of paramount importance to recognize that the Total Harmonic Distortion (THD) embedded in the output current undergoes a rigid and consistent surveillance, ensuring optimal functioning of the system. This operational framework assures that the Proportional and Integral (PI) parameters, intrinsic to PI controllers, are aptly optimized within the D–Q scheme, a task that is proficiently realized through the adept

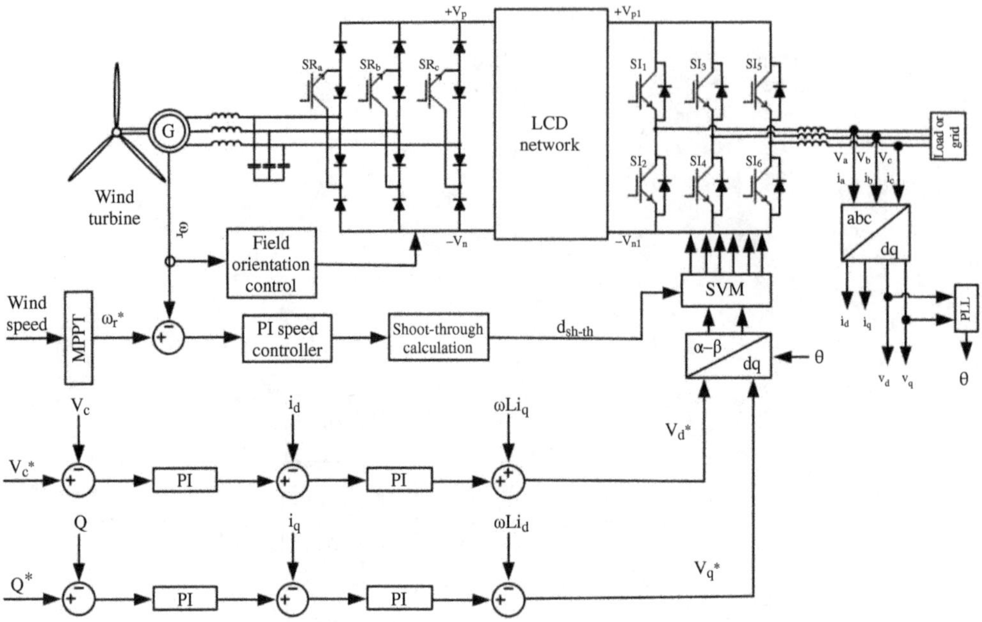

Figure 3.11 Systematic arrangement of wind-energy system encompassing PMSG, matrix converter, and LCD network [23]. Source: Karaman et.al. (2013) / with permission of IEEE.

application of the PSO method. Here, it is vital to underscore that this optimization activity is judiciously conducted while simultaneously maintaining a clear and unwavering understanding of the intricate interrelations existing among miscellaneous elements and the aforementioned PI parameters. Essentially, the PSO-PI control strategy meticulously and tactically uplifts the quality of the stator currents within the motor, most notably in scenarios characterized by unbalanced and visibly distorted input voltage, while concurrently and diligently attenuating the THD perceptible within the output current, thereby safeguarding the operational integrity and functionality of the motor under various operating conditions.

3.4.2 Simulation Results

Figure 3.10 visually presents the simulation outcomes pertinent to the QZS-IMC-based induction motor drive system, showcasing both its stable steady-state and adept dynamic performance. A detailed account of the system parameters can be referred to in Table 3.1, which meticulously delineates the relevant values and configurations that were utilized during the simulation process to evaluate the performance capabilities of the QZS-IMC.

Initiating from an entirely stationary state, the motor drive system begins its operation, methodically accelerating with the aim to achieve a specifically pre-established rotor speed, set at a precise 1500 rpm, all while no load is present. Once this initial phase is navigated, a meticulously calculated 25 Nm step alteration in the load torque is inserted, taking place following a pause of 1.5 s. In a subsequent phase, spanning from the 2 to 3 s interval, there is a strategic diminution of the targeted rotor speed, reducing it to a lower level of 900 rpm. An illustrative display of the resultant data from simulations, wherein the boosting operation is purposely excluded by allocating values $m = 1$ and $D = 0$, is portrayed in Figure 3.13. Within this illustration, the DC-link's peak voltage, accurately

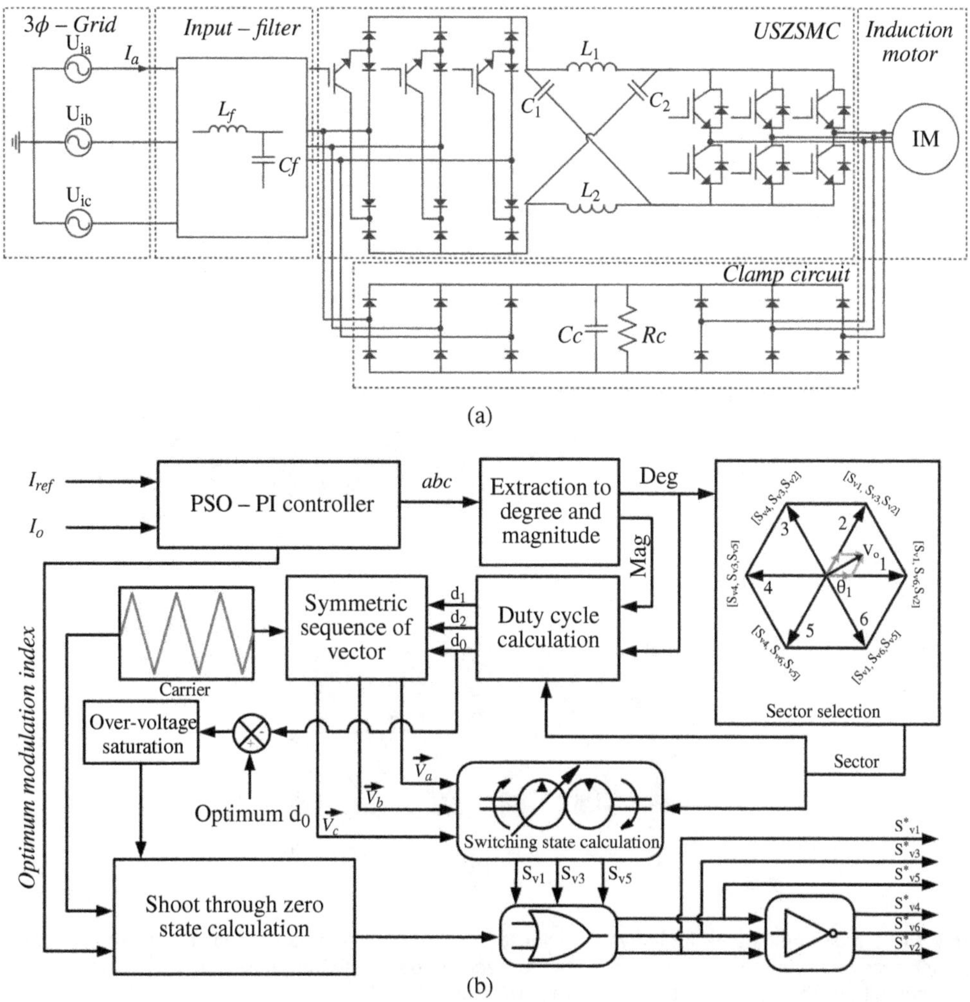

Figure 3.12 Methodology of particle swarm optimization for enhancing motor stator currents under conditions of distorted input voltages. (a) The USZS-MC topology driving an IM and (b) particle swarm optimization control strategy algorithm [24]. Source: Sebtahmadi et.al. (2018)/with permission of IEEE.

Table 3.1 System parameters [16].

Items	Value
Input AC source	380 V/50 Hz
Inductance L_1 and L_2	0.44 mH
Capacitance C_1 and C_2	100 μF
IM's rated power	4 kW
IM's rated speed	1430 rpm
IM's rated current	8 A

Source: Liu et.al. (2013)/with permission of IEEE.

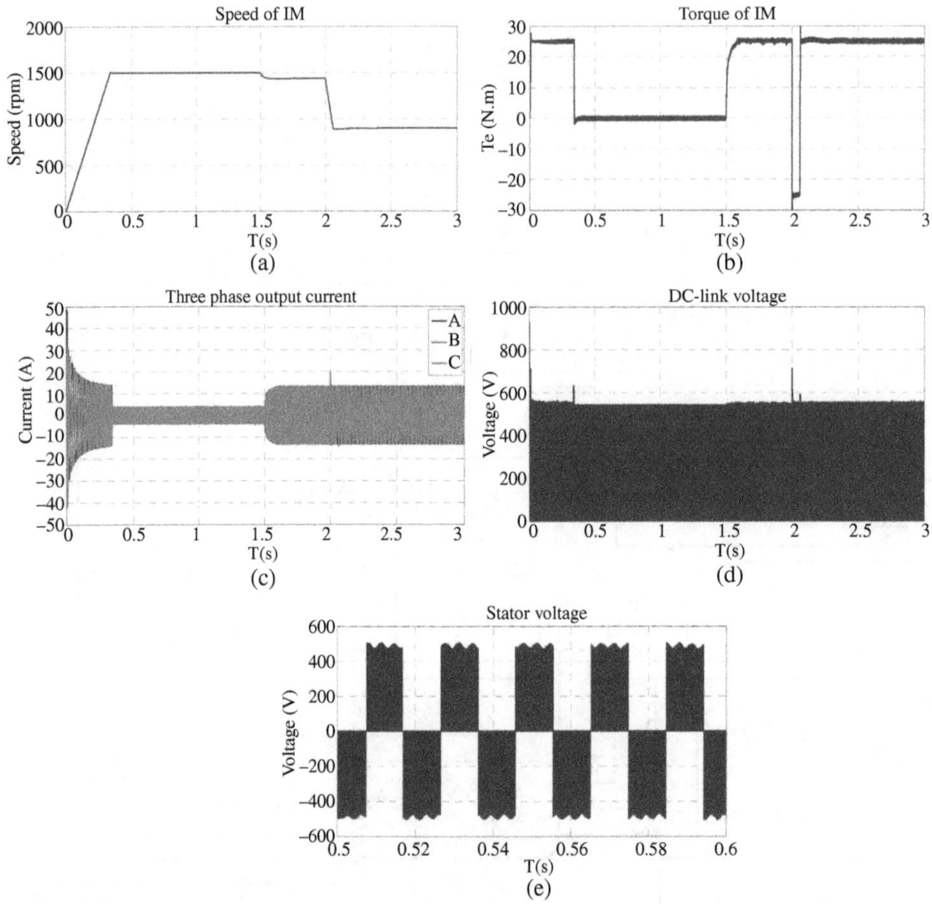

Figure 3.13 Simulation data excluding voltage boost ($m = 1$, $D = 0$): (a) motor speed; (b) electromagnetic torque; (c) three phase output currents; (d) DC-link voltage; (e) output single-phase voltage directed to stator winding [16]. Source: Liu et.al. (2013)/with permission of IEEE.

measured to be 540 V, is distinctly delineated in Figure 3.13d. When placed in situations where the QZS-IMC operates in a manner analogous to a standard IMC—notably absent of any shoot-through operations—its buck mode propels the motor drive to function below a speed of 1500 rpm when at the predetermined rated torque.

Utilizing the QZS-IMC's boost mode, and judiciously designating a value of 0.1 to D, a methodical strategy unfolds with an aim to counteract the voltage gain impediment, notably embedded within the conventional IMC. The intricate computation, facilitated by the application of equation (3.5), yields a subsequent voltage boost factor, herein denoted as B, and precisely quantified as 1.125. Figure 3.14, serving as a graphical reflection, elucidates the simulation results, tailoring them to this explicit case. Particularly, Figure 3.14d meticulously delineates the achievement of an output voltage measuring 604 V, a product elicited from the operations within the QZS network. Upon closer inspection, it is discerned that the QZS network harbors the capacity to amplify the DC-link voltage, thereby endowing the QZS-IMC, as proposed, with the proficiency to secure a voltage gain that transcends one and, in subsequent stages, rectifies the intrinsic limitation in voltage gain observed in the traditional IMC. In a resultant action, the heightened voltage adeptly drives the motor rotor

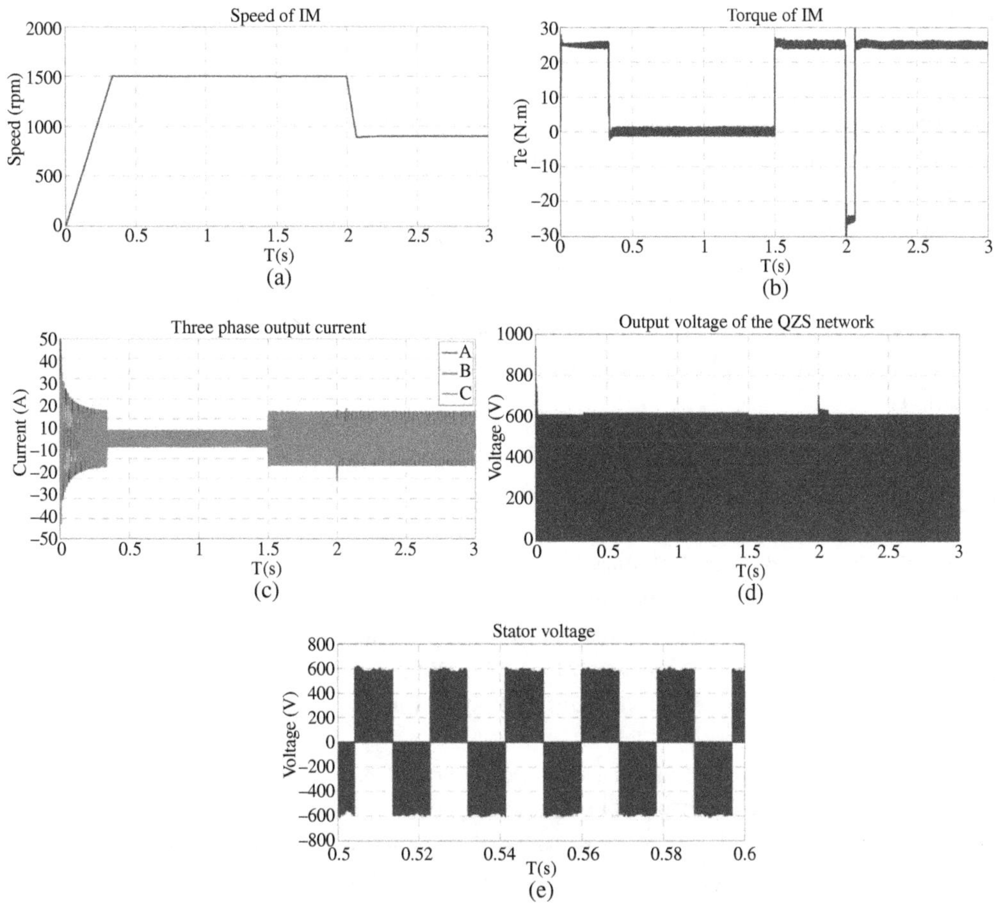

Figure 3.14 Resultant simulation data with an incorporated voltage boost ($m = 0.9, D = 0.1$): (a) speed of the IM; (b) electromagnetic torque; (c) three phase output currents; (d) DC-link voltage behavior; (e) single-phase voltage output to stator winding, cited from [16]. Source: Liu et.al. (2013)/with permission of IEEE.

speed to synchronize with the 1500 rpm benchmark, while simultaneously ensuring accommodation of the rated load torque.

3.5 Conclusion

This chapter meticulously unveils an all-encompassing overview, methodically sketching the topologies, operative principles, and modulation strategies, alongside a diverse spectrum of applications specifically related to the ZS-IMC, which is particularly distinguished by its abstention from utilizing an all-silicon solution. The investigative lens has been cast broadly over numerous Z-source and Quasi-Z-source indirect matrix converters. These explorations fundamentally pivot upon embedding the Z-source network squarely between the respective rectifier and inverter segments inherent to a conventional IMC. This astute fusion of QZS-IMC topologies strategically maneuvers around the voltage gain limitations synonymous with traditional IMCs. This tactful integration thereby orchestrates a conducive environment for facilitating both buck and boost operations, while simultaneously mitigating the need for an extensive switch array. This in turn substantiates claims of enhanced cost-effectiveness, elevated efficiency, and heightened reliability when juxtaposed with the back-to-back converter paradigm. An in-depth exploration of the rudimentary modulation methodology of ZS-IMC has been presented herein. Furthermore, numerous optimized modulation methodologies have been addressed, some of which are distinctly devised to mitigate common-mode voltage or attenuate inductor current ripple. The applications of ZS-IMC, including its ability to drive induction motors and its implementation within PMSG wind energy systems, have been meticulously developed and discussed. Simulation results, particularly those concerning a QZS-IMC driving an induction motor, have been illustrated and analyzed. Additionally, these advancements in technology and methodology pave the way for further exploration and subsequent adoption within additional IMC industrial applications.

References

1 P. W. Wheeler, J. Rodriguez, J. C. Clare, and L. Empringham, "Matrix converter: A technology review," *IEEE Trans. Ind. Electron.,* vol. 49, no. 2, pp. 276–288, Apr. 2002.

2 Y. D. Yoon and S. K. Sul, "Carrier-based modulation technique for matrix converter," *IEEE Trans. Power Electron.*, vol. 21, no. 6, pp. 1691–1703, Nov. 2006.

3 F. Z. Peng, "Z-source inverter", *IEEE Trans. Ind. Appl.*, vol. 39, no. 2, pp. 504–510, Mar./Apr. 2003.

4 J. Anderson and F.Z. Peng, "Four quasi-Z-source inverters", in *Proc. IEEE Power Electronics Specialists Conf. (PESC)*, pp. 2743–2749, Jun. 15–19, 2008.

5 S. Zhang, K. J. Tseng, and T. D. Nguyen, "Novel three-phase AC-AC Z-Source converters using matrix converter theory", in *Proc. IEEE Energy Conversion Congr. Expo. (ECCE)*, pp. 3063–3070, Sept. 20–24, 2009.

6 W. Song and Y. Zhong, "A study of Z-source matrix converter with high voltage transfer ratio", in *Proc. IEEE Vehicle Power and Propulsion Conf. (VPPC)*, pp. 1–6, Sept. 3–5, 2008.

7 K. Park and K. Lee, "A novel sparse matrix converter with a Z-source network", in *Proc. IEEE 35th Annu. Conf. Industrial Electronics (IECON)*, pp. 4487–4492, Nov. 3–5, 2009.

8 K. Park and K. Lee, "A Z-source sparse matrix converter under a voltage sag condition", in *Proc. IEEE Energy Conversion Congr. Expos. (ECCE)*, pp. 2893–2898, Sept. 12–16, 2010.

9 E. Karaman, M. Farasat, and A. M. Trzynadlowski,"A 3Φ-3Φ quasi Z-source matrix converter for residential wind energy systems", in *Proc. IEEE Energy Conversion Congr. Expo. (ECCE)*, pp. 240–246, Sept. 15–20, 2012.

10 E. Karaman, F. Niu, and A. M. Trzynadlowski, "Three-phase switched-inductor Z-source matrix converter", in *Proc. 27th Annu. IEEE Applied Power Electronics Conf. Expo. (APEC)*, pp. 1449–1454, Feb. 5–9, 2012.

11 E. Karaman, M. Farasat, and A. M. Trzynadlowski, "Indirect matrix converters as generator–grid interfaces for wind energy systems", *IEEE J. Emerg. Select. Top. Power Electron.*, vol. 2, no. 4, pp. 776–783, Dec. 2014.

12 E. Karaman, M. Farasat, and A.M. Trzynadlowski, "A comparative study of series and cascaded Z-source matrix converters", *IEEE Trans. Ind. Electron.*, vol. 61, no. 10, pp. 5164–5173, Oct. 2014.

13 F. B. Effah, A. J. Watson, P. W. Wheeler, J. C. Clare, and L. De-Lillo, "Space-vector-modulated three-level Z-source hybrid direct AC-AC power converter", in *Proc. 15th European Conf. Power Electronics and Applications (EPE)*, pp. 1–10, Sept. 2–6, 2013.
14 X. Liu, P. C. Loh, F. Z. Peng, P. Wang, and F. Gao, "Modulation of three-level Z-source indirect matrix converter", in *Proc. IEEE Energy Conversion Congr. Expo. (ECCE)*, pp. 3195–3201, Sept. 12–16, 2010.
15 E. Karaman, M. Farasat, F. Niu, and A. M. Trzynadlowski, "Three-phase to single-phase super-sparse matrix converters", in *Proc. 27th Annu. IEEE Applied Power Electronics Conf. Expo. (APEC)*, pp. 1061–1066, Feb. 5–9, 2012.
16 S. Liu, B. Ge, H. Abu-Rub, X. Jiang, and F.Z. Peng, "A novel indirect quasi-Z-source matrix converter applied to induction motor drives", in *Proc. IEEE Energy Conversion Cong. Expo. (ECCE)*, pp. 2440–2444, Sept. 15–19, 2013.
17 C. N. El-Khoury, H. Y. Kanaan, I. Mougharbel, and K. Al-Haddad, "Optimized modulation technique for series Z-source very sparse matrix converter," *IEEE 12th Int. Conf. Compatibility, Power Electronics and Power Engineering (CPE-POWERENG 2018)*, Doha, Qatar, 2018.
18 M. Bozorgi and M. Farasat, "Improved design and space vector modulation of a Z-source ultra-sparse matrix converter: Analysis, implementation, and performance evaluation," *IEEE Trans. Ind. Appl.*, vol. 54, no. 4, pp. 3737–3748, Jul.–Aug. 2018.
19 M. Bozorgi and M. Farasat, "An in-depth investigation of a Z-source ultrasparse matrix converter in buck and boost modes of operation," *IEEE Trans. Ind. Electron.*, vol. 65, no. 6, pp. 5177–5187, Jun. 2018.
20 M. Bozorgi, A. Hakemi, M. Farasat, and M. Monfared, "Modulation techniques for common-mode voltage reduction in the Z-source ultra sparse matrix converters," *IEEE Trans. Power Electron.*, vol. 34, no. 1, pp. 958–970, Jan. 2019.
21 V. Siva, M. Raghuram, A. Singh, S. K. Singh, and Y. P. Siwakoti, "Switching strategy to reduce inductor current ripple and common mode voltage in quasi Z-source ultra sparse matrix converter," *IEEE J. Emerg. Select. Top. Ind. Electron.*, vol. 4, no. 4, pp. 1159–1169, Oct. 2023.
22 S. Sousa, S. Pinto, F. Silva, and J. Maia, "Extended voltage range AC drive using a Z source indirect matrix converter", in *Proc. Int. Conf. Electrical Machines (ICEM)*, pp. 953–958, Sept. 2–5, 2012.
23 E. Karaman, M. Farasat, and A.M. Trzynadlowski, "Permanent-magnet synchronous-generator wind-energy systems with boost matrix converters," in *Proc. IEEE Energy Conversion Congr. Expo. (ECCE)*, pp. 2977–2981, Sept. 15–19, 2013.
24 S. S. Sebtahmadi, H. B. Azad, S. H. A. Kaboli, M. D. Islam, and S. Mekhilef, "A PSO-DQ current control scheme for performance enhancement of Z-source matrix converter to drive IM fed by abnormal voltage," *IEEE Trans. Power Electron.*, vol. 33, no. 2, pp. 1666–1681, Feb. 2018.
25 E. Karaman, M. Farasat, and A. M. Trzynadlowski, "Matrix converter with a series Z-source", in *Proc. 38th Annu. Conf. IEEE Industrial Electronics Society (IECON)*, pp. 6093–6098, Oct. 25–28, 2012.

4

Z-Source/Quasi-Z-Source Indirect Matrix Converter (All SiC Solution)

4.1 Introduction

In the chapter that precedes, we embarked on an insightful journey through the varied landscape of ZS-IMC and its meticulously crafted design strategies. These strategies, quintessentially, revolve around the strategic placement of the ZS/QZS network within the complex framework of the system. In instances where the ZS/QZS network is embedded within the DC-link, consequently forming a bridge between the rectifier and the inverter of the Indirect Matrix Converter (IMC), a configuration emerges that does not strictly adhere to a silicon-based structure. The embodiment of this specific configuration is intrinsically tied to the inclusion of robust inductors and capacitors within the DC-link bus, ushering in an elevation in both the physical size and weight of the overall system. In stark contrast, the interposition of the ZS/QZS network between the alternating current (AC) source and the rectifier enables the crafting of a system architecture that leans entirely on a silicon-based foundation. This ensuing chapter seeks to delve deeper, exploring further into the caverns of ZS-IMC, with a discerning focus predominantly oriented toward designs that leverage the utility of silicon as their foundational base.

4.2 Topologies and Operating Principle

4.2.1 Topologies

As depicted in Figure 4.1, the design intricacies of the ZS/QZS-IMC, emphasizing an all-silicon strategy as noted in references [1–3, 5], unfold. The system seamlessly integrates five cardinal elements. The initial component is a tri-phase AC source, vital for ensuring stable power input. Following this, the ZS/QZS network emerges, acting as the system's power regulation backbone. After power modulation, the current proceeds through an introductory rectifier, responsible for converting the incoming alternating current. The subsequent inverter then readies the power for final consumption, and the terminal component is the AC load, where the processed power is ultimately utilized. Delving deeper, the ZS configuration is marked by its trio of inductors labeled L_a, L_b, and L_c. Accompanying these are capacitors designated as C_a, C_b, and C_c, and a set of three bidirectional switches, namely S_a, S_b, and S_c. In contrast, the QZS setup is more elaborate. It houses double the number of inductors and capacitors—specifically, L_{a1}, L_{a2}, L_{b1}, L_{b2}, L_{c1}, L_{c2} for inductors and C_{a1}, C_{a2}, C_{b1}, C_{b2}, C_{c1}, C_{c2} for capacitors. Despite the increased components, it maintains the same three bidirectional switches (S_a, S_b, S_c). All these switches share a hallmark of consistent

Impedance Source Matrix Converters and Control, First Edition.
Yushan Liu, Xiao Li, Baoming Ge, Haitham Abu-Rub, and Frede Blaabjerg.

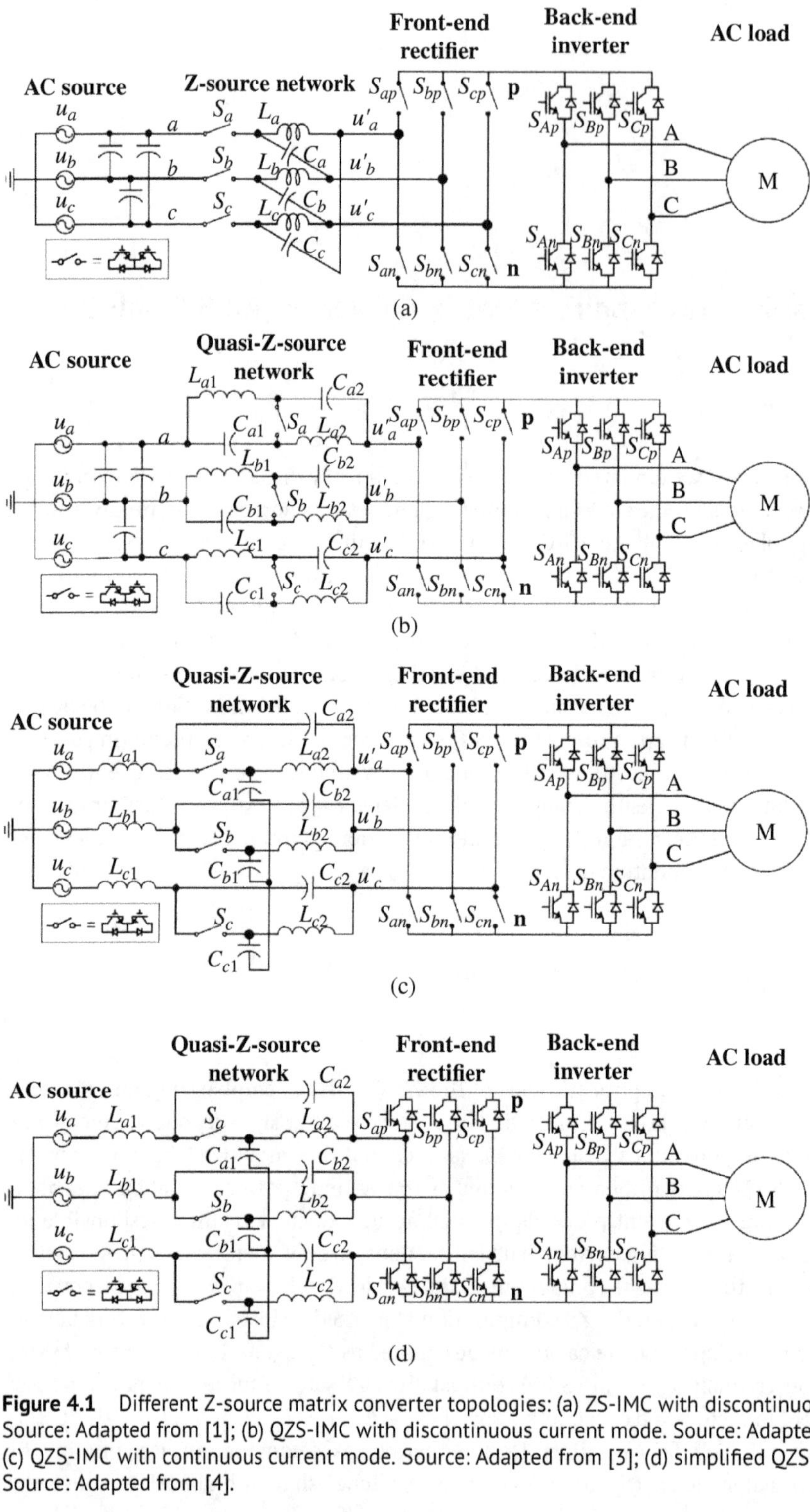

Figure 4.1 Different Z-source matrix converter topologies: (a) ZS-IMC with discontinuous current mode. Source: Adapted from [1]; (b) QZS-IMC with discontinuous current mode. Source: Adapted from [2]; (c) QZS-IMC with continuous current mode. Source: Adapted from [3]; (d) simplified QZS-IMC. Source: Adapted from [4].

switching characteristics, efficiently governed by a singular gate signal, aptly named S_x. The strategic and meticulous design of the ZS/QZS framework ensures that the ZS/QZS-IMC operates proficiently in both buck and boost mechanisms. One notable distinction, as per Figure 4.1a, is the ZS-IMC's leaner design, housing three inductors and capacitors, whereas the alternative QZS-IMC doubles this count. While the ZS-IMC topology, constrained by a modest voltage boost ratio of merely 1.15, manifests inherent phase shifts from the Z-network, yielding less precise control, and experiences discontinuous current at the Z-source network's forefront, necessitating supplementary input filters to mitigate input current Total Harmonic Distortion (THD), [3] it concurrently elevates the system's overall volume and expense, and attenuates efficiency. In Figure 4.1b, a nuanced portrayal of a variant, discontinuous quasi-Z-source Indirect Matrix Converter (QZS-IMC) topology is presented, further elaborated in the document [2]. This distinct topology amplifies its foundational setup by tactically integrating an auxiliary set of three inductors and capacitors, distinguishing it from the conventional ZS-IMC topology. This methodological integration enables the attainment of voltage gains, which are noticeably in the range of four to five times, and, under specific circumstances related to the switches' voltage rating, potentially even surpassing this range. Notably, this is achieved while meticulously avoiding the induction of phase shift from the quasi-Z (QZ) network. Despite its adept capabilities, it still perpetuates the need for the incorporation of additional input filters to ensure optimal functionality. Conversely, the continuous QZS-IMC topology, distinctly illustrated in Figure 4.1c and originally introduced in [3], adeptly obviates the requirement for input filters. It achieves this by judiciously integrating a continuous quasi-Z source network with the LC filter, presenting a cohesive and efficient unit. A thorough and scrupulous comparison of the three ZS/QZS-IMC topologies, vividly illustrated in Figure 4.1, has been systematically executed [4]. This encompasses probing into various critical facets, including voltage gain, the oscillations present in current and voltage, the stresses experienced by inductor current and capacitor voltage, the current and voltage stresses encountered by ZS/qZS switch, filtering capabilities, the THD in input and output voltage, and operational efficiency. A succinct and comprehensive evaluation, encapsulating the characteristic features and performance metrics of the triad of ZS/QZS-IMC topologies, is encapsulated meticulously in Table 4.1.

Table 4.1 Comparison of three ZS/QZS-IMC (all-silicon solution) AC/AC converter topologies [1].

	ZS-IMC	**Discontinuous QZS-IMC**	**Continuous QZS-IMC**
Additional input filter	Yes	Yes	No
Voltage gain	Low	High	High
Current ripples	High	Low	Low
Voltage ripples	Low	High	Middle
Inductor current stress	Low	High	Middle
Capacitor voltage stress	Middle	Low	High
Switch current stress	Low	High	High
Switch voltage stress	Low	Low	High
Input current THD	Middle	High	Low
Output voltage THD	High	Middle	Low
Efficiency	High	Low	High

Source: Adapted from Liu et.al. (2015)/IEEE.

In [5], a thorough examination of the common-mode voltage (CMV) challenge within the QZS-IMC framework was presented, especially when the system operates under a continuous current regime. This research led to the formulation of three innovative modulation techniques tailored for the QZS-IMC. The primary objective of these techniques was to address the CMV concerns, and the results were promising. A noteworthy reduction in the peak value of CMV was achieved, registering a decrease by a significant 42%.

As delineated in [4], Figure 4.1d visually unfolds a representation of the adeptly modified QZS-IMC, which emanates from the precise structural amendments implemented into the original topology, succinctly depicted in Figure 4.1c. This judiciously reconfigured setup necessitates a minimized assembly of power switches, precisely totaling 12, thereby facilitating a discernible decrement in the overarching system's fiscal expenditure. By way of contrast, the configuration encapsulated in Figure 4.1c mandates the inclusion of 18 switches, mimicking the architecture of the back-to-back pulse-width modulation converter, notably excluding the DC link capacitor. An additional meritorious advantage conferred by the QZS network incorporates the inclusion of the LC filtering facet, obviating the prerequisite for a distinct input filter. Nonetheless, an indispensable contemplation pertaining to this streamlined QZS-IMC resides in a circumscribed performance. Explicitly, the system's input current demonstrates a tendency toward exhibiting amplified harmonics, barring the incorporation of a strategy to adjust the input current in a closed-loop manner. Consequently, the actuation of input current closed-loop regulation emerges as quintessential to mitigating the perceptible harmonics in the input current.

4.2.2 Operating Principle

Figure 4.2 conveys a visual representation of the analogous circuit pertaining to the QZS-IMC topology, accentuating conditions wherein a continuous input current persists through both the Shoot-Through (ST) and Non-Shoot-Through (NST) operational states. Within the confines of the

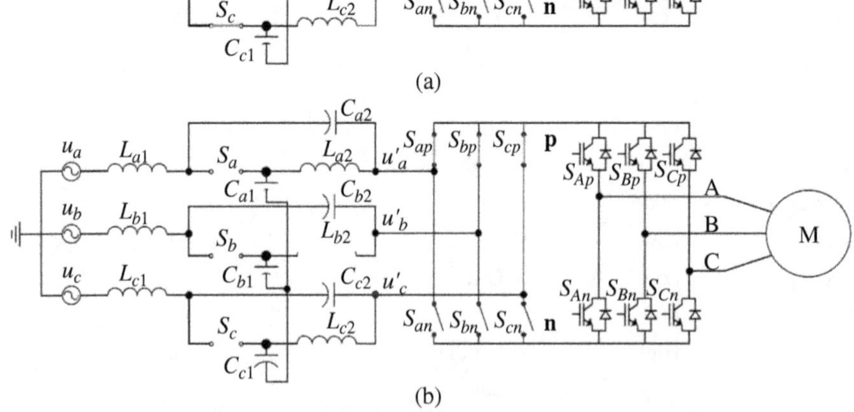

Figure 4.2 Equivalent circuit of the QZS-IMC; (a) non-shoot-through state and (b) shoot-through state.

NST state, vividly depicted in Figure 4.2a, activation of the switches, duly noted as S_x (here "x" substitutes for identifiers a, b, or c), is observable ($S = 1$). This proactive engagement facilitates a targeted discharge of the inductors into the capacitors, consequently orchestrating each output phase voltage of the QZS network to materialize as the amalgamated sum of two distinct capacitor voltages. Alternatively, Figure 4.2b succinctly outlines the ST state, during which the switches, S_x, are consciously disengaged ($S = 0$). This judicious disconnection precipitates the input portion of the front-end rectifier transitioning into a short-circuited state, thereby initiating the inductors' charging mechanism.

During the NST state, from Figure 4.2a, the following voltage and current equations can be obtained:

$$\begin{bmatrix} u_{Ca2} \\ u_{Cb2} \\ u_{Cc2} \end{bmatrix} = -\begin{bmatrix} u_{La2} \\ u_{Lb2} \\ u_{Lc2} \end{bmatrix} \tag{4.1}$$

$$\begin{bmatrix} u_{La2} \\ u_{Lb2} \\ u_{Lc2} \end{bmatrix} = \begin{bmatrix} u_a \\ u_b \\ u_c \end{bmatrix} - \begin{bmatrix} u_{Ca1} \\ u_{Cb1} \\ u_{Cc1} \end{bmatrix} \tag{4.2}$$

$$\begin{bmatrix} u_{La2} \\ u_{Lb2} \\ u_{Lc2} \end{bmatrix} = \begin{bmatrix} u_a \\ u_b \\ u_c \end{bmatrix} + \begin{bmatrix} u_{Ca2} \\ u_{Cb2} \\ u_{Cc2} \end{bmatrix} - \begin{bmatrix} u'_a \\ u'_b \\ u'_c \end{bmatrix} \tag{4.3}$$

$$\begin{bmatrix} i_{La1} \\ i_{Lb1} \\ i_{Lc1} \end{bmatrix} - \begin{bmatrix} i_{La2} \\ i_{Lb2} \\ i_{Lc2} \end{bmatrix} = \begin{bmatrix} i_{Ca1} \\ i_{Cb1} \\ i_{Cc1} \end{bmatrix} - \begin{bmatrix} i_{Ca2} \\ i_{Cb2} \\ i_{Cc2} \end{bmatrix} \tag{4.4}$$

Upon examination of the ST state, as depicted in Figure 4.2b, one can derive the subsequent voltage and current equations:

$$\begin{bmatrix} u_a \\ u_b \\ u_c \end{bmatrix} + \begin{bmatrix} u_{Ca2} \\ u_{Cb2} \\ u_{Cc2} \end{bmatrix} = \begin{bmatrix} u_{La1} \\ u_{Lb1} \\ u_{Lc1} \end{bmatrix} \tag{4.5}$$

$$\begin{bmatrix} u_{Ca1} \\ u_{Cb1} \\ u_{Cc1} \end{bmatrix} = \begin{bmatrix} u_{La2} \\ u_{Lb2} \\ u_{Lc2} \end{bmatrix} \tag{4.6}$$

$$\begin{bmatrix} i_{La1} \\ i_{Lb1} \\ i_{Lc1} \end{bmatrix} = -\begin{bmatrix} i_{Ca2} \\ i_{Cb2} \\ i_{Cc2} \end{bmatrix}, \quad \begin{bmatrix} i_{La2} \\ i_{Lb2} \\ i_{Lc2} \end{bmatrix} = -\begin{bmatrix} i_{Ca1} \\ i_{Cb1} \\ i_{Cc1} \end{bmatrix} \tag{4.7}$$

In this context, "u" represents voltage and "i" symbolizes current, where the subscripts 1 and 2 pertain to capacitors 1 and 2 of phase x, respectively. Likewise, the subscripts 1 and 2 correspond to inductors 1 and 2 of phase x, where x equates to a, b, or c.

Over the duration of a single switching cycle, denoted as T_s, and assuming the time interval during the Shoot-Through (ST) state is symbolized as T_0, the ST duty cycle can consequently be defined by the relation $D = T_0/T_s$. The average voltages across the inductors and the mean currents through the capacitors ought to be nullified over one switching period under stable conditions. By employing equations (4.1) through (4.7), derivations can be attained as follows:

$$\begin{bmatrix} i_{La1} \\ i_{Lb1} \\ i_{Lc1} \end{bmatrix} = \begin{bmatrix} i_{La2} \\ i_{Lb2} \\ i_{Lc2} \end{bmatrix} = \begin{bmatrix} i_a \\ i_b \\ i_c \end{bmatrix} \tag{4.8}$$

$$\begin{bmatrix} u_{Ca2} \\ u_{Cb2} \\ u_{Cc2} \end{bmatrix} = \frac{D}{1-D} \begin{bmatrix} u_{Ca1} \\ u_{Cb1} \\ u_{Cc1} \end{bmatrix} \tag{4.9}$$

$$\begin{bmatrix} u_{Ca1} \\ u_{Cb1} \\ u_{Cc1} \end{bmatrix} = \frac{1-D}{1-2D} \begin{bmatrix} u_a \\ u_b \\ u_c \end{bmatrix} \tag{4.10}$$

Upon amalgamating equations (4.3) and (4.10), resultant equations can subsequently be derived as follows:

$$\begin{bmatrix} u'_a \\ u'_b \\ u'_c \end{bmatrix} = \frac{1-D}{1-2D} \begin{bmatrix} u_a \\ u_b \\ u_c \end{bmatrix} \tag{4.11}$$

One can articulate the factor, denoted as B, which serves to augment the voltage, in the manner illustrated below:

$$B = \frac{u_o}{u_i} = \frac{1}{1-2D} \tag{4.12}$$

In the field of electronic systems, where the parameter u_i serves as an indicator for the magnitude of our input voltage source and, concurrently, u_o stands as a measurement of the amplitude of the output voltage originating from the QZS network, a nuanced clarification becomes imperative. Specifically, the voltage gain G, which is integrally tied to the design and functioning of the QZS-IMC under consideration, has a definable mathematical relationship. This relationship, when analyzed and broken down, can be expressed in a structured format, which will be presented as follows:

$$G = Bm \tag{4.13}$$

In the instance where $m = m_i\, m_o$ denotes the modulation index pertinent to the IMC, m_i indicates the modulation index intrinsic to the front-end rectifier, and m_o symbolizes that corresponding to the back-end inverter, Figure 4.3 serves to elucidate the linkage between the QZS-IMC's voltage gain and modulation index. A discernible characteristic from the illustration is the capability of the voltage gain to surpass 1, under the premise that both the modulation index and the ST duty cycle are astutely chosen, with the latter (symbolized as D) being under 0.5, heralding the activation of the boost mode. Conversely, by opting for a lower modulation index, realization of the buck mode becomes achievable.

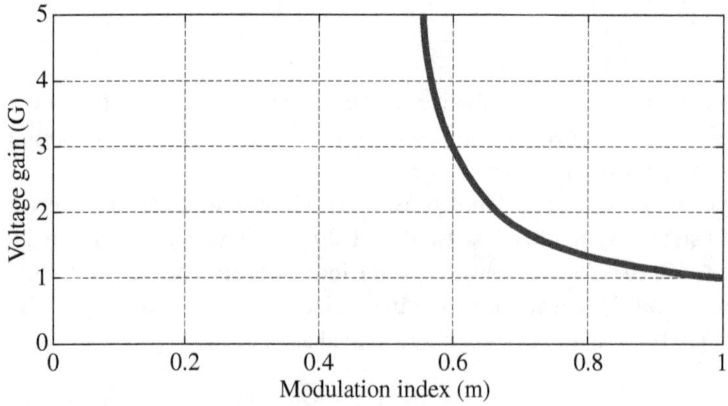

Figure 4.3 Voltage gain versus modulation index of the QZS-IMC [3]. Source: Adapted from Liu et.al. (2013)/Emerald Publishing Limited.

4.2.3 Parameters Design of the QZS-Network

Within the QZS, the inductor is allocated the essential duty of limiting the high-frequency current ripple, and concurrently, the capacitor is engineered to temper the high-frequency voltage ripple. Individually, they each substantiate a cardinal role in fortifying the stability of their corresponding electrical parameters.

Figure 4.2a reveals that the voltage across inductor L_{x1}, where "x" denotes a, b, or c, corresponds to the discrepancy between the input voltage and that across capacitor C_{x1}. In a parallel manner, the voltage passing through inductor L_{x2} maintains equivalency with the voltage spanning across capacitor C_{x2}. Such relational dynamics are applicable in a non-shoot-through state, a conclusion drawn from reference [5].

$$\begin{cases} L_{x1} = \dfrac{|u_i - u_{Cx1}|(1-D)T_s}{\Delta i_i} \\ L_{x2} = \dfrac{|u_{Cx2}|(1-D)T_s}{\Delta i_{L_{x2}}} \end{cases} \tag{4.14}$$

Given that the variables $\Delta i_i = r_L\ I_s$ and $\Delta i_{Lx2} = r_L\ I_{Lx2}$ hold particular values, the ensuing equations can be logically derived as follows:

$$\begin{cases} L_{x1} \geq \dfrac{|u_{i_{\max}} - u_{C_{x1\max}}|(1-D)T_s}{r_L I_s} = \dfrac{\sqrt{2}V_s D(1-D)T_s}{(1-2D)r_L I_s} \\ L_{x2} \geq \dfrac{|u_{C_{x2\max}}|(1-D)T_s}{r_L I_{L_{x2}}} = \dfrac{\sqrt{2}V_s D(1-D)T_s}{(1-2D)r_L I_{L_{x2}}} \end{cases} \tag{4.15}$$

In this context, I_s, I_{Lx2}, and V_s signify the root mean square (RMS) values of the input current and the inductor L current, and the input voltage, respectively. Notably, r_L is utilized as the designated ripple coefficient specifically for inductor currents.

During the Shoot-Through (ST) state, illustrated in Figure 4.2b, the current traversing the capacitor C_{x1} is equivalent to that of the inductor L_{x2}, and similarly, the current through the capacitor equates to the input current. These relationships allow certain derivations to be obtained, as indicated in [5].

$$\begin{cases} C_{x1} = \dfrac{i_{L_{x2}} DT_s}{\Delta u_{C_{x1}}} \\ C_{x2} = \dfrac{i_i DT_s}{\Delta u_{C_{x2}}} \end{cases} \tag{4.16}$$

Given that $\Delta u_{Cx1} = r_C\ u_{Cx1}$ and $\Delta u_{Cx2} = r_C\ u_{Cx2}$ are identified, it becomes feasible to derive the subsequent equations:

$$\begin{cases} C_{x1} \geq \dfrac{i_{L_{x2\max}} DT_s}{\Delta u_{C_{x1}}} = \dfrac{\sqrt{2}I_s D(1-2D)T_s}{(1-D)r_C V_s} \\ C_{x2} \geq \dfrac{i_{i_{\max}} DT_s}{\Delta u_{C_{x2}}} = \dfrac{\sqrt{2}I_s(1-2D)T_s}{r_C V_s} \end{cases} \tag{4.17}$$

In this context, u_{Cx1} and u_{Cx2} represent the RMS values of the voltages across capacitor C_{x1} and capacitor C_{x2}, respectively. The pre-determined ripple coefficient for the voltages of the capacitors is denoted as r_C, while I_s signifies the RMS values pertinent to the input current.

4.3 Modulation Methods

4.3.1 Conventional Space Vector Modulation Method

At the onset of this chapter, there's an in-depth exploration of the conventional dual SVM method. This methodology, which thoughtfully integrates zero vectors, has been effectively implemented in the rectifier phase of the QZS-IMC system, as indicated by references [3, 5]. Originating from the foundational principles of the dual pulse-width modulation (PWM) modulation framework, dual SVM is characterized by its indirect modulation strategy. A closer look reveals that its operational logic is mirrored from the structured AC-DC-AC conversion process. This said process encompasses two key phases: a virtual rectifier in the initial phase, followed by a virtual inverter in the succeeding phase. These phases collectively reflect the mechanics of the standard matrix converter setup.

Analyzing from a structural or topological standpoint, it becomes evident that the design of the indirect matrix converter is distinctly partitioned. Specifically, it divides into two primary sections, namely the rectification phase and the inversion phase. Such an inherent division suggests that any modulation strategy tailored for the indirect matrix converter should inherently focus on these two separate segments, each catering to the unique operational attributes of the rectifier and inverter, respectively. Of the various techniques available, the dual SVM emerges as a front-runner, offering precision and efficiency, making it especially fitting for the indirect matrix converter's requirements.

4.3.1.1 Rectifier-stage SVM

Employing the input current space vector as its foundation, the space vector modulation within the rectifier-stage strategically targets the realization of a unity power factor on the input flank.

i) Interval division

Presuming that the voltage source originates from a three-phase input phase:

$$\begin{aligned} u_a &= U_{im}\cos(\omega_i t) \\ u_b &= U_{im}\cos(\omega_i t - 2\pi/3) \\ u_c &= U_{im}\cos(\omega_i t + 2\pi/3) \end{aligned} \tag{4.18}$$

In the electrical system being discussed, U_{im} accurately symbolizes the amplitude of the input phase voltage spanning across all three distinct phases, while its angular frequency finds representation through ω_i. A crucial endeavor of the modulation strategy applied to the bidirectional switch rectifier stage involves not merely producing a discernibly positive DC voltage on the DC side but also scrupulously maximizing the rate at which the voltage is utilized, ensuring an optimal and efficient operational paradigm. Through a methodical analysis, the cycle of the input phase voltage is carefully partitioned into six well-defined intervals. Within this categorization, the meticulous division and apt selection of the input voltage interval, in tandem with the effective current vector, are invariably influenced by the zero-crossing point of the input phase voltage, a relationship that finds its visual representation in Figure 4.4. Importantly, each of these identified intervals distinctively features one-phase voltage that sustains the highest absolute value, while the polarities of the ancillary two-phase voltages are meticulously structured to be antithetical to it, ensuring a balanced and stable electrical environment throughout the system's operation.

ii) Space vector diagram

During the operation of the rectifier stage, a complex interplay emerges from six combinations of bidirectional switches, culminating in the generation of three uniquely characterized types of

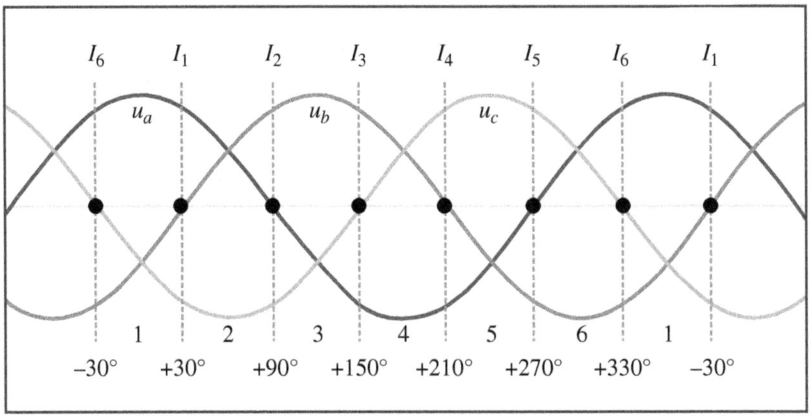

Figure 4.4 Rectifier-stage sector division diagram.

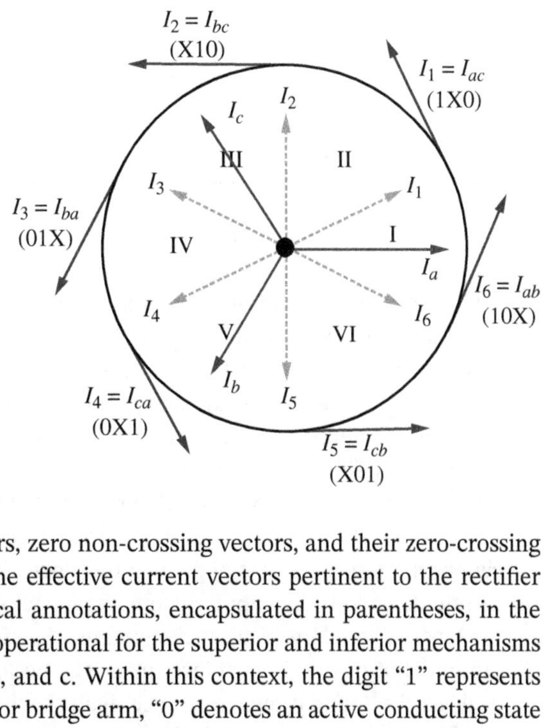

Figure 4.5 Space vector diagram of rectifier-stage current.

space vectors: the effective current vectors, zero non-crossing vectors, and their zero-crossing counterparts. A visual demarcation of the effective current vectors pertinent to the rectifier stage is provided in Figure 4.5. Numerical annotations, encapsulated in parentheses, in the depiction symbolize the switching state operational for the superior and inferior mechanisms located within input bridges labeled a, b, and c. Within this context, the digit "1" represents conduction occurring through the superior bridge arm, "0" denotes an active conducting state through the inferior bridge arm, and "X" communicates a state where both the upper and lower bridge arms are non-conducting. For example, the annotation "X10" illustrates a scenario where non-conduction prevails through both bridge arms of phase a, while conduction is facilitated through the superior bridge arm of phase b and through the inferior bridge arm of phase c. It is perceptible that each pairing of superior and inferior bridge arms enables simultaneous conduction through two disparate phase bridge arms, thus signifying that a total of two bridges, not localized within an identical phase, are actively conducting.

iii) Space vector table

During operations where the current vector classifies as an active or non-zero crossing entity, the S_x switch, located in the quasi-Z source structure, consistently remains in its conductive

Table 4.2 Space vector table for rectifier-stage current.

N	U_{dc}	S_{ap}, S_{bp}, S_{cp}	S_{an}, S_{bn}, S_{cn}	S_x	Current vector	Description
1	u_{ac}	1, 0, 0	0, 0, 1	1	I_1	Effective current vector
2	u_{bc}	0, 1, 0	0, 0, 1	1	I_2	Effective current vector
3	$-u_{ab}$	0, 1, 0	1, 0, 0	1	I_3	Effective current vector
4	$-u_{ac}$	0, 0, 1	1, 0, 0	1	I_4	Effective current vector
5	$-u_{bc}$	0, 0, 1	0, 1, 0	1	I_5	Effective current vector
6	u_{ab}	1, 0, 0	0, 1, 0	1	I_6	Effective current vector
7	0	1, 0, 0	1, 0, 0	1	I_7	Non-ST zero vector
8	0	0, 1, 0	0, 1, 0	1	I_8	Non-ST zero vector
9	0	0, 0, 1	0, 0, 1	1	I_9	Non-ST zero vector
		1, 1, 1	1, 0, 0	0	I_{10}	
10	0	1, 1, 1	0, 1, 0	0	I_{11}	Shoot-through zero vector
		1, 1, 1	0, 0, 1	0	I_{12}	
		1, 0, 0	1, 1, 1	0	I_{13}	
11	0	0, 1, 0	1, 1, 1	0	I_{14}	Shoot-through zero vector
		0, 0, 1	1, 1, 1	0	I_{15}	

phase. On the flip side, in instances where every upper bridge arm (or its counterpart, the lower bridge arms) in the rectifier's tri-phase stage actively conducts simultaneously, there's a shift in the status of the S_x switch within the quasi-Z source layout. Specifically, it transitions to an inactive or off mode, correlating with the moment the current vector shifts to a zero-crossing nature. For a more detailed understanding, Table 4.2 systematically showcases a complete array of potential switch state permutations.

iv) Vector synthesis and duty cycle

Presented in Figure 4.6 is a detailed visual representation that thoughtfully showcases the array of current vectors directly associated with the operations of the rectifier stage. Carefully analyzed in this depiction, the input current vector, which is denoted by the label I_{ref}, undergoes a specific counterclockwise motion. This motion is of pivotal importance, primarily because it establishes itself as the foundational reference against which other vectors can be compared. As we delve deeper, it becomes apparent that the comprehensive vector, termed I_{ref}, is the product of a meticulous combination of several vectors. These include two adjacent effective space vectors, aptly labeled I_α and I_β. Additionally, we recognize the presence of a distinct non-zero crossing vector named I_{0r}, and yet another vector recognized as a zero-crossing entity, I_{st}. Each of these vectors, it's worth noting, holds a unique position, making its mark within a designated sector of the operational space. To facilitate a more structured understanding, it becomes essential to introduce certain duty ratios. These ratios, symbolized as d_α, d_β, d_{0r}, and d_{st}, have been tailored to match the characteristics of I_α, I_β, I_{0r}, and I_{st} vectors, respectively. By analyzing the visual data presented in the figure and correlating it with the underlying principles, one can systematically derive expressions that

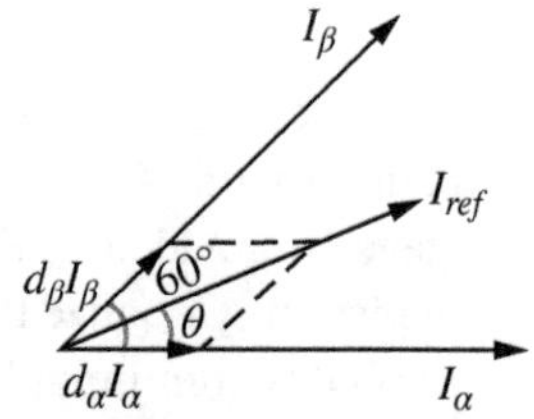

Figure 4.6 Vector synthesis diagram of rectifier-stage current.

detail the dynamics of these duty ratios:

$$\begin{cases} d_\alpha = m_i \sin(60° - \theta_i) \\ d_\beta = m_i \sin \theta_i \\ d_{0r} = 1 - d_\alpha - d_\beta - d_{st} \\ d_{st} = D = const \end{cases} \tag{4.19}$$

Ensuring that $0° \leq \theta_j \leq 60°$, which signifies the angle lying between the reference vector and its adjacent effective vector within each respective sector, $0 \leq m_i = \frac{I_{im}}{I_{dc}} \leq 1$ exemplifies the modulation degree pertaining to the rectifier stage's current vector. Moreover, I_{im} represents the amplitude of the current within the three-phase input phase, while I_{dc} illustrates the amplitude of the direct current circulating through the DC bus.

4.3.1.2 Inverter-stage SVM

The inverter stage's space vector modulation bears resemblance to traditional inverters. In this setup, the objective is to leverage the output voltage space vector for effective control over the voltage on the output side.

i) Segmentation of intervals

Let's postulate that the voltage sources for the three-phase output phase are represented as:

$$\begin{aligned} u_A &= U_{om} \cos(\omega_o t) \\ u_B &= U_{om} \cos(\omega_o t - 2\pi/3) \\ u_C &= U_{om} \cos(\omega_o t + 2\pi/3) \end{aligned} \tag{4.20}$$

Here, U_{om} stands for the magnitude associated with the three-phase output phase voltage, while ω_o signifies its angular frequency.

The cycle of the output line voltage can be segregated into six distinct intervals. Based on the points where the output line voltage crosses zero, these intervals are determined and the appropriate effective voltage vector is chosen. This segmentation and vector selection can be visualized in Figure 4.7.

ii) Space vector diagram

Within the inverter stage, which embodies a total ensemble of six switches, its architectural design finds an analogous counterpart in traditional inverter systems, adeptly synthesizing six impactful space vectors in addition to two zero vectors, all intrinsically intertwined with the output line voltage. Figure 4.8 provides a visual elucidation of the effective voltage vectors that are inherently associated with the output voltage. Numbers, encapsulated within parentheses and annotated within the graphical representation, elucidate the operational statuses of the upper and lower switches spread across the output phases, termed as a, b, and c, respectively. In this structural context, the notation "1" symbolizes the activation of the upper bridge arm,

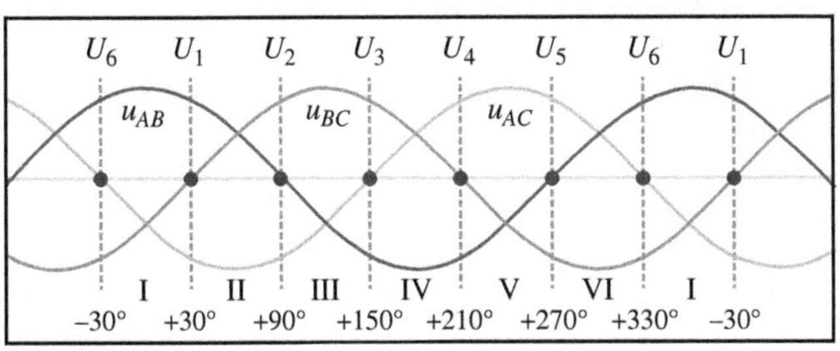

Figure 4.7 Inverter-stage sector division diagram.

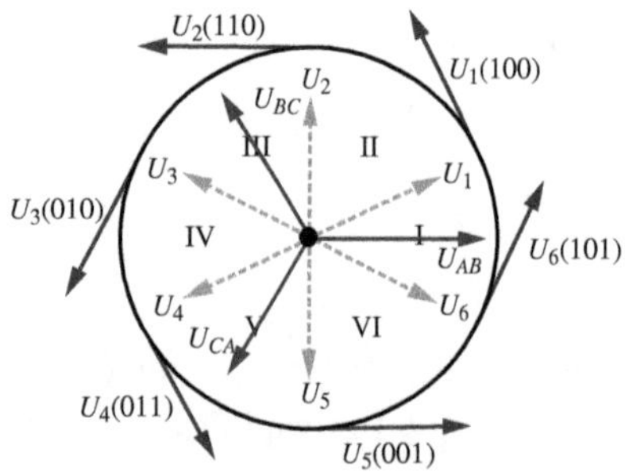

Figure 4.8 Space vector diagram of Inverter-stage voltage.

Table 4.3 Space vector table for inverter-stage voltage.

N	i_{dc}	S_{Ap}, S_{Bp}, S_{Cp}	S_{An}, S_{Bn}, S_{Cn}	S_x	Voltage vector	Description
1	i_A	1, 0, 0	0, 1, 1	1	U_1	Effective vector
2	$-i_C$	1, 1, 0	0, 0, 1	1	U_2	Effective vector
3	i_B	0, 1, 0	1, 0, 1	1	U_3	Effective vector
4	$-i_A$	0, 1, 1	1, 0, 0	1	U_4	Effective vector
5	i_C	0, 0, 1	1, 1, 0	1	U_5	Effective vector
6	$-i_B$	1, 0, 1	0, 1, 0	1	U_6	Effective vector
7	0	1, 1, 1	0, 0, 0	1/0	U_7	Zero vector
8	0	0, 0, 0	1, 1, 1	1/0	U_8	Zero vector

while "0" highlights the engagement of the lower bridge arm. A point warranting observation emerges in the sporadic instances where bridge arms, emanating from divergent phases, undergo simultaneous activation. To be more precise, a synchronized activation is perceptible in three bridges, impacting both the upper and lower bridge arms, albeit without simultaneous occurrence within a singular phase context.

iii) Space vector table

In drawing parallels with the traditional inverter stage, for the situations where the output voltage vector encompasses both an effective vector and a zero vector, a comprehensive list of all possible switch combination states can be referenced in Table 4.3.

iv) Vector synthesis and duty cycle

In a manner akin to the rectifier stage, the assembled output voltage vector, in relation to a reference current vector, rotates in a counterclockwise direction. The reference voltage U_{ref} is a culmination of two neighboring effective space vectors, U_μ and U_ν, along with the zero vector U_{0i}, present in its designated sector. Figure 4.9 visually represents the duty cycles, termed as d_μ, d_ν, and d_{0i}, corresponding to vectors U_μ, U_ν, and U_{0i}, respectively. Based on this illustration, the duty cycle can be mathematically expressed as:

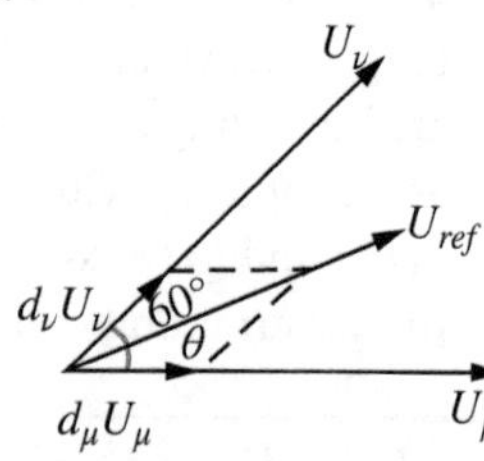

Figure 4.9 Vector synthesis diagram of inverter-stage voltage.

$$\begin{cases} d_\mu = m_o \sin(60° - \theta_o) \\ d_\nu = m_o \sin\theta_o \\ d_{0i} = 1 - d_\mu - d_\nu \end{cases} \tag{4.21}$$

4.3.1.3 Coordination of dual SVM

In the architecture of the indirect matrix converter, the absence of a substantial energy storage intermediary means the rectifier and inverter stages are interdependent. Thus, when the rectifier stage is in operation, a synergistic approach with the inverter stage is essential to ensure the system functions seamlessly.

i) Synergistic operating principles
   Using the dual space vector modulation approach, during a single switch cycle, the rectifier stage provides two distinct non-zero voltages and a singular zero voltage on the DC side, a relationship depicted in Figure 4.10. The modulatory actions of the inverter stage, in harmony with the rectifier stage, are as follows:
   a) In the time frame where the rectifier stage emits a zero voltage, the inverter stage is aligned to output a zero vector.
   b) During the moments when the rectifier stage is executing a non-through effective vector action, the inverter stage seamlessly carries out a respective space vector modulation.

ii) Unification of dual space duty cycle
   In the endeavor to establish a cohesive modulation methodology, particularly for the indirect matrix converter, an assimilation of two distinct processes is paramount. The first, concerning the rectifier stage, involves the meticulous modulation of the input phase current space vector. Simultaneously, the second process navigates through the modulation of the output line voltage space vector within the inverter stage. This inclusive and integrative strategy not only heralds a comprehensive modulation framework for the indirect matrix converter but also enables precise control over the sinusoidal waveform. Furthermore, it effectively manages the power factor at disparate points of contact—specifically at the input and output terminals. Through a thoughtful consolidation of duty cycles from dual stages, the resultant synthesis of these modulation strategies can be systematically articulated and is elucidated as follows:

$$\begin{cases} d_{\alpha\mu} = d_\alpha \, d_\mu \\ d_{\beta\mu} = d_\beta \, d_\mu \\ d_{\alpha v} = d_\alpha \, d_v \\ d_{\beta v} = d_\beta \, d_v \\ d_{\alpha 0i} = d_\alpha \, d_{0i} \\ d_{\beta 0i} = d_\beta \, d_{0i} \end{cases} \tag{4.22}$$

iii) Switching timing of dual space coordination
   Adhering to the fundamental coordination principle inherent between the two-stage space vectors, we explore a particular scenario in which the input current space vector at the rectifier

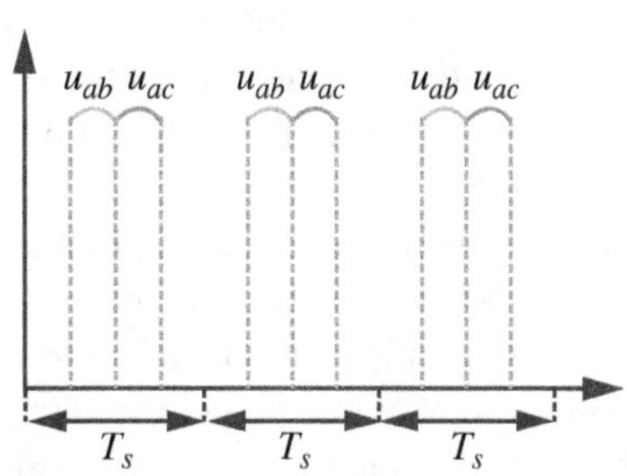

Figure 4.10 DC side voltage diagram of rectifier stage.

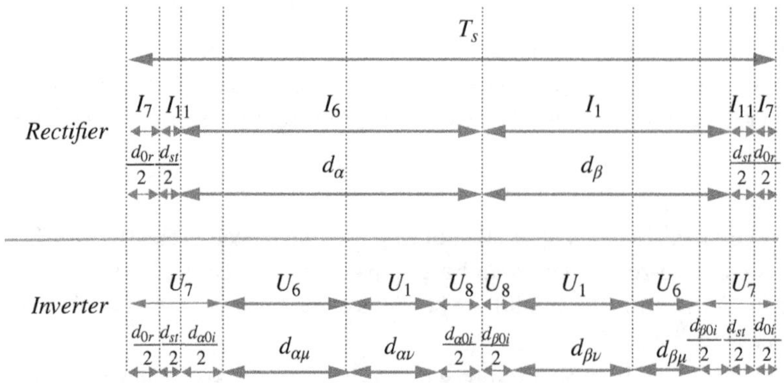

Figure 4.11 Switching sequence diagram of dual space vector.

stage and the output voltage space vector within the inverter stage engage in simultaneous operation, specifically residing within sector 1. Within the confines of a singular PWM cycle, Figure 4.11 provides a graphical representation that elucidates the synchronized timing, in conjunction with the switching sequences occurring amidst the two stages. A key observation from this depiction is the inverter stage consistently maintaining a zero current vector condition whenever the rectifier stage undergoes a vector switch. This operational alignment ensures zero-current transitions in the rectifier stage, significantly diminishing conversion-related losses. Simultaneously, the inverter stage follows a dead-time transition approach akin to traditional inverters. It's noteworthy that when the rectifier stage operates in a zero-vector condition, leading to a zero DC bus voltage, the inverter stage mirrors this with its own zero-vector operation, enhancing voltage utilization efficiency.

4.3.2 Modulation Methods with Common-mode Voltage Reduction

In PWM-regulated power converters, the challenge of CMV is pervasive. This phenomenon introduces several complications within electrical infrastructures [6, 7]. A notable concern is the leakage current: it traverses through an unintended path, specifically the parasitic capacitance found between the stator core and winding of motors. This not only accelerates the degradation of winding insulation but also heightens the risk of motor insulation failures. Moreover, it becomes a source of broad-spectrum electromagnetic interference (EMI) [8–10]. Given these ramifications, mitigating CMV in power converters becomes a crucial endeavor.

To address the CMV challenges in traditional converters and inverters, experts primarily focus on two predominant solutions. Firstly, incorporating supplementary hardware elements offers one approach, such as the integration of passive filters, active cancellers, or filters specifically designed for electromagnetic interference, as indicated in [11]. For instance, a unique common-mode filter based on the inductor–capacitor–inductor (LCL) configuration has been discussed in [12]. This design not only offers an alternate route for ground currents but also limits the amplitude of EMI disruptions channeled into the electrical grid. Another noteworthy technique is showcased in [13], where an adaptive common-mode filter has been designed to counterbalance output CMV fluctuations. Furthermore, in the studies [14, 15], distinct output filter setups were introduced, aiming to mitigate both CMV and rapid voltage changes at the terminals of PWM-driven motor windings. Yet, one significant limitation of these strategies is that they tend to augment the converter's dimensions, weight, and overall expenditure.

The refinement of modulation strategies has emerged as a cost-efficient approach to mitigating CMV issues, as documented in several studies [16–20]. One such method presented in [16] specifically aims at decreasing both the "virtual DC-link" voltage and the accompanying CMV. In another instance, a predictive control approach, as discussed in [17], focuses on the CMV reduction. Furthermore, an innovative space vector PWM design strategy is highlighted in [18], which minimizes common-mode disturbances, achieving CMV reductions ranging between 11% and 33%, albeit introducing two additional switching transitions in each PWM cycle. A study in [19] proposes a duo of CMV minimization techniques: the initial method integrates four active states and omits the zero state in the inverter phase for CMV attenuation, while the subsequent technique excludes zero vectors but incorporates two active vectors in the rectifier phase. Additionally, it leverages two zero vectors in the inverter stage for the creation of output voltage references. Both techniques, however, couldn't attain the zero DC-link current transition. Instead, a four-step voltage transition method was utilized, as presented in [19]. Another method described in [20] involves selecting the nearest three active inverter stage states, excluding zero states, with an objective to diminish the IMC's CMV, while upholding a maximum voltage gain of 0.866. Yet, achieving zero DC-link current transition remains elusive. It's noteworthy to mention that the majority of the above-illustrated methods are primarily designed for conventional inverters, DMCs, or IMCs. The prevalent modulation techniques for Z-source/quasi-Z source matrix converters are chiefly oriented toward executing fundamental voltage amplification functionalities along with adaptable frequency and voltage outputs. Conventional space vector modulations, as well as PWM, have been devised for such converters, as cited in [1–5, 21, 22]. A synergistic method combining PWM and pulse amplitude modulation is introduced in [23] targeting the quasi-Z source DMC to curtail switching losses. In the subsequent segments, we delve into the CMV challenges pertaining to QZS-IMC, harnessing equivalent circuitry and established modulation methodologies. This exploration enables us to discern the primary contributors to heightened CMV, paving the way for the development of two tailored modulation techniques to attenuate the CMV in QZS-IMC.

4.3.2.1 Common-mode Voltage

In the context of a three-phase AC load, Figure 4.12 illustrates the pathway of leakage current within the quasi-Z source IMC. The CMV, symbolized as u_{ng}, is characterized by the voltage disparity between the neutral point, designated "n," and the power supply's ground point, marked "g" [24, 25]. Concurrently, the parameter Z_L is indicative of the leakage impedance.

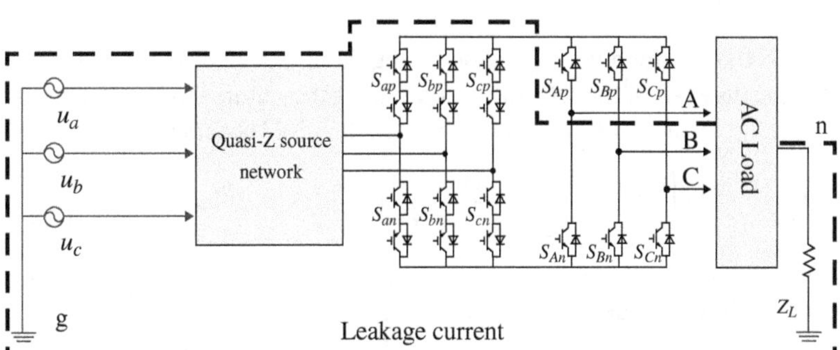

Figure 4.12 Leakage current path of ZS/QZS-IMC [5]. Source: Liu et.al. (2015)/with permission of John Wiley & Sons.

In a situation where R and L represent the per-phase load resistance and inductance, respectively, the following relationships can be observed

$$\begin{cases} u_A - u_{ng} = Ri_A + L\dfrac{di_A}{dt} \\ u_B - u_{ng} = Ri_B + L\dfrac{di_B}{dt} \\ u_C - u_{ng} = Ri_C + L\dfrac{di_C}{dt} \end{cases} \tag{4.23}$$

Considering the variables u_A, u_B, and u_C as the converter output voltages relative to the ground, and recognizing i_A, i_B, and i_C as corresponding output currents, we can establish certain relationships. With the foundational assumption that the collective sum of these output currents, i.e., i_A+i_B+i_C, equals zero and referring to equation (4.23), the derivation of CMV can be articulated as follows:

$$u_{ng} = \frac{u_A + u_B + u_C}{3} \tag{4.24}$$

In instances where the AC load establishes a direct connection to the source, the CMV is observed to be zero. Conversely, in scenarios wherein the AC load links to the source utilizing a quasi-Z source IMC, a deviation is noticed as the aggregate of the converter output voltages doesn't retain a zero value, influenced by the ensuing CMV.

The evaluation of the CMV unveils four potential outcomes, anchored in the previously discussed modulation technique. Illustratively, when the reference for the input current vector is positioned within sector 1, and concurrently, the reference for the output voltage vector resides in sector 1 with a boost factor denoted as B, the subsequent CMV computations unfold as follows:

i) In the situation where the modulating vector for the inverter stage is determined as V_7, each of the three-phase outputs interfaces directly with the positive terminal of the DC-link bus. This setup consequently yields the specific CMV.

$$u_{ng} = \frac{u_A + u_B + u_C}{3} = Bu_i \sin \omega_i t \tag{4.25}$$

Given that the peak input voltage for the IMC is articulated as Bu_i, it follows that the utmost value of the CMV is

$$u_{ng}\Big|_{max} = Bu_i \tag{4.26}$$

In this context, the term "u_i" refers specifically to the amplitude of the voltage in the input phase.

ii) When V_1 is chosen to serve as the modulating vector during the inverter stage, the output of phase-a is linked to the positive rail of the DC-link bus, whereas the outputs for both phase-b and phase-c are connected to the negative rail. Consequently, the CMV becomes

$$u_{ng} = \frac{u_A + u_B + u_C}{3} = \frac{Bu_i \left[\sin \omega_i t + 2 \sin\left(\omega_i t + \frac{2\pi}{3}\right)\right]}{3} \tag{4.27}$$

Differentiating the function "u_{ng}" with respect to the variable $\omega_i t$ and setting $\omega_i t$ to zero, one can discern the point of maximum value, which is derived from

$$u_{ng} = \frac{Bu_i \left[\sin 0 + 2 \sin \frac{2\pi}{3}\right]}{3} = \frac{Bu_i}{\sqrt{3}} \tag{4.28}$$

iii) In instances where V_2 is chosen to act as the modulating vector within the inverter stage, the outputs of both the a-phase and b-phase find themselves connected to the positive rail of the DC-link bus. Conversely, the output of the c-phase is aligned with the negative rail. Under such circumstances, the resultant CMV can be described as

$$u_{ng} = \frac{u_A + u_B + u_C}{3} = \frac{Bu_i\left[2\sin\omega_i t + \sin\left(\omega_i t + \frac{2\pi}{3}\right)\right]}{3} \tag{4.29}$$

At $\omega_i t = \pi/3$, the maximum value can be obtained by

$$u_{ng} = \frac{Bu_i\left[2\sin\frac{\pi}{3} + \sin\pi\right]}{3} = \frac{Bu_i}{\sqrt{3}} \tag{4.30}$$

iv) Selecting V_0 as the modulating vector for the inverter stage implies that the outputs across all three phases are uniformly connected to the negative rail of the DC-link bus. Subsequently, the CMV becomes

$$u_{ng} = \frac{u_A + u_B + u_C}{3} = Bu_i \sin\left(\omega_i t + \frac{2\pi}{3}\right) \tag{4.31}$$

At $\omega_i t = 0$, the maximum value can be obtained by

$$u_{ng} = Bu_i \sin\frac{2\pi}{3} = \frac{\sqrt{3}}{2}Bu_i \tag{4.32}$$

The preceding analysis elucidates that the magnitude of the common-mode voltage notably correlates with various factors such as the IMC switch combination, the level of input voltage, and the applied boost factor. Furthermore, a varied spectrum of peak values of CMV, originating from diverse sectors and switch combinations, exhibits distinctions, all of which are systematically cataloged in Table 4.4.

Table 4.4 The maximum value of common-mode voltage in the different switch vector.

Sector of rectifier-stage modulation vector	Switch vector of inverter stage (S_{Ap}, S_{Bp}, S_{Cp})	CMV peak values
1, 3, 5	100, 010, 001	$\frac{Bu_i}{\sqrt{3}}$
	110, 101, 011	$\frac{Bu_i}{\sqrt{3}}$
	000	$\frac{\sqrt{3}}{2}Bu_i$
	111	Bu_i
2, 4, 6	100, 010, 001	$-\frac{Bu_i}{\sqrt{3}}$
	110, 101, 011	$-\frac{Bu_i}{\sqrt{3}}$
	000	$-Bu_i$
	111	$-\frac{\sqrt{3}}{2}Bu_i$

4.3.2.2 Common-Mode Voltage Reduction Method I

The technique of space vector modulation, as discussed previously, meticulously integrates a combination of two adjacent voltage vectors, and is further supplemented by the incorporation of two zero vectors, thereby meticulously crafting the PWM outputs essential for the operational phase of the inverter. According to the insights provided in Table 4.4, and as corroborated by reference [5, 19], it is paramount that the switching patterns executed during the inverter stage assiduously abstain from implementing zero-switching states, an approach that strategically minimizes CMV within the confines of a quasi-Z source IMC framework. By astutely employing four geographically contiguous active vectors to synthesize the output voltage reference, as exemplified in Figure 4.13—with the emphasis on sector 1 for the sake of illustration—the derivation of the output voltage vector reference is modulated by the spatially adjacent vectors: V_1, V_2, V_3, and V_6. It merits attention that vectors V_3 and V_6 are judiciously substituted in lieu of the zero vectors V_7 and V_0, respectively. Moving forward, Figure 4.14 provides a detailed visualization and elucidation.

$$V_{ref} = d_6V_6 + d_1V_1 + d_2V_2 + d_3V_3 \tag{4.33}$$

The duty cycles are calculated as follows:

$$\begin{cases} d_1 = \dfrac{T_1}{T_s} = m_o \sin\left(\dfrac{\pi}{3} - \theta_o\right) \\ d_2 = \dfrac{T_2}{T_s} = m_o \sin\theta_o \\ d_3 = d_6 = \dfrac{1}{2}(1 - d_1 - d_2) \end{cases} \tag{4.34}$$

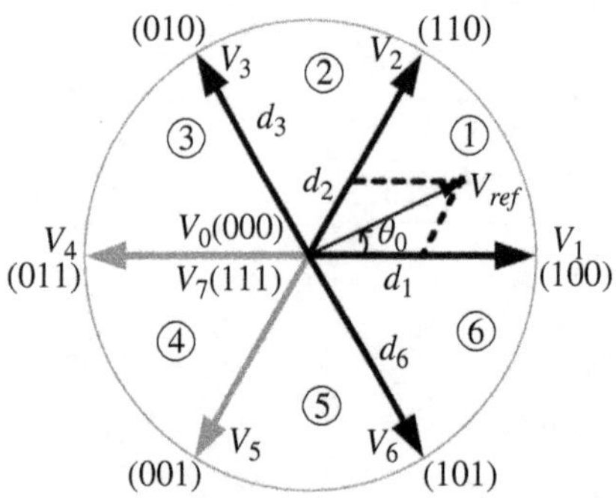

Figure 4.13 Space vector diagram of back-end inverter stage at Method I [5]. Source: Liu et.al. (2015)/with permission of John Wiley & Sons.

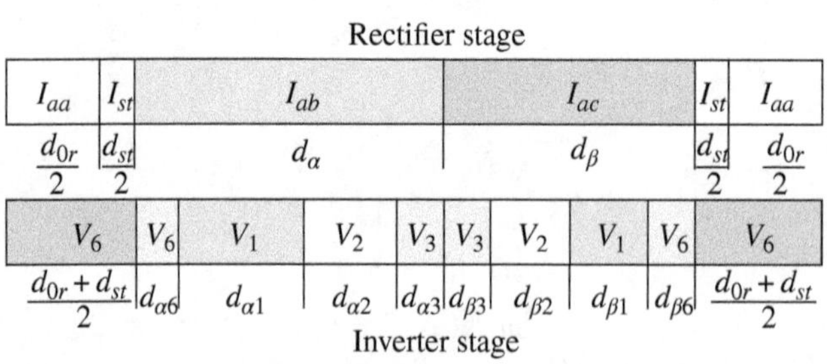

Figure 4.14 The switching pattern of SVM in rectifier stage and inverter stage [5]. Source: Liu et.al. (2015)/with permission of John Wiley & Sons.

In Figure 4.14, the switching patterns are exemplified with specific focus on sector 1, where the reference for both input current vector and output voltage vector, denoted as V_{ref}, is positioned. Given the rectifier stage's ability to yield two distinct DC voltage values, the overall output in terms of line-to-line voltage is segmented into two portions during a single PWM period. A particular sequencing of voltage space vectors is adopted: V_6-V_1-V_2-V_3 followed by V_3-V_2-V_1-V_6. Depending on the active current vector, the inverter stage makes its selection. Specifically, when I_{ab} serves as the current vector in the rectifier stage, the first part of the sequence, V_6-V_1-V_2-V_3, is prioritized. Conversely, if I_{ac} is the dominating current vector, the inverter opts for the latter part, V_3-V_2-V_1-V_6. Regarding duty cycles, they are

$$\begin{aligned}
d_{\alpha 1} &= d_\alpha \cdot d_1 \\
d_{\alpha 2} &= d_\alpha \cdot d_2 \\
d_{\alpha 3} &= d_{\alpha 6} = \frac{1}{2}(1 - d_1 - d_2)d_\alpha \\
d_{\beta 1} &= d_\beta \cdot d_1 \\
d_{\beta 2} &= d_\beta \cdot d_2 \\
d_{\beta 3} &= d_{\beta 6} = \frac{1}{2}(1 - d_1 - d_2)d_\beta
\end{aligned} \tag{4.35}$$

As depicted in Figure 4.15, the switch timing sequence has been structured, in alignment with what's presented in Figure 4.14. The introduction of four active vectors for the creation of the

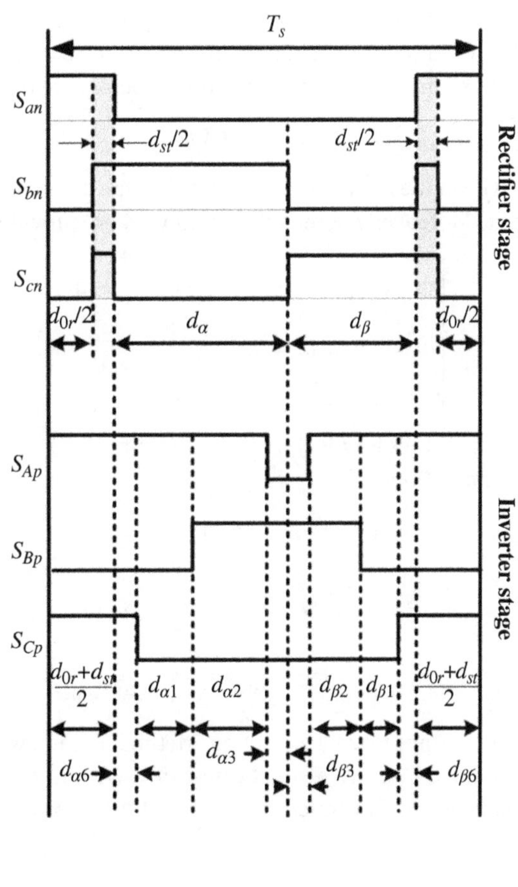

Figure 4.15 Switches timing sequence to reduce common-mode voltage in QZS-IMC [5]. Source: Liu et.al. (2015)/with permission of John Wiley & Sons.

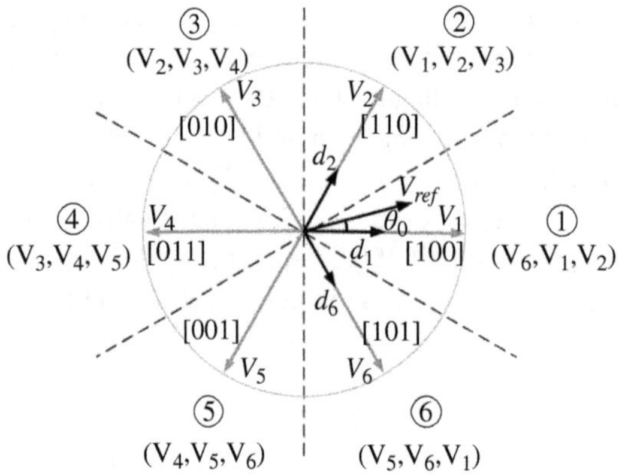

Figure 4.16 Space vector diagram of back-end inverter stage at method II [5]. Source: Liu et.al. (2015)/with permission of John Wiley & Sons.

output vector reference results in a diminished peak magnitude for the CMV, which now reduces to $Bu_i/\sqrt{3}$, when benchmarked against Bu_i.

4.3.2.3 Common-Mode Voltage Reduction Method II

As indicated in the study presented in [20], for the inverter stage of the quasi-Z source IMC, the output voltage vector reference is constructed using a trio of the nearest active vectors. In order to derive new voltage vector references, the conventional space vector modulation's six sectors undergo a rotational shift of 30°. This adjustment is visually represented in Figure 4.16.

As an illustrative instance, consider the situation where both the targeted output voltage vector and the space vector of the input current reside in sector 1. In this scenario, the active vectors V_6, V_1, and V_2 collaboratively shape the reference for the output voltage, denoted as V_{ref}. The corresponding duty cycles for these vectors are represented as d_6, d_1, and d_2, respectively.

$$V_{ref} = d_1 V_1 + d_2 V_2 + d_6 V_6 \tag{4.36}$$

Given the nature of the duty cycles, the cumulative sum of d_6, d_1, and d_2 is expected to equate to one. Moreover, referencing equation (4.36) provides additional insight into this relationship.

$$\begin{cases} m_o \cos\alpha_0 = \dfrac{2}{\sqrt{3}}d_1 + \dfrac{1}{\sqrt{3}}d_2 + \dfrac{1}{\sqrt{3}}d_6 \\ m_o \sin\alpha_0 = d_2 - d_6 \\ 1 = d_1 + d_2 + d_6 \end{cases} \tag{4.37}$$

The values for the duty cycles, specifically d_6, d_1, and d_2, can be derived based on the following computations:

$$d_1 = -1 + \sqrt{3}m_o \cos(\alpha_0)$$

$$d_2 = 1 - \frac{\sqrt{3}}{2}m_o \cos(\alpha_0) + \frac{1}{2}m_o \sin(\alpha_0)$$

$$d_6 = 1 - \frac{\sqrt{3}}{2}m_o \cos(\alpha_0) - \frac{1}{2}m_o \sin(\alpha_0) \tag{4.38}$$

In an approach resembling Method I, the switching patterns for the inverter stage can be classified into two distinct categories. Within this division, voltage vectors adhere to a particular sequence as illustrated in Figure 4.17: beginning with V_6, then progressing to V_1, followed by

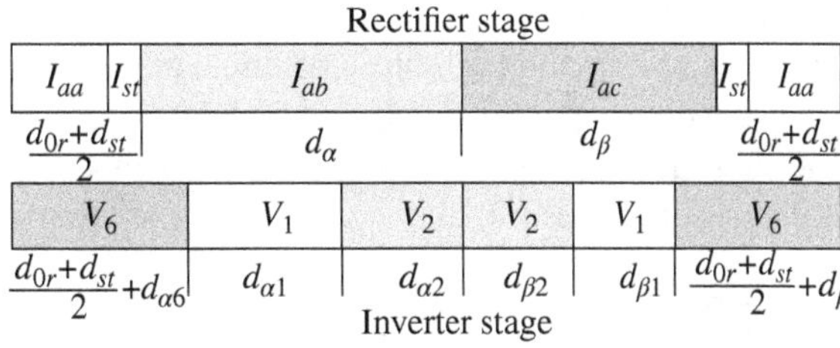

Figure 4.17 The switching pattern of SVM in rectifier stage and inverter stage at method II [5]. Source: Liu et.al. (2015)/with permission of John Wiley & Sons.

V_2, repeating V_2, reverting to V_1, and concluding with V_6. With this arrangement in place, the corresponding duty cycles can be ascertained through the subsequent formulae.

$$\begin{cases} d_{\alpha 1} = d_\alpha d_1 \\ d_{\alpha 2} = d_\alpha d_2 \\ d_{\alpha 3} = d_\alpha d_3 \end{cases}, \quad \begin{cases} d_{\beta 1} = d_\beta d_1 \\ d_{\beta 2} = d_\beta d_2 \\ d_{\beta 3} = d_\beta d_3 \end{cases} \tag{4.39}$$

Figure 4.18 shows the specific timing sequence of switches corresponding to Figure 4.17.

In the approach outlined as Method II, the configuration is such that it harnesses the capabilities of three active vectors, directing them toward the formulation of the output voltage vector reference. By doing so, the intervention of zero space vectors during the inverter operation is successfully circumvented. A noteworthy outcome of this methodology is the substantial decline in the

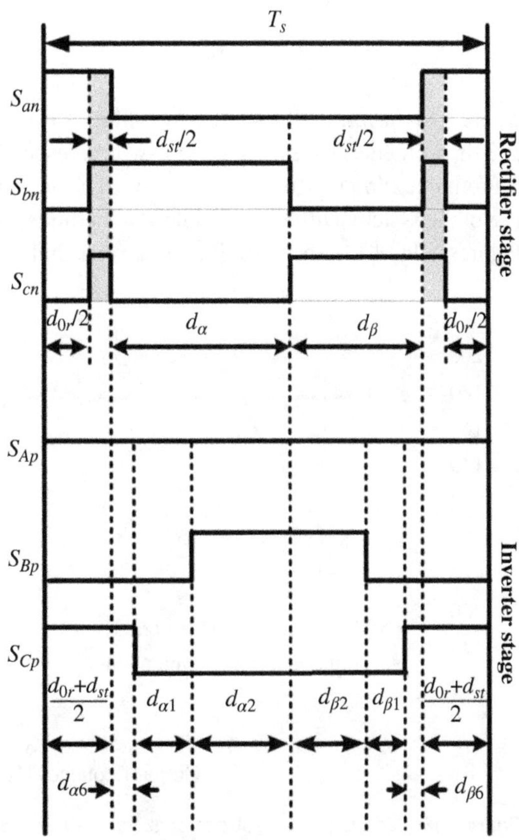

Figure 4.18 Switches timing sequence to reduce common-mode voltage in QZS-IMC at method II [5]. Source: Liu et.al. (2015)/with permission of John Wiley & Sons.

CMV peak value. More precisely, when juxtaposed against the prevalent space vector modulation techniques, this method exhibits a potential to slash the CMV peak by a remarkable 42%.

Drawing upon equation (4.38), it is imperative for the modulation index linked with the inverter stage to adhere to the stipulations outlined in equation (4.40). This compliance ensures the sustenance of the positive duty cycles when operating with the trio of active vectors.

$$\frac{2}{3} \leq m_o \leq 1 \tag{4.40}$$

Based on insights derived from equation (4.40), the modulation degree linked to the inverter stage must lie within a specific range. Specifically, it should exceed 0.667 but remain below 1. This inherent limitation of the modulation approach presents a constraint, potentially constricting the effective operational range for the output voltage. It's worth noting that the apex common-mode voltage for the ZS/QZS-IMC is influenced by factors such as the magnitude of the input voltage, the designated duty cycle of the through zero vector, and the chosen modulation scheme of the inverter stage. Interestingly, it remains unaffected by fluctuations in input/output frequency. An illustrative correlation between the peak common-mode voltage and the output frequency, considering various modulation approaches, is demonstrated in Figure 4.19.

Referencing Figure 4.19, we notice that through the enhancement of the modulation scheme used for the inverter stage and the subsequent replacement of the SVM's zero vector with a more effective vector, the peak common-mode voltage equates to $1/\sqrt{3}$ when compared to the original value. A comprehensive assessment from [26] sheds light on the power loss associated with the inverter stage's power switch tube during a single control cycle. It's quantified as:

$$P_{sw\_loss} = \frac{6}{T}\sum_{1}^{N}(E_{ON} + E_{OFF})_i \tag{4.41}$$

where N is the number of switch switching, T is the control cycle, E_{ON} and E_{OFF} are the conduction and turn-off energy losses of power switch, respectively.

From equation (4.40), it can be seen that the number of switch switching affects the switching loss of the system. Different modulation methods have different numbers of switch switching. From Figures 4.11, 4.15, and 4.18, it can be seen that within one control cycle, the conventional SVM

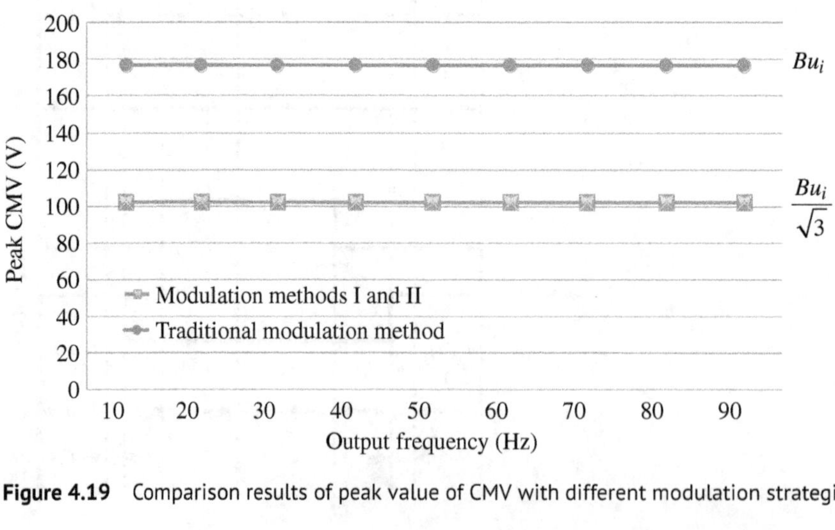

Figure 4.19 Comparison results of peak value of CMV with different modulation strategies.

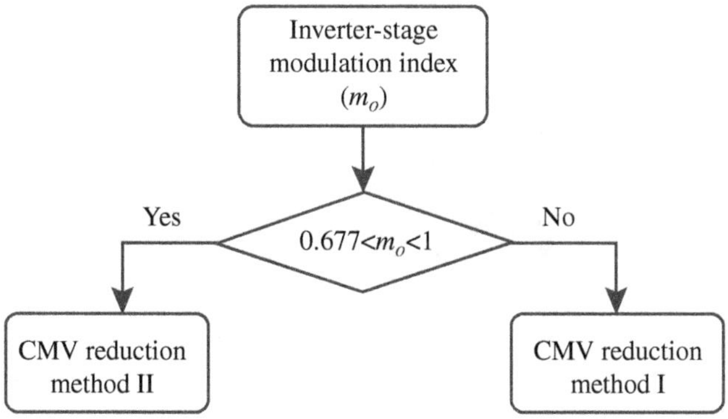

Figure 4.20 Hybrid modulation algorithms for common-mode voltage reduction of ZS/QZS-IMC.

modulation method has 6 switch switching times for the inverter stage. For the modulation method using common-mode voltage suppression method one, the switch switching times for the inverter stage are also 6. While using common-mode voltage suppression method two, because there is one-phase bridge arm in a sector no switch switching is performed, thereby reducing the number of switch switching. Its switch switching times for the inverter stage are 4, and the switch loss is lowest.

In order to suppress common-mode voltage while minimizing switch loss and satisfying a wider range of adjustable voltage gain, two common-mode voltage suppression methods need to coordinate and divide work, as shown in Figure 4.20.

4.4 Simulation and Experimental Results

Drawing upon the comprehensive exploration delineated in Section 4.3, the common-mode voltage affiliated with the Z-source/quasi-Z-source indirect matrix converter undergoes a methodical simulation alongside an experimental scrutiny.

The system under examination boasts specific parameters, which include an AC source characterized by a 50-V line-to-line peak voltage and an input frequency benchmarked at 50 Hz. The shoot-through duty cycle, symbolized as (D), stands at 0.15. Components of the quasi-Z source network include inductors L_1 and L_2, both valued at 2 mH, capacitors C_1 and C_2 with a capacitance of 100 μF each, and a switching rhythm, or frequency (f_s), set to resonate at 10 kHz. When examining the converter's output metrics, it delivers a frequency of 30 Hz and serves an AC load characterized by specifications $R = 5\ \Omega$ and $L = 3\ m$H. The modulation index of the inverter, indicated as (m_0), is pegged at 0.85. Delving deeper into graphical representations, both Figures 4.21 and 4.24 showcase waveforms—both simulated and practical—of the inverter's reference vector. This vector is an outcome of a synthesis process involving two zero vectors and two effective vectors, all functioning at an output frequency of 30 Hz. Shifting focus, Figures 4.22 and 4.25 offer insights into both simulated and actual waveform outputs of the inverter's reference vector, curated by intertwining four effective vectors under the umbrella of the first common-mode suppression technique. Rounding off the visual aids, Figures 4.23 and 4.26 display waveforms, both theoretical and empirical, of the inverter reference vector; these are the result of the strategic integration of three effective vectors, coupled with a nuanced sector division, under the guidelines of the second common-mode suppression strategy.

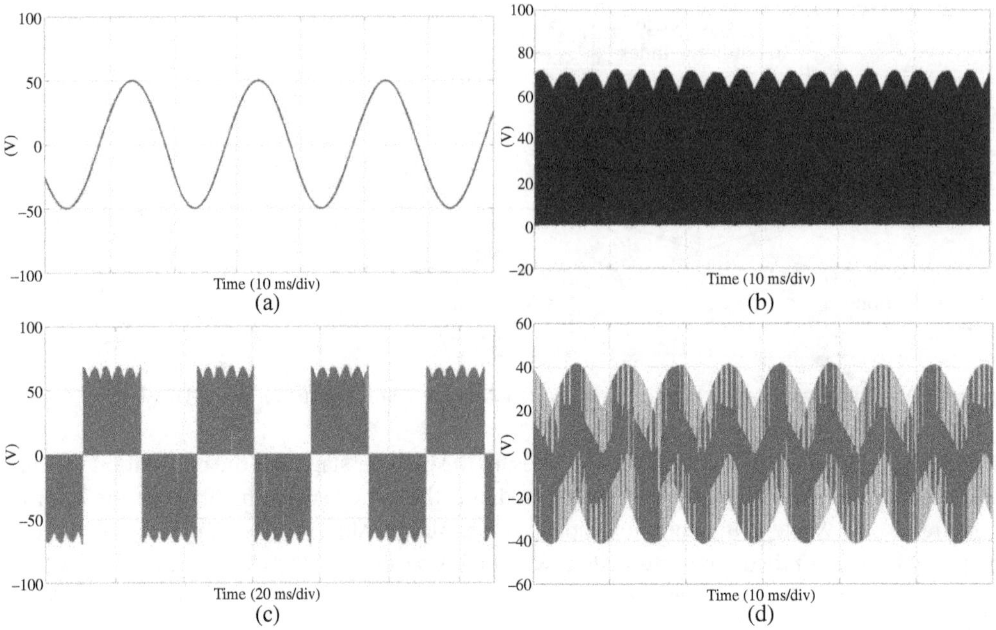

Figure 4.21 Simulation results of conventional SVM (a) input voltage; (b) DC-link voltage; (c) output voltage; (d) common-mode voltage [27].

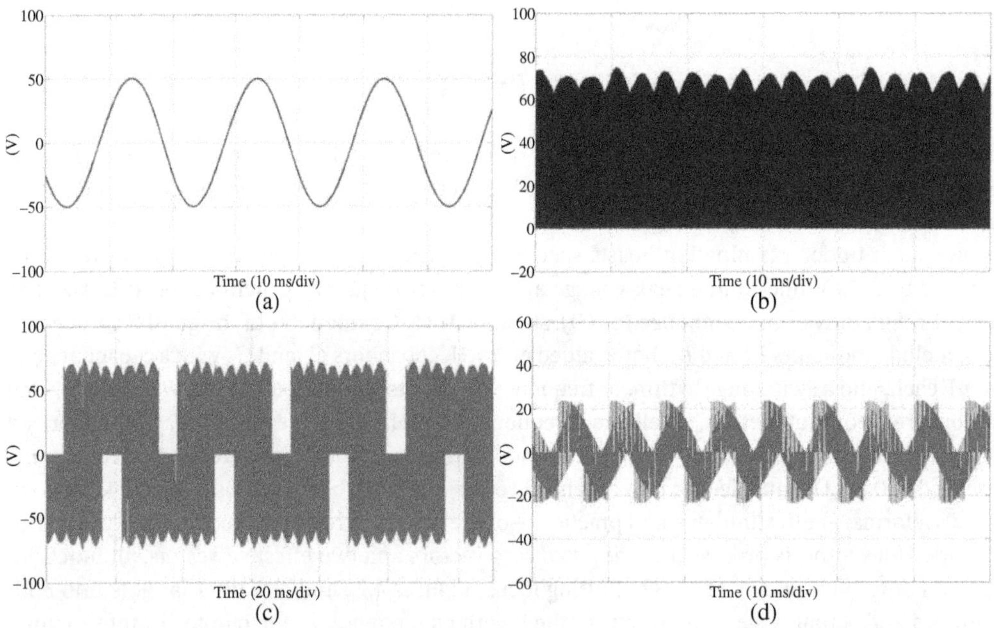

Figure 4.22 Simulation results of common-mode voltage reduction method I (a) input voltage; (b) DC-link voltage; (c) output voltage; (d) common-mode voltage [27].

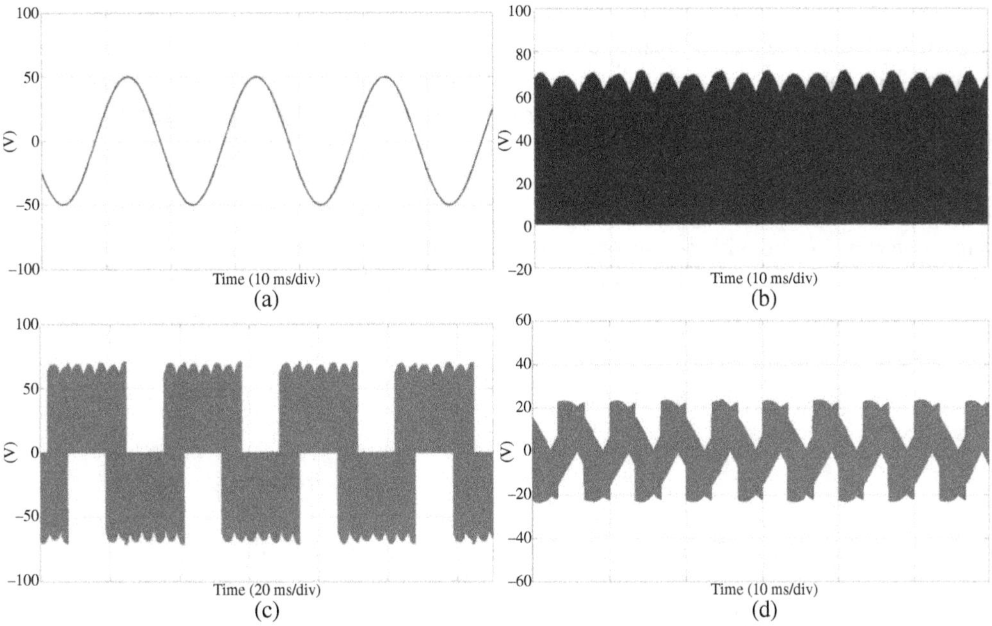

Figure 4.23 Simulation results of common-mode voltage reduction method II (a) input voltage; (b) DC-link voltage; (c) output voltage; (d) common-mode voltage [27].

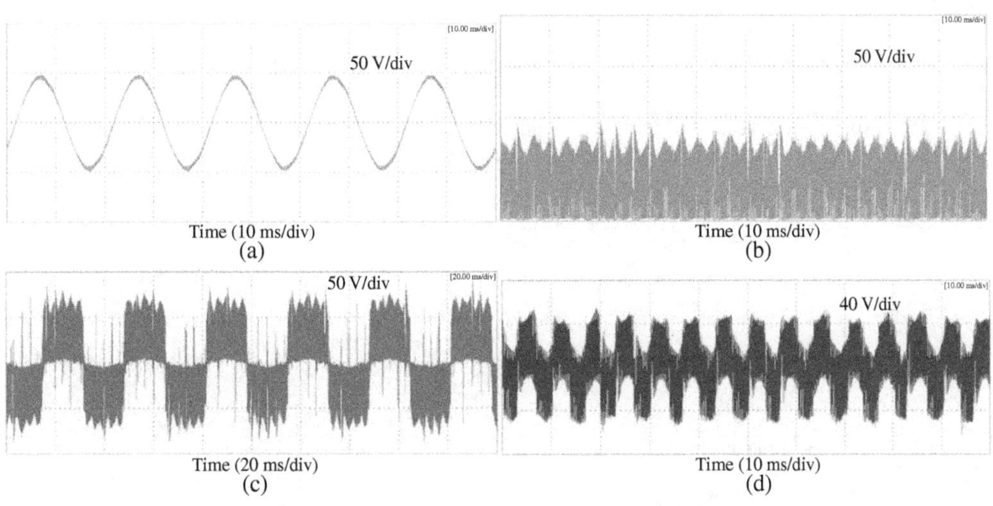

Figure 4.24 Experimental results of conventional SVM (a) input voltage; (b) DC-link voltage; (c) output voltage; (d) common-mode voltage [27].

In the simulation executed, the source's voltage exhibited a maximum of 50 V. The duty cycle associated with the through-zero vector, symbolized by D, was consistently marked at 0.15. By leveraging equation (4.12), the boost factor, denoted as B, was determined to be 1.4286. This calculation set the bus voltage peak value for the indirect matrix converter at approximately 71 V. From the insights of Figure 4.24, it's evident that the conventional modulation approach, which crafts the reference vector by integrating zero vectors and two other distinct vectors, gives rise to a peak common-mode voltage of around 40 V. Intriguingly, this value parallels the apex input phase voltage identified in the indirect matrix converter. By shifting to an alternate modulation strategy, where all non-through zero vectors are supplanted by their corresponding effective vectors and where the reference vector is crafted through the amalgamation of four effective vectors following the first common-mode suppression technique, Figure 4.25 highlights a substantial reduction in the peak common-mode voltage to a mere 24 V. Furthermore, adjustments to the sectors within the inverter stage and the choice to utilize three effective vectors to construct the reference vector, in line with the second common-mode suppression methodology, underscore that the peak common-mode voltage is refined to approximately 25 V, as portrayed in Figure 4.26.

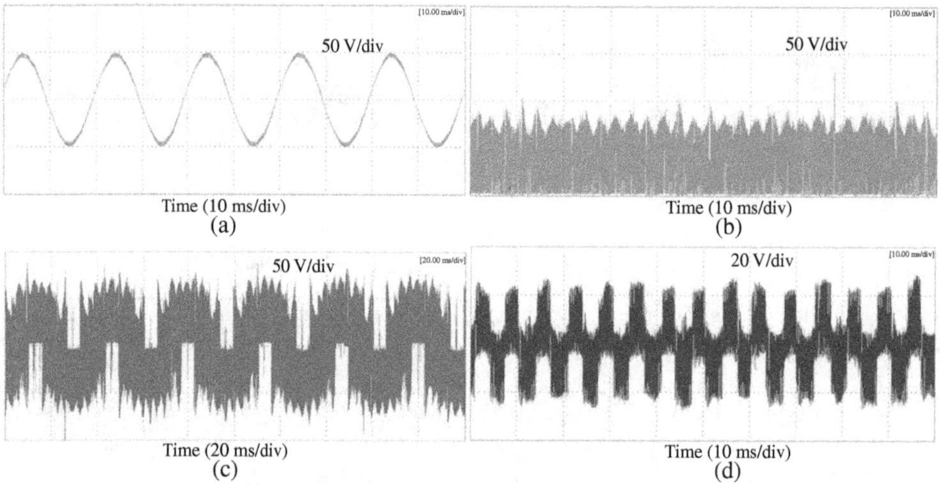

Figure 4.25 Experimental results of common-mode voltage reduction method I (a) input voltage; (b) DC-link voltage; (c) output voltage; (d) common-mode voltage [27].

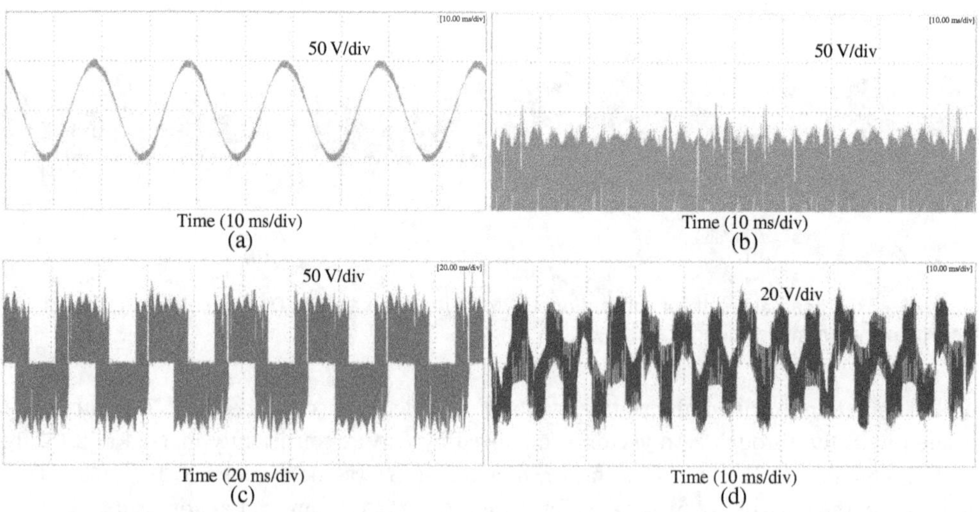

Figure 4.26 Experimental results of common-mode voltage reduction method II (a) input voltage; (b) DC-link voltage; (c) output voltage; (d) common-mode voltage [27].

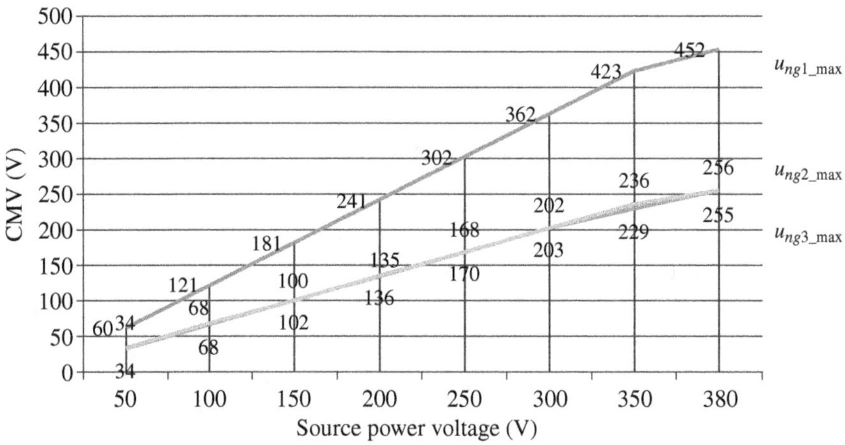

Figure 4.27 Common-mode voltage of different input voltages [27].

In the context of empirical tests, the chosen parameters paralleled those designated for earlier simulations. As highlighted in Figure 4.24, the application of a classic modulation method led to the realization of a common-mode voltage peaking at 44 V. However, when the two common-mode voltage reduction techniques, as elucidated in this study, were implemented, the common-mode voltage was trimmed down to a value of 27 V. This decrement is visually captured in Figures 4.25 and 4.26. An intriguing observation is the efficacy of the second suppression approach, which remarkably reduced the frequency of switch activations within one cycle, leading to a tangible reduction in energy losses. The symmetry between simulated and actual results is strikingly clear. Further, it's crucial to understand that any fine-tuning in the voltage on the source side of the Z-source/quasi-Z-source indirect matrix converter introduces proportional fluctuations in the system's common-mode voltage.

Figure 4.27 visually conveys the behavior of peak common-mode voltage under diverse source voltage conditions, alongside different modulation approaches. In this context, the term u_{ng1_max} represents the highest attainable common-mode voltage when the chosen modulation strategy synthesizes the reference vector by merging zero vectors with a pair of effective vectors. On the other hand, when the modulation leverages four effective vectors for the reference vector construction, the pinnacle common-mode voltage is denoted by u_{ng2_max}. Lastly, when the modulation framework integrates the reference vector by blending three effective vectors, the resultant peak value of the common-mode voltage is symbolized as u_{ng3_max}.

4.5 Conclusion

1) The creation of the input current reference vector in the rectifier stage, when utilizing a Z-source/quasi-Z-source indirect matrix converter, can be accomplished through three distinctive methodologies, each presenting a unique set of merits and limitations. When the objective is to achieve a minimized harmonic amplitude, the third methodology emerges as the superior choice, albeit at the expense of an increase in switch occurrences and, consequently, heightened switch losses within a singular cycle. On the other hand, if the augmentation of system efficiency and the minimization of switch losses are prioritized, the first methodology proves to be optimal, boasting the minimal switch occurrences and losses among the three.

Moreover, considering it from an algorithmic perspective, the first methodology also presents itself as the most straightforward to put into practice. Consequently, the research encapsulated within this chapter opts for the first methodology as the modulating technique for the rectifier stage.

2) To mitigate the frequency of switch activations and subsequently minimize switch losses, adherence to specific principles concerning the positioning of zero vectors, efficacious vectors of current in the rectifier stage, and effective voltage vectors in the inverter stage is imperative. Firstly, the zero vectors pertinent to the rectifier stage must be strategically positioned at each extremity of every modulation cycle. Secondly, throughout the vector transition process in the rectifier stage, it is incumbent upon the inverter stage to operate invariably in a zero-vector state, thereby preventing any current from traversing the DC bus and safeguarding the zero-current transformation within the rectifier stage. Lastly, aiming to curtail the switch instances within the inverter stage, the sequence for its output voltage vector should adhere to the following order: $V_0 \rightarrow V_m \rightarrow V_n \rightarrow V_7 \rightarrow V_n \rightarrow V_m \rightarrow V_0$.
3) Within the realm of Z-source and quasi-Z-source indirect matrix converters, it's noteworthy to mention the unique characteristics of the common-mode voltage. Primarily, its determination hinges majorly on the source voltage, coupled with the duty cycle associated with the zero vector. Intriguingly, its dependency remains unswayed by variations in the output voltage frequency. Analyzing its amplitude, it's evident that the maximum magnitude this common-mode voltage can reach is congruent with the value denoted as Bu_i.
4) A viable approach to mitigate the peak common-mode voltage involves judiciously selecting two effective vectors for the inverter stage as substitutes for zero vectors. Further, by identifying and utilizing four effective vectors, those in closest proximity to the sector harboring the reference vector, for the synthesis of the reference vector, one can employ the common-mode suppression methodology - the first approach, to attain a reduction in the common-mode voltage, approximating 42%.
5) By meticulously redefining the vector sectors tied to the inverter stage, one can harness the potential of three potent vectors to craft the desired reference vector. This structured alignment not only paves the way for a notable reduction of about 42% in the common-mode voltage but also contributes to mitigating switch-related energy losses. However, it's pivotal to comprehend that the extent of this method's success remains tethered to the specific modulation degree inherent to the inverter stage.

References

1 S. Liu, B. Ge, X. Jiang, and H. Abu-Rub, F. Z. Peng, "Comparative evaluation of three Z-source/quasi-Z-source indirect matrix converters", *IEEE Trans. Ind. Electron.*, vol. 62, no. 2, pp. 692–701, Feb. 2015.

2 S. Liu, B. Ge, X. Jiang, H. Abu-Rub, and F. Z. Peng, "Modeling, analysis, and motor drive application of quasi-Z-source indirect matrix converter", *Int. J. Comput. Math. Electr. Electron. Eng.*, vol. 33, no. 1/2, pp. 298–319, 2014.

3 S. Liu, B. Ge, X. Jiang, H. Abu-Rub, and F. Z. Peng, "A novel quasi-Z-source indirect matrix converter", *Int. J. Circ. Theo. Appl.*, vol. 43, pp. 438–454, 2013.

4 S. Liu, B. Ge, X. Jiang, H. Abu-Rub, and F. Z. Peng, "Simplified quasi-Z source indirect matrix converter", *Int. J. Circ. Theo. Appl.*, vol. 43, pp. 1775–1793, Oct. 2014.

5 X. You, B. Ge, S. Liu, N. Nie, X. Jiang, and H. Abu-Rub, "Common mode voltage reduction of quasi-Z source indirect matri Janx converter", *Int. J. Circ. Theo. Appl.*, vol. 44, no. 1, pp. 162–184, Jan. 2016.

6 J. M. Erdman, R. J. Kerkman, D. W. Schlegel, and G. L. Skibinski, "Space-vector PWM with reduced common-mode voltage for five-phase induction motor drives," *IEEE Trans. Ind. Electron.*, vol. 60, no. 10, pp. 4159–4168, 2013.

7 C. Hou, C. Shih, P. Cheng, and A. M. Hava, "Common-mode voltage reduction pulsewidth modulation techniques for three-phase grid-connected converters," *IEEE Trans. Power Electron.*, vol. 28, no. 4, pp. 1971–1979, 2013.

8 J. Shang, Y. W. Li, N. R. Zargari, and Z. Cheng, "PWM strategies for common-mode voltage reduction in current source drives," *IEEE Trans. Power Electron.*, vol. 29, no. 10, pp. 5431–5445, 2014.

9 N. O. Cetin and A. M. Hava, "Topology and PWM method dependency of high frequency leakage current characteristics of voltage source inverter driven AC motor drives," *IEEE Energy Conver. Congr. Expos.*, vol. 2012, pp. 3430–3437, 2012.

10 N. Zhu, D. Xu, B. Wu, N. R. Zargari, M. Kazerani, and F. Liu, "Common-mode voltage reduction methods for current-source converters in medium-voltage drives," *IEEE Trans. Power Electron.*, vol. 28, no. 2, pp. 995–1006, 2013.

11 F. Luo, D. Dong, D. Boroyevich, P. Mattavelli, and S. Wang, "Improving high-frequency performance of an input common mode EMI filter using an impedance-mismatching filter," *IEEE Trans. Power Electron.*, vol. 29, no. 10, pp. 5111–5115, 2014.

12 M. H. Hedayati, A. B. Acharya, and V. John, "Common-mode filter design for PWM rectifier-based motor drives," *IEEE Trans. Power Electron.*, vol. 28, no. 11, pp. 5364–5371, 2013.

13 D. Barater, G. Buticchi, E. Lorenzani, and C. Concari, "Active common-mode filter for ground leakage current reduction in grid-connected PV converters operating with arbitrary power factor," *IEEE Trans. Ind. Electron.*, vol. 61, no. 8, pp. 3940–3950, 2014.

14 A. Hirofumi and T. Doumoto, "An approach to eliminating high-frequency shaft voltage and ground leakage current from an inverter-driven motor," *IEEE Trans. Ind. Appl.*, vol. 40, no. 4, pp. 1162–1169, 2004.

15 X. Chen, D. Xu, F. Liu, and J. Zhang, "A novel inverter-output passive filter for reducing both differential- and common-mode dv/dt at the motor terminals in PWM drive systems," *IEEE Trans. Ind. Electron.*, vol. 54, no. 1, pp. 419–426, 2007.

16 R. Peña, R. Cárdenas, E. Reyes, J. Clare, and P. Wheeler, "Control of a doubly fed induction generator via an indirect matrix converter with changing DC voltage," *IEEE Trans. Ind. Electron.*, vol. 58, no. 10, pp. 4664–4674, 2011.

17 R. Vargas, U. Ammann, J. Rodríguez, and J. Pontt, "Predictive strategy to control common-mode voltage in loads fed by matrix converters," *IEEE Trans. Ind. Electron.*, vol. 55, no. 12, pp. 4372–4380, 2008.

18 J. Espina, C. Ortega, A. Arias, and J. Balcells, "Space vector modulation strategy to reduce the common mode perturbations in matrix converters," *IEICE Electron. Expo.*, vol. 7, no. 4, pp. 281–287, 2010.

19 T.D. Nguyen and H.-H. Lee, "Modulation strategies to reduce common-mode voltage for indirect matrix converters," *IEEE Trans. Ind. Electron.*, vol. 59, no. 1, pp. 129–140, 2012.

20 T.D. Nguyen and H. H. Lee, "A new SVM method for an indirect matrix converter with common-mode voltage reduction," *IEEE Trans. Industr. Inform.*, vol. 10, no. 1, pp. 61–72, 2014.

21 S. Liu, B. Ge, H. Abu-Rub, F. Z. Peng, and Y. Liu, "Quasi-Z-source matrix converter based induction motor drives," in *IECON 2012–38th Annual Conference on IEEE Industrial Electronics Society*, 2012, pp. 5303–5307.

22 S. Liu, B. Ge, X. Jiang, H. Abu-Rub, and F. Peng, "Modeling, analysis, and motor drive application of quasi-Z-source indirect matrix converter," *Int. J. Comput. Math. Electr. Electron. Eng.*, vol. 33, no. 1/2, pp. 298–319, 2013.

23 L. Qin, F. Z. Peng, and B. Ge, "Pulse-width-amplitude-modulated voltage-fed quasi-Z-source direct matrix converter with maximum constant boost," in *Twenty-Seventh Annual IEEE Applied Power Electronics Conference and Exposition (APEC)*, pp. 641–646, 2012.

24 T. Shi, Q. Huang, Y. Yan, and C. Xia, "Suppression of common mode voltage for matrix converter based on improved double line voltage synthesis strategy," *IET Power Electron.*, vol. 7, no. 6, pp. 1384–1395, 2014.

25 H. H. Lee and H. M. Nguyen, "An effective direct-SVM method for matrix converters operating with low-voltage transfer ratio," *IEEE Trans. Power Electron.*, vol. 28, no. 2, pp. 920–929, 2013.

26 R. Peña *et al*, "Control of a doubly fed induction generator via an indirect matrix converter with changing DC voltage," *IEEE Trans. Ind. Electron.*, vol. 58, no. 10, pp. 4664–4674, 2011.

27 S. Liu, "Research on Z-source/quasi-Z-source Indirect Matrix Converter," Ph.D. dissertation, School of Electrical Engineering, Beijing Jiaotong Univ., 2014. (In Chinese).

5

Comparison of Typical Z-Source/Quasi-Z-Source Matrix Converters

5.1 Introduction

Matrix Converters (MCs) are revolutionary devices that create a seamless bridge between an AC source and its corresponding AC load, eliminating the traditional requirement of intermediate energy storage mechanisms. The salient features of MCs lie in their adoption of highly efficient all-silicon power converter cells. This adoption not only negates the need for DC-link capacitors, but also ensures the presentation of an output voltage with dynamic amplitude and frequency variations. Another noteworthy aspect is the sinusoidal nature of both their input and output currents. Moreover, these converters are designed to operate with a consistent unity power factor on the input side, optimizing power usage. While exploring the intricacies of MC topologies, two dominant designs can be identified: the Direct MC (DMC) [1, 2] and the Indirect MC (IMC) [3, 4]. Though both these designs exhibit similar performance markers, such as the harmonic content in the current and a voltage gain restricted to 0.866, they are commonly paired with input filters, as highlighted in Figure 5.1. A distinguishing feature of the IMC is its ability to adeptly address the current commutation problems that the DMC is known to encounter. Consequently, the IMC has garnered significant attention and preference across numerous application domains. The forthcoming sections of this chapter will delve deeper into the specifics and advantages of the IMC topology.

In the realm of MCs, addressing the persistent challenges related to voltage gain has been a focal point of academic research. To navigate these limitations, there's been a significant thrust toward over-modulation strategies, the primary objective being the amplification of the output voltage [5, 6]. This pioneering approach, though laudable for its innovation, is not without its pitfalls. It inadvertently introduces low-order harmonics, which cast a shadow on the pristine nature of both the output voltage and input current. Venturing further, another innovative proposition that has garnered attention is the seamless integration of a step-up transformer within the MC framework. This design choice ensures a direct and efficient liaison with the power supply. However, while the straightforwardness of this solution makes it attractive, it's not without compromises. Parameters such as physical size, weight, economic feasibility, and overall operational efficiency are drawn into question [7]. Taking a different trajectory, experts have looked at the potential of merging an AC chopper with the MC. This avant garde system, known in academic circles as the matrix-reactance frequency converter (MRFC), holds the promise of achieving a voltage gain that transcends the number one [8]. But, as with most innovative solutions, it comes with its own set of complexities. The system's control architecture becomes notably intricate, and it brings into play an added power conversion stage. Additionally, there's a pronounced reliance of the voltage transfer ratio on the unique interplay of circuit parameters and the specifics of the load in question [7].

Impedance Source Matrix Converters and Control, First Edition.
Yushan Liu, Xiao Li, Baoming Ge, Haitham Abu-Rub, and Frede Blaabjerg.

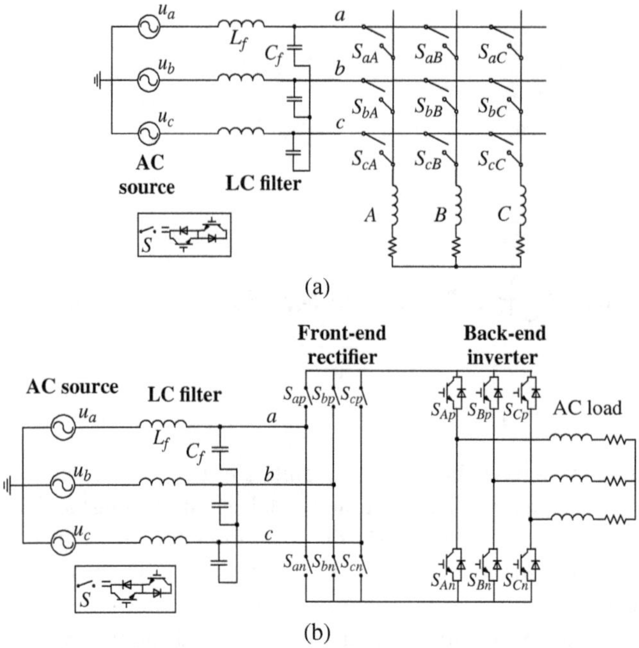

Figure 5.1 Topologies of conventional (a) direct matrix converter and (b) indirect matrix converter.

Facing the continuous voltage challenges exhibited by the buck-derived three-phase VSI, advancements in inverter technology yielded the development of the Z-source (ZS) and quasi-Z-source (qZS) inverters. These evolutionary designs are distinguished by features such as their capability for single-stage power conversion, robustness in safeguarding against unexpected short-circuits, and the flexibility to accommodate a versatile spectrum of input voltage ranges [9–11]. One particularly notable progression involves the deliberate integration of the ZS/qZS architectures into the intermediate DC link of the IMC structure [12–14]. Another is the refinement of the MC, which now boasts a lesser number of rectifier-side switches, thus enhancing its efficiency [15, 16]. As visually represented in Figure 5.2a,b, both the ZS and qZS iterations of the IMC are embedded with an impedance network specifically within the DC link. This unique incorporation seeks to directly remedy the existing limitations on the MC's voltage gain potential. Drawing a comparative analysis, these configurations, when juxtaposed against an IMC supplemented with a step-up transformer, emerge superior in aspects of weight, spatial efficiency due to a reduced volume, and minimized power dissipation. Intriguingly, the introduction of bidirectional switches in the impedance network, which might be perceived as an added complexity, doesn't tarnish these attributes. By making subtle adjustments in the modulation strategy, the operational integrity of the ZS/QZS-IMC is preserved, ensuring that the quality of the MC output remains consistent. The output presents a stable voltage gain, which remarkably remains independent of any load variations. This eliminates the traditional need for additional power conversion stages, representing a groundbreaking departure from methodologies like the MRFC. Nevertheless, a critical observation reveals that embedding the qZS within the intermediary DC link somewhat compromises the IMC's renowned distinguished all-silicon MC cell feature.

Explorations have been made into integrating ZS/QZS constructs directly on the AC input side, as can be observed from designs like the ZS-DMC [17], the QZS-IMC [18], and the ZS-IMC [19].

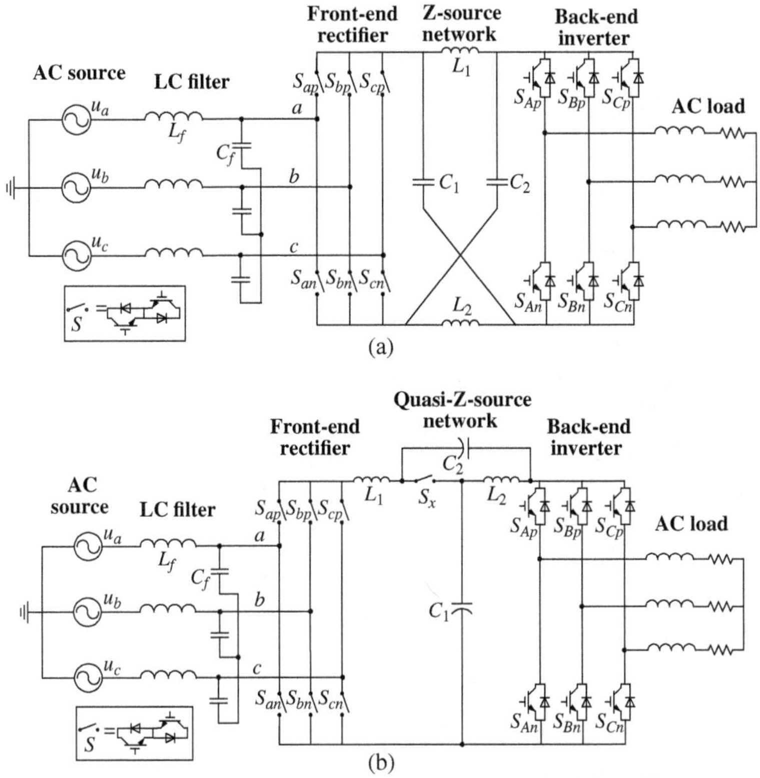

Figure 5.2 ZS/qZS indirect matrix converter with impedance source network in DC link. (a) Z-source indirect matrix converter; (b) quasi-Z-source indirect matrix converter.

Illustrated in Figure 5.3 panels (a) through (c), these design innovations aim to amplify the voltage gain while still upholding the revered all-silicon MC cell signature. However, they face an inherent challenge: the input currents display a discontinuous pattern, leading to a surge in total harmonic distortion (THD). This elevates the necessity for the inclusion of an input filter, like the second-order LC filter showcased in Figure 5.3, to mitigate harmonic interferences and optimize power quality, especially when interfaced with the main grid.

Within the comprehensive scope of the research presented in [19], a distinct QZS-IMC configuration integrated with an LC filter emerges, as portrayed in Figure 5.4. This innovative approach stands as a compelling alternative to the traditional input filters often observed in the quintessential ZS/QZS-IMC, depicted in Figure 5.3. At its core, this design fosters a seamless interface, enabling the qZS module to establish a direct linkage to the AC power grid, thereby facilitating the culmination of the IMC structure. Unveiling this iteration of the QZS-IMC has shed light on a couple of seminal advantages. On the one hand, there's the realization of a significant boost in voltage, effectively shattering the conventional limitations inherent to classical MCs. Simultaneously, the design boasts an intrinsic second-order filtration mechanism, rendering superfluous the necessity for external LC filter appendages. This construct faithfully encapsulates the all-silicon MC architecture, ensuring that the elemental strengths synonymous with both the orthodox MC and qZS inverters are undiminished. A point of prominence lies in its prowess to consistently generate unbroken sinusoidal grid currents, nullifying the requirement for added LC filtering elements. This efficiency translates into tangible benefits: a palpable reduction in

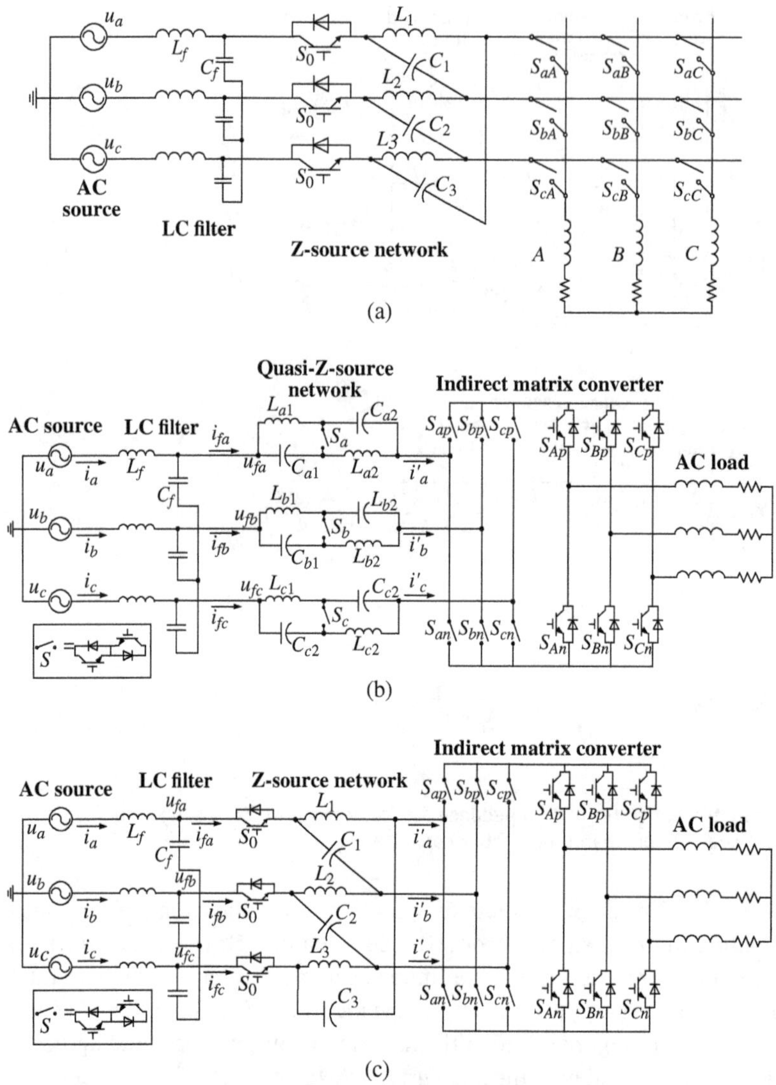

Figure 5.3 ZS/qZS matrix converters with impedance source network at AC input side. (a) ZS-DMC. Source: Adapted from [17]; (b) QZS-IMC. Source: Adapted from [18]; (c) ZS-IMC. Source: Adapted from [19].

operational expenditures, a pronounced leaning toward space optimization, marked strides in energy conservation, and a decidedly lightweight construction. A detailed analytical discourse, encompassing a myriad of facets like the dynamics of voltage surge, nuanced oscillations in voltage and current, potential stress points on both passive and active components, and a plethora of other considerations, juxtaposing this evolved QZS-IMC with the emblematic QZS-IMC portrayed in Figure 5.3b and the ZS-IMC in Figure 5.3c, is meticulously catalogued in [19]. From a holistic perspective, drawing from the granular insights and comparative delineations set forth in [19], the inference is unambiguous: the reimagined QZS-IMC, in its avant garde avatar, holds a discernibly superior position, making it the choice par excellence for applications in the domain of AC motor drives, especially when weighed against its traditional ZS/QZS brethren.

In the ensuing discourse of this chapter, we embark on the meticulous task of architecting a model that resonates with the LC-filter-integrated QZS-IMC [20]. Our investigative lens widens, accommodating an in-depth theoretical probe, both into the realm of voltage magnification potentialities and the nuanced mechanics underpinning the LC filtration mechanics. Further, we illuminate the systematic strategies underlying impedance crafting. The soundness of the model we present, coupled with the foundational theoretical scaffolds and the articulated design methodologies, gains credence from a confluence of simulation results and empirical experimental observations.

5.2 Operation Analysis of Novel QZS-IMC

5.2.1 Discussed Topology

Figure 5.4 delineates the architecture of the LC-filter-integrated QZS-IMC, which fundamentally breaks down into three principal elements: the intrinsic qZS mesh, the front-end rectifier, followed by the terminal inverter. Within the qZS matrix, one can observe a unique amalgamation of three paired inductors: L_{a1} paired with L_{a2}, L_{b1} in conjunction with L_{b2}, and L_{c1} linked to L_{c2}. Complementing this, the design features an ensemble of six capacitors, catalogued as C_{a1}, C_{a2}, C_{b1}, C_{b2}, C_{c1}, and C_{c2}. Furthermore, three bidirectional switches, labeled S_a, S_b, and S_c, are seamlessly integrated. Their congruent switching behaviors permit unified control via a single gate pulse, aptly represented by the symbol S_x.

5.2.2 Buck Operation

When operating in the buck regime, switches labeled S_a, S_b, and S_c come into action. The front-end rectifier and the downstream inverter mimic the behavior of a conventional IMC, striving to generate the designated output voltage levels.

5.2.3 Boost Operation

In Figure 5.5, the functioning of the boost mode can be delineated into two separate phases: the phase involving shoot-through and its counterpart, the non-shoot-through phase. Adjusting the duration of the shoot-through period is key to realizing the targeted voltage amplification levels.

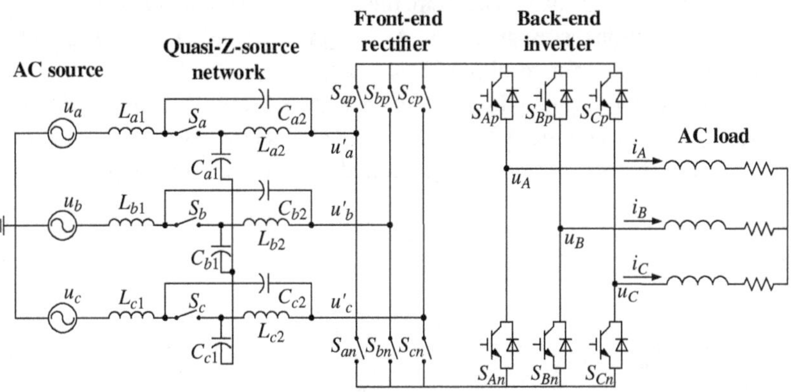

Figure 5.4 LC-filter-integrated quasi-Z-source indirect matrix converter in [19, 20]. Source: Adapted from Liu et.al. (2015)/IEEE.

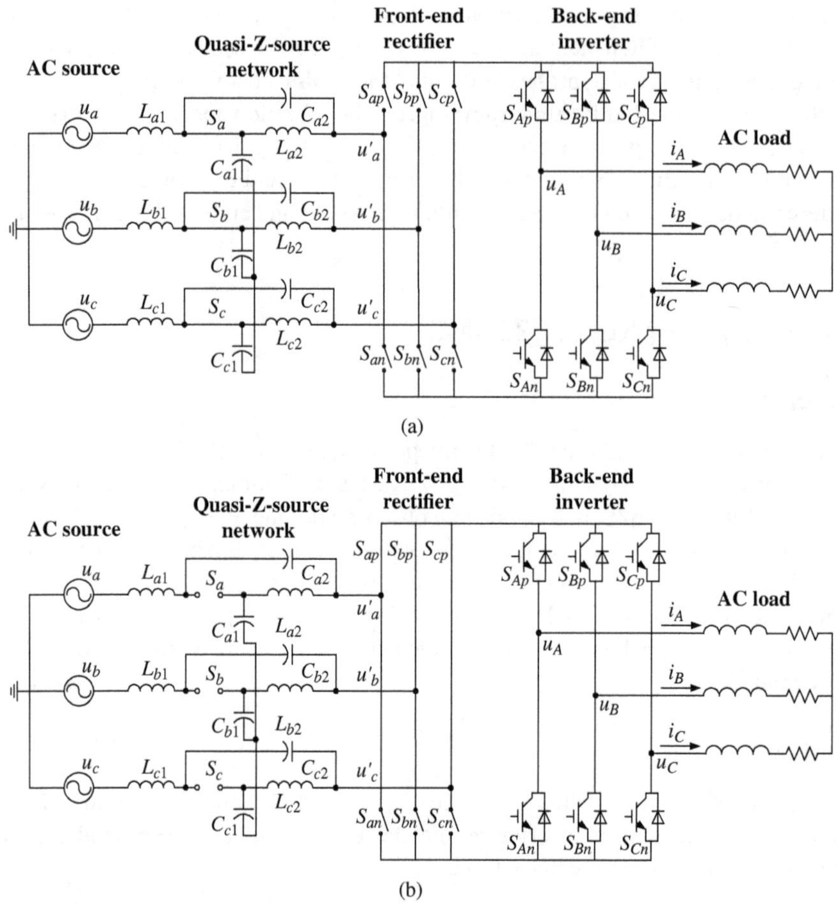

Figure 5.5 Working states of the LC-filter-integrated QZS-IMC. (a) Non-shoot-through state; (b) shoot-through state.

5.2.3.1 Non-shoot-through state

During the phase when no shoot-through occurs, Figure 5.5a offers a visual depiction wherein switches S_a, S_b, and S_c find themselves engaged. In tandem, the preliminary rectifier displays operational characteristics mirroring traditional configurations. As this phase unfolds, inductors transfer energy to their associated capacitors, and the resulting phase voltage output from the qZS network becomes the cumulative sum of the voltages spanning two capacitors. Assuming the conditions: L_{a1} equals L_{a2}, L_{b1} matches L_{b2}, L_{c1} is the same as L_{c2} (collectively represented as L), and C_{a1} parallels C_{a2}, C_{b1} aligns with C_{b2}, C_{c1} mirrors C_{c2} (collectively denoted as C), the inferences drawn from Figure 5.5a suggest:

$$Lp\mathbf{I}_{L1} = \mathbf{u}_{\mathrm{i}} - \mathbf{u}_{C1} - R_L\mathbf{I}_{L1} \tag{5.1}$$

$$Lp\mathbf{I}_{L2} = -\mathbf{u}_{C2} - R_L\mathbf{I}_{L2} \tag{5.2}$$

$$Lp\mathbf{I}_{L1} = \mathbf{u}_{\mathrm{i}} + \mathbf{u}_{C2} - \mathbf{u}' - R_L\mathbf{I}_{L1} \tag{5.3}$$

$$Cp\mathbf{u}_{C1} = \mathbf{I}_{L1} - \mathbf{I}' \tag{5.4}$$

$$Cp\mathbf{u}_{C2} = \mathbf{I}_{L2} - \mathbf{I}' \tag{5.5}$$

where $p = \frac{d}{dt}$; the qZS inductor-L_1 current $\mathbf{I}_{L1} = [i_{La1}\ i_{Lb1}\ i_{Lc1}]^T$, i.e., the grid current; the qZS inductor-L_2 current $\mathbf{I}_{L2} = \left[i_{La2}\ i_{Lb2}\ i_{Lc2}\right]^T$; the qZS capacitor-$C_1$ voltage $\mathbf{u}_{C1} = \left[u_{Ca1}\ u_{Cb1}\ u_{Cc1}\right]^T$; the qZS capacitor-$C_2$ voltage $\mathbf{u}_{C2} = \left[u_{Ca2}\ u_{Cb2}\ u_{Cc2}\right]^T$; the grid voltage $\mathbf{u}_\text{i} = \left[u_a\ u_b\ u_c\right]^T$; the qZS network output voltage $\mathbf{u}' = \left[u'_a\ u'_b\ u'_c\right]^T$; the qZS network output current $\mathbf{I}' = \left[i'_a\ i'_b\ i'_c\right]^T$; R_L is the stray resistance of inductors.

5.2.3.2 Shoot-through state

During the shoot-through state, all switches—S_a, S_b, and S_c—cease operation simultaneously. This condition induces a short circuit at the input region of the preliminary rectifier, leading the inductors to initiate their charging process. The representation of this condition can be observed in Figure 5.5b. Here, the switches within the qZS network maintain an OFF status, while the topmost switches of the initial rectifier get engaged. A closer examination of Figure 5.5b provides the understanding that:

$$Lp\mathbf{I}_{L1} = \mathbf{u}_\text{i} + \mathbf{u}_{C2} - R_L\mathbf{I}_{L1} \tag{5.6}$$

$$Lp\mathbf{I}_{L2} = \mathbf{u}_{C1} - R_L\mathbf{I}_{L2} \tag{5.7}$$

$$Cp\mathbf{u}_{C2} = -\mathbf{I}_{L1}, Cp\mathbf{u}_{C1} = -\mathbf{I}_{L2} \tag{5.8}$$

Within a given switching cycle T_s, if we denote the shoot-through duration as T_0, then the shoot-through duty cycle is represented by the ratio $D = T_0/T_s$. By altering this duty cycle D, it becomes feasible to elevate the QZS network's output voltage, denoted as u′, to the desired level.

5.3 Small-Signal Modeling of QZS-IMC

From (5.1)–(5.8), we obtain the average state-space equations as

$$\begin{cases} Lp\mathbf{I}_{L1} = (1-D)(\mathbf{u}_i + \mathbf{u}_{C2} - \mathbf{u}' - R_L\mathbf{I}_{L1}) + D(\mathbf{u}_i + \mathbf{u}_{C2} - R_L\mathbf{I}_{L1}) \\ Lp\mathbf{I}_{L2} = (1-D)(-\mathbf{u}_{C2} - R_L\mathbf{I}_{L2}) + D(\mathbf{u}_{C1} - R_L\mathbf{I}_{L2}) \\ Cp\mathbf{u}_{C1} = (1-D)(\mathbf{I}_{L1} - \mathbf{I}') - D\mathbf{I}_{L2} \\ Cp\mathbf{u}_{C2} = (1-D)(\mathbf{I}_{L2} - \mathbf{I}') - D\mathbf{I}_{L1} \end{cases} \tag{5.9}$$

It's important to highlight that within the context of both equations (5.9) and subsequent parts of this chapter, by substituting the vectors $\mathbf{I}_{L1}$, $\mathbf{I}_{L2}$, $\mathbf{u}_{C1}$, $\mathbf{u}_{C2}$, $\mathbf{u}'$, and $\mathbf{u}_\text{i}$ with i_{L1}, i_{L2}, u_{C1}, u_{C2}, u', and u_i, it leads us to the derivation of equations corresponding to a single phase as opposed to the combined three phases. This transformation is clearly illustrated in (5.10), representing the average state-space equations for one phase. Given that $L_{a1} = L_{a2} = L_{b1} = L_{b2} = L_{c1} = L_{c2} = L$ and $C_{a1} = C_{a2} = C_{b1} = C_{b2} = C_{c1} = C_{c2} = C$, it's deduced that $i_{L1} = i_{L2} = i_L$ for both (5.9) and (5.10). Moreover, equation (5.10) is an offshoot derived directly from (5.9).

$$\begin{bmatrix} L & 0 & 0 \\ 0 & C_1 & 0 \\ 0 & 0 & C_2 \end{bmatrix} p \begin{bmatrix} i_L \\ u_{C1} \\ u_{C2} \end{bmatrix} = \begin{bmatrix} -R_L & D-1 & D \\ 1-2D & 0 & 0 \\ 1-2D & 0 & 0 \end{bmatrix} \begin{bmatrix} i_L \\ u_{C1} \\ u_{C2} \end{bmatrix} + \begin{bmatrix} 1 & 0 \\ 0 & D-1 \\ 0 & D-1 \end{bmatrix} \begin{bmatrix} u_i \\ i' \end{bmatrix} \tag{5.10}$$

When introducing small-signal perturbations to parameters like the input voltage, the qZS network's output current, and the shoot-through duty cycle, the respective variables become denoted as $u_i = \overline{u}_i + \hat{u}_i$, $i' = \overline{i'} + \hat{i'}$, and $d = D + \hat{d}$. Leveraging the equation (5.10), it allows us to derive the corresponding small-signal state equations, which are:

$$\begin{bmatrix} Lp\hat{i}_L \\ C_1p\hat{u}_{C1} \\ C_2p\hat{u}_{C2} \end{bmatrix} = \begin{bmatrix} -R_L & D-1 & D \\ 1-2D & 0 & 0 \\ 1-2D & 0 & 0 \end{bmatrix} \begin{bmatrix} \hat{i}_L \\ \hat{u}_{C1} \\ \hat{u}_{C2} \end{bmatrix} + \begin{bmatrix} 1 & 0 & u_{C1}+u_{C2} \\ 0 & D-1 & i'-2i_L \\ 0 & D-1 & i'-2i_L \end{bmatrix} \begin{bmatrix} \hat{u}_i \\ \hat{i'} \\ d \end{bmatrix} \tag{5.11}$$

Within the s-plane, representations for small-signal variants of capacitor voltages and currents across the inductors can be articulated as:

$$\begin{cases} \hat{u}_{C1}(s) = G_{C1ui}(s)\hat{u}_i(s) + G_{C1ii}(s)\hat{i'}(s) + G_{C1d}(s)\hat{d}(s) \\ \hat{u}_{C2}(s) = G_{C2ui}(s)\hat{u}_i(s) + G_{C2ii}(s)\hat{i'}(s) + G_{C2d}(s)\hat{d}(s) \\ \hat{i}_L(s) = G_{Lui}(s)\hat{u}_i(s) + G_{Lii}(s)\hat{i'}(s) + G_{Ld}(s)\hat{d}(s) \end{cases} \tag{5.12}$$

where

$$G_{C1ui}(s) = \frac{C_2(1-2D)}{s^2C_2C_1L + sR_LC_2C_1 + (1-D)(1-2D)C_2 - DC_1(1-2D)} \tag{5.13}$$

$$G_{C1ii}(s) = \frac{(D-1)C_2(sL+R_L)}{s^2C_1C_2L + sR_LC_1C_2 + (1-D)(1-2D)C_2 - DC_1(1-2D)} \tag{5.14}$$

$$G_{C1d}(s) = \frac{(1-2D)C_2(u_{C1}+u_{C2}) + C_2(sL+R_L)(i'-2i_L)}{s^2C_1C_2L + sR_LC_1C_2 + (1-D)(1-2D)C_2 - DC_1(1-2D)} \tag{5.15}$$

$$G_{C2ui}(s) = \frac{C_1(1-2D)}{s^2C_1C_2L + sR_LC_1C_2 + (1-D)(1-2D)C_2 - DC_1(1-2D)} \tag{5.16}$$

$$G_{C2ii}(s) = \frac{(D-1)C_1(sL+R_L)}{s^2C_1C_2L + sR_LC_1C_2 + (1-D)(1-2D)C_2 - DC_1(1-2D)} \tag{5.17}$$

$$G_{C2d}(s) = \frac{(1-2D)C_1(u_{C1}+u_{C2}) + C_1(sL+R_L)(i'-2i_L)}{s^2C_1C_2L + sR_LC_1C_2 + (1-D)(1-2D)C_2 - DC_1(1-2D)} \tag{5.18}$$

$$G_{Lui}(s) = \frac{C_1C_2s}{s^2C_1C_2L + sR_LC_1C_2 + (1-D)(1-2D)C_2 - DC_1(1-2D)} \tag{5.19}$$

$$G_{Lii}(s) = \frac{(D-1)^2C_2 + D(D-1)C_1}{s^2C_1C_2L + sR_LC_1C_2 + (1-D)(1-2D)C_2 - DC_1(1-2D)} \tag{5.20}$$

$$G_{Ld}(s) = \frac{sC_1C_2(u_{C1}+u_{C2}) + (D-1)C_2(i'-2i_L) + DC_1(i'-2i_L)}{s^2C_1C_2L + sR_LC_1C_2 + (1-D)(1-2D)C_2 - DC_1(1-2D)} \tag{5.21}$$

5.4 Voltage Gain Investigation

5.4.1 Modeling IMC

The system with an integrated LC-filter QZS-IMC strategically adopts the space vector modulation (SVM) technique for current input within its leading front-end rectifier. Conversely, it harnesses the SVM technique focusing on output voltage when considering its trailing back-end inverter [19]. Delving deeper into the working modes of the said front-end rectifier, they distinctly bifurcate into active and zero vector states. It's imperative to emphasize that, during the stretches of these zero vector phases, there emerges a scenario of the shoot-through zero vector getting inserted. This particular vector induces a short-circuit scenario throughout the tri-phase source, subsequently amplifying the voltage. To understand these dynamics better, the method for calculating vector duty cycles is outlined subsequently.

For the leading front-end rectifier, the modulation index adopts the symbol m_i. Within its realm, the angular position of the input current vector is designated by θ_i. To further refine the understanding, the specific duty cycles attributed to distinct current vectors within an isolated switching period are signified as d_α, d_β, d_s, and d_{or} in (5.22). Transitioning our focus to the latter part of the system, the back-end inverter embodies characteristics reminiscent of a conventional voltage source inverter. Distinguishing this inverter is its integration of six dynamic vectors and an accompanying pair of zero vectors. The delineation and derivation of duty cycles for each of these vectors are detailed in the ensuing sections.

$$\begin{cases} d_\alpha = m_i \sin\left(\frac{\pi}{3} - \theta_i\right) \\ d_\beta = m_i \sin\theta_i \\ d_s = const \ (d_s < 1 - d_\alpha - d_\beta) \\ d_{or} = 1 - d_\alpha - d_\beta - d_s \end{cases} \tag{5.22}$$

$$\begin{cases} d_\mu = m_o \sin\left(\frac{\pi}{3} - \theta_o\right) \\ d_v = m_o \sin\theta_o \\ d_{oi} = 1 - d_\mu - d_v \end{cases} \tag{5.23}$$

Within the architecture of the back-end inverter, a specific term, m_o, represents the modulation index in (5.23). Furthermore, when observing the output voltage vector, its corresponding angle is designated by θ_o. Active vectors that are relevant to this output voltage come with duty cycles labeled as d_μ and d_v. In contrast, the zero vector's duty cycle is referenced as d_{oi}. To ensure a harmonized relationship between the fluctuations of input currents and the ensuing voltages during the modulation phase, it becomes essential to weave in a modulation strategy. This strategy ought to consider the diverse combinations stemming from the operational states of both rectifying and inverting switches. A graphical representation of this switch sequence can be consulted in Figure 5.6. The duty cycles associated with these sequences can be deduced from the following mathematical derivations:

$$\begin{cases} d_{\mu\alpha} = d_\mu \cdot d_\alpha \\ d_{\mu\beta} = d_\mu \cdot d_\beta \\ d_{v\alpha} = d_v \cdot d_\alpha \\ d_{v\beta} = d_v \cdot d_\beta \\ d_{oi\alpha} = d_{oi} \cdot d_\alpha \\ d_{oi\beta} = d_{oi} \cdot d_\beta \end{cases} \tag{5.24}$$

With T_s denoting the period for switching, the duty cycles associated with various output voltage vectors within this period are represented as $d_{\mu\alpha}$, $d_{\mu\beta}$, $d_{v\alpha}$, $d_{v\beta}$, $d_{oi\alpha}$, and $d_{oi\beta}$.

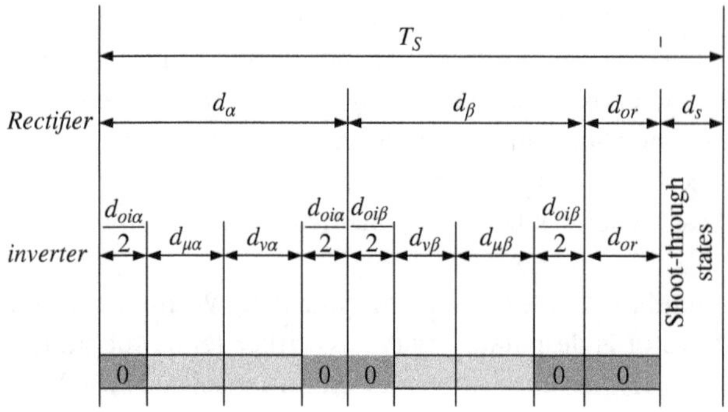

Figure 5.6 Switching sequence for the LC-filter-integrated quasi-Z-source indirect matrix converter.

The switching signals that are generated serve a crucial role in actuating the power switches within the QZS-IMC system. Specifically, on the rectifier's side, a set of six bidirectional switches are incorporated, which contrasts with the inverter section where six insulated gate bipolar transistors (IGBTs) are predominantly used. A unique feature of this configuration is the existence of a virtual DC link. This link provides a connection between the two sides but does not possess a capacitor or any DC/DC conversion mechanism. A comprehensive elucidation regarding the switching strategy for each individual switch spanning the two bridges can be described as follows:

$$S_{xy} = \begin{cases} 1, & \text{Switch } S_{xy} \text{ closed} \\ 0, & \text{Switch } S_{xy} \text{ open} \end{cases}, \quad x = a, b, c, A, B, C; \; y = p, n \tag{5.25}$$

Regarding the rectifier segment, the voltage u_{DC} across the DC-link is determined by the state of the rectifier switches and the qZS network's resultant output voltages, denoted as $\mathbf{u}'$. This relationship can be expressed as:

$$u_{DC} = \begin{bmatrix} S_{ap} - S_{an} & S_{bp} - S_{bn} & S_{cp} - S_{cn} \end{bmatrix} \cdot \mathbf{u}' \tag{5.26}$$

For the qZS network, the resultant output currents, represented as $\mathbf{I}'$, are influenced by the operational states of the rectifier switches and are inherently tied to the current i_{DC} traversing the DC-link. This can be articulated as:

$$\mathbf{I}' = \begin{bmatrix} S_{ap} - S_{an} \\ S_{bp} - S_{bn} \\ S_{cp} - S_{cn} \end{bmatrix} \cdot i_{DC} \tag{5.27}$$

On the inverter's end, the current flowing through the DC link, denoted as i_{DC}, is inherently associated with the operational conditions of the inverter switches. This relationship is further shaped by the output currents labeled as i_A, i_B, and i_C. They are described as:

$$i_{DC} = \begin{bmatrix} S_{Ap} & S_{Bp} & S_{Cp} \end{bmatrix} \cdot \begin{bmatrix} i_A \\ i_B \\ i_C \end{bmatrix} = \begin{bmatrix} S_{Ap} & S_{Bp} & S_{Cp} \end{bmatrix} \cdot \mathbf{I}_o \tag{5.28}$$

In the final phase, the inverter stage's output voltages are formulated, taking into account the switching actions of the inverter switches. The DC-link voltage, u_{DC}, further plays a pivotal role in this formulation. Specifically, they can be articulated as:

$$\mathbf{u}_o = \begin{bmatrix} u_A \\ u_B \\ u_C \end{bmatrix} = \begin{bmatrix} S_{Ap} - S_{An} \\ S_{Bp} - S_{Bn} \\ S_{Cp} - S_{Cn} \end{bmatrix} \cdot u_{DC} \tag{5.29}$$

Within the IMC framework, the switch-driven dynamics, spanning both the rectifier and inverter stages, are articulated mathematically through equations (5.25) to (5.29). An integral facet of ensuring smooth IMC operation hinges on the ability to sustain a DC-link voltage that remains non-negative, expressly articulated as u_{DC} being greater than or equal to zero.

5.4.2 Voltage Gain Analysis

In conditions characterized by a steady state, both the mean voltage traversing the inductor and the mean current coursing through the capacitor equate to zero within one complete switch cycle. Drawing insights from equations (5.9), we can deduce that:

$$\begin{cases} u_{C1} = \dfrac{(1-D)u_i - R_L i_L}{1-2D} \\ u_{C2} = \dfrac{Du_i - R_L i_L}{1-2D} \\ u' = \dfrac{u_i - 2R_L i_L}{1-2D} \\ i_{L1} = i_{L2} = i_L \\ i' = \dfrac{1-2D}{1-D} i_L \end{cases} \tag{5.30}$$

As inferred from equations (5.30), disregarding the resistance of the inductor, the factor denoted as B, responsible for boosting the voltage, can be formulated as follows:

$$B = \frac{u'}{u_i} = \frac{1}{1-2D} \tag{5.31}$$

Based on the set of equations from (5.22) to (5.29), we can derive the average DC-link current, represented as I_{DC}, and the peak virtual DC-link voltage, denoted as V_{DC}, associated with the IMC.

$$I_{DC} = \frac{1}{1-D} I_{om} m_o \cos\theta_{out},\ V_{DC} = \frac{3}{2} V_{im} m_i \cos\theta_{ind} \tag{5.32}$$

In this context, I_{om} signifies the magnitude of the inverter's output current, namely $\boldsymbol{I}_o$. The term θ_{out} corresponds to the phase offset or displacement of the output, while V_{im} symbolizes the magnitude of the rectifier's input voltage, represented as $\boldsymbol{u'}$. Furthermore, θ_{ind} is used to indicate the phase displacement at the input side. Given the principle of power equilibrium on the inverter's end, it can be concluded that

$$(1-D) V_{DC} I_{DC} = \frac{3}{2} V_{om} I_{om} \cos\theta_{out} \tag{5.33}$$

In this formulation, V_{om} signifies the magnitude of the voltage output, labeled as $\mathbf{u}_o$, from the inverter. Figure 5.4 illustrates the arrangement of the input current vector, $\mathbf{I}_{L1}$, in alignment with the input voltage vector, $\mathbf{u}_i$, ensuring adherence to a unity power factor as mentioned in reference [20]. Drawing insights from equations (5.32) and (5.33), there exists a relationship between the amplitude of the inverter's output voltage and that of the input voltage.

$$V_{om} = V_{im} m_i m_o \tag{5.34}$$

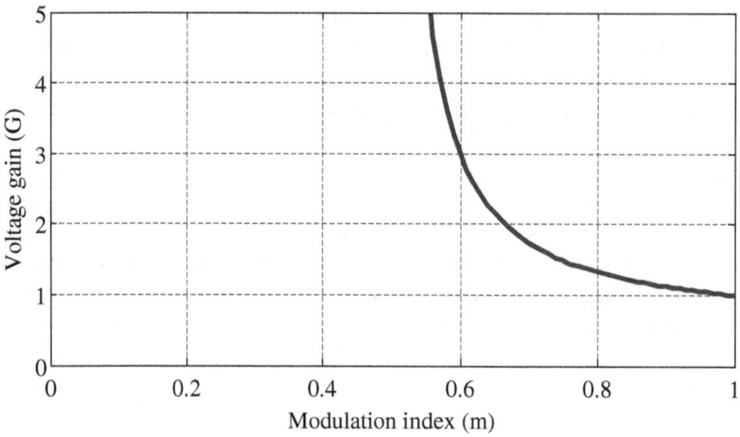

Figure 5.7 Voltage gain versus modulation index of the novel QZS-IMC.

Subsequently, by utilizing the information presented in equations (5.31) and (5.34), the voltage gain, represented by G, associated with the QZS-IMC under consideration can be calculated as follows:

$$G = \frac{V_{om}}{U_m} = \frac{m_i m_o}{1 - 2D} = Bm \tag{5.35}$$

In our present discussion, the grid voltage magnitude is symbolized by U_m. The term modulation index is designated by m, holding equivalence with $m_i m_o$. As illustrated in Figure 5.7, a clear relation exists between the modulation index (m) and voltage gain. With the judicious choice of the modulation index (m) combined with the shoot-through duty ratio (D), the LC-filter-integrated QZS-IMC has the potential to achieve a voltage gain surpassing 0.866.

5.5 QZS Network's Filtering Function Investigation

An in-depth study was conducted on the qZS network's filtering attributes using both large signal analysis within the circuit framework and s-domain analysis for minor signal variations.

5.5.1 Circuit Large Signal Analysis

Figure 5.4 depicts the continuity of the grid current $\mathbf{I}_{L1}$, a result of the qZS inductors L_{a1}, L_{b1}, and L_{c1}. This innovative QZS-IMC configuration negates the need for supplementary input filters, while retaining both the voltage boost capability and the features of the standard IMC. When contrasting the typical ZS/QZS MCs from Figure 5.3 with the LC-filter-integrated QZS-IMC of Figure 5.4, it's apparent that the latter seamlessly draws a consistent AC current from the grid without necessitating extra input filters. This demonstrates the integrated filtering capability of the LC-filter-combined QZS system. On the contrary, the converter designs in Figure 5.3 depend on added LC filters to ensure a stable grid current, as the absence of such filters would result in irregular AC currents. Furthermore, the erratic input currents observed in the ZS and QZS designs of Figure 5.3 peak significantly higher than those in Figure 5.4, leading to increased current stress.

In the traditional ZS/QZS MCs, illustrated by Figure 5.3, the inclusion of an LC filter becomes necessary to eliminate the high-frequency elements present in the ZS/QZS network's input currents. Taking Figure 5.3b as a reference, the input currents—i_{fa}, i_{fb}, and i_{fc}—within the QZS network are characterized by their discontinuity and pronounced high-frequency elements, making them unsuitable for direct grid integration due to their heightened harmonic levels. This prompts the integration of an auxiliary LC filter between the grid and the QZS network in Figure 5.3b. Consequently, the resultant grid currents, namely i_a, i_b, and i_c in Figure 5.3b, manifest as continuous with minimal harmonic distortions. When examining the phase "a" of Figure 5.3b, this LC filter operates as a second-order system and is represented as:

$$\begin{cases} L_f p i_a = (u_a - u_{fa}) - R_f i_a \\ C_f p u_{fa} = i_a - i_{fa} \end{cases} \Rightarrow (L_f C_f p^2 + R_f C_f p + 1)\ i_a = i_{fa} + C_f p u_a \tag{5.36}$$

Referring to Figure 5.3b, the symbols L_f, C_f, and R_f delineate the inductance, capacitance, and damping resistance of the filter, respectively. In the QZS network, u_{fa} and i_{fa} signify the input voltage and current of phase-a, whereas the grid's phase-a voltage and current are captured by u_a and i_a.

Building on the innovative QZS-IMC, by referencing the expansive signal state-space equation denoted as (5.9), we deduce the following:

$$\begin{cases} Lp(i_{L1} + i_{L2}) = u_i - (1 - 2D)u' - R_L(i_{L1} + i_{L2}) \\ Cp(u_{C1} + u_{C2}) = (1 - 2D)(i_{L1} + i_{L2}) - 2(1 - D)i' \end{cases} \tag{5.37}$$

Using equation (5.30), the output voltage of the qZS network in Figure 5.4 is determined to be $u' = u_{C1} + u_{C2}$. Subsequently, by applying equation (5.37), we derive the second-order system of the qZS network illustrated in Figure 5.4, as follows:

$$[LCp^2 + R_L Cp + (1 - 2D)^2]\ (i_{L1} + i_{L2}) = 2(1 - D)(1 - 2D)\ i' + Cpu_i$$

As a result of factor $i_{L1} = i_{L2}$, it leads to

$$[LCp^2 + R_L Cp + (1 - 2D)^2]\ i_{L1} = (1 - D)(1 - 2D)\ i' + 0.5\ Cpu_i \tag{5.38}$$

Equation (5.38) mirrors the expression found in (5.36), albeit with differing coefficients. This underlines the inherent filtering capability of the qZS network in association with the IMC bridges, as depicted in Figure 5.4.

5.5.2 S-Domain Small-Signal Analysis

In Figure 5.4, the grid currents, denoted as I_{L1}, are characterized by the small-signal model in equation (5.12). When visualized through a signal flow diagram, this is represented in Figure 5.8 [20].

The transfer function denoted as $G_{Lui}(s)$ is associated with the grid voltage $\boldsymbol{u}_i$, and its characterization is uninfluenced by the qZS network's output current $\boldsymbol{I}'$. Conversely, $G_{Lii}(s)$ denotes a transfer function associated with $\boldsymbol{I}'$, which operates independently from the grid voltage $\boldsymbol{u}_i$. Another significant function is $G_{Ld}(s)$, which corresponds to the shoot-through duty ratio D. The relationship between the current $\boldsymbol{I}'$ and the grid currents $\boldsymbol{I}_{L1}$ is articulated by $G_{Lii}(s)$. As

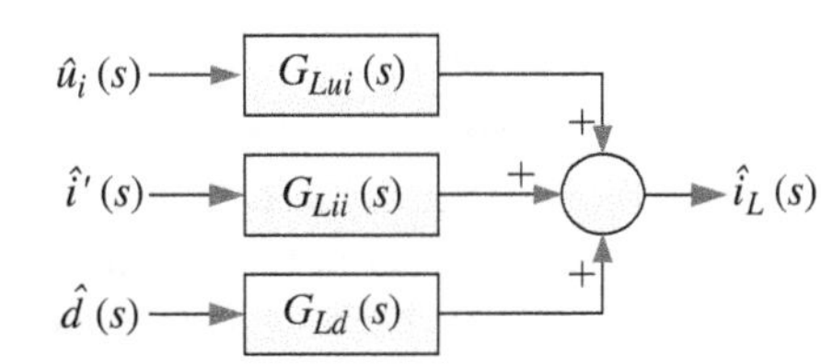

Figure 5.8 Signal flow diagram of small-signal model of the grid currents I_{L1} [20]. Source: Liu et.al. (2016)/with permission of IEEE.

evidenced in expression (5.20), this relationship mirrors that of a standard LC filter. This indicates the inherent filtering capability of the novel QZS-IMC's qZS network via $G_{Lii}(s)$. When making an assumption that L_1 equals L_2 (both represented as L) and C_1 equals C_2 (denoted as C) in relation to (5.20), the characteristic frequency, ω_n, and the damping factor, ζ, for $G_{Lii}(s)$ are as follows:

$$\begin{cases} \omega_n = \dfrac{1-2D}{\sqrt{LC}} \\ \zeta = \dfrac{(R_L + R_C)C}{2(1-2D)\sqrt{LC}} \end{cases} \tag{5.39}$$

The behavior of $G_{Lii}(s)$ reveals a direct influence from the qZS network's capacitance, its inductance, and the duty cycle during shoot-through phases. Turning our attention to Figure 5.9, which illustrates the Bode plots for the transfer function $G_{Lii}(s)$, a few observations become clear. Specifically, as the inductance and capacitance of qZS, as well as the shoot-through duty ratio, experience an increment, both the resonant and crossover frequencies exhibited by $G_{Lii}(s)$ tend to reduce. This decline tends to foster enhanced stability within the system. Moreover, an escalated qZS inductance and capacitance lead to a pronounced shift in the amplitude–frequency curve, resulting in a higher quality factor. In contrast, a surge in the shoot-through duty ratio introduces a more subtle variation in the amplitude–frequency response, leading to a reduction in the quality factor.

5.6 Parameters Design of QZS Network

Drawing insights from the prior filtration functionality analysis, it becomes evident that the inductor within the qZS setup plays a pivotal role in moderating the current ripples at switching frequencies. Such moderation ensures that the fluctuations present in the grid current remain confined within stipulated bounds. To put this into perspective, consider the guidelines set by IEEE 519 [21], which mandate that the THD of the grid current not exceed 5% of the foundational current. Concurrently, the capacitor of the qZS configuration is tasked with curtailing the ripple effects observed in the network's output voltage, denoted as $\mathbf{u}'$. Furthermore, it's imperative to maintain the grid current in a manner that aligns with the unity power factor. This means that the deviation angle at the grid's foundational frequency, f_0, should approximate zero.

5.6.1 Switching Frequency Ripple Limit

Referring to Figure 5.5a and taking phase a into consideration for our calculations, we can determine the ripple present in the inductor current as follows:

$$\Delta i_L = \frac{|u_a - u_{Ca1}|(1-D)}{f_s L_1} \tag{5.40}$$

Upon examining Figure 5.5b, we can derive the ripple associated with the capacitor voltage in the following manner:

$$\Delta u_C = \frac{i_{La2} D T_s}{C_2} \tag{5.41}$$

Let's postulate that the acceptable ripple for inductor current falls within $k_1 I_{\text{rated}}$, while for the capacitor voltage it should not exceed $k_2 V_{C,\text{rated}}$. Here, k_1 and k_2 represent specific coefficients. Furthermore, I_{rated} corresponds to the standard grid current, and $V_{C,\text{rated}}$ signifies the rated voltage

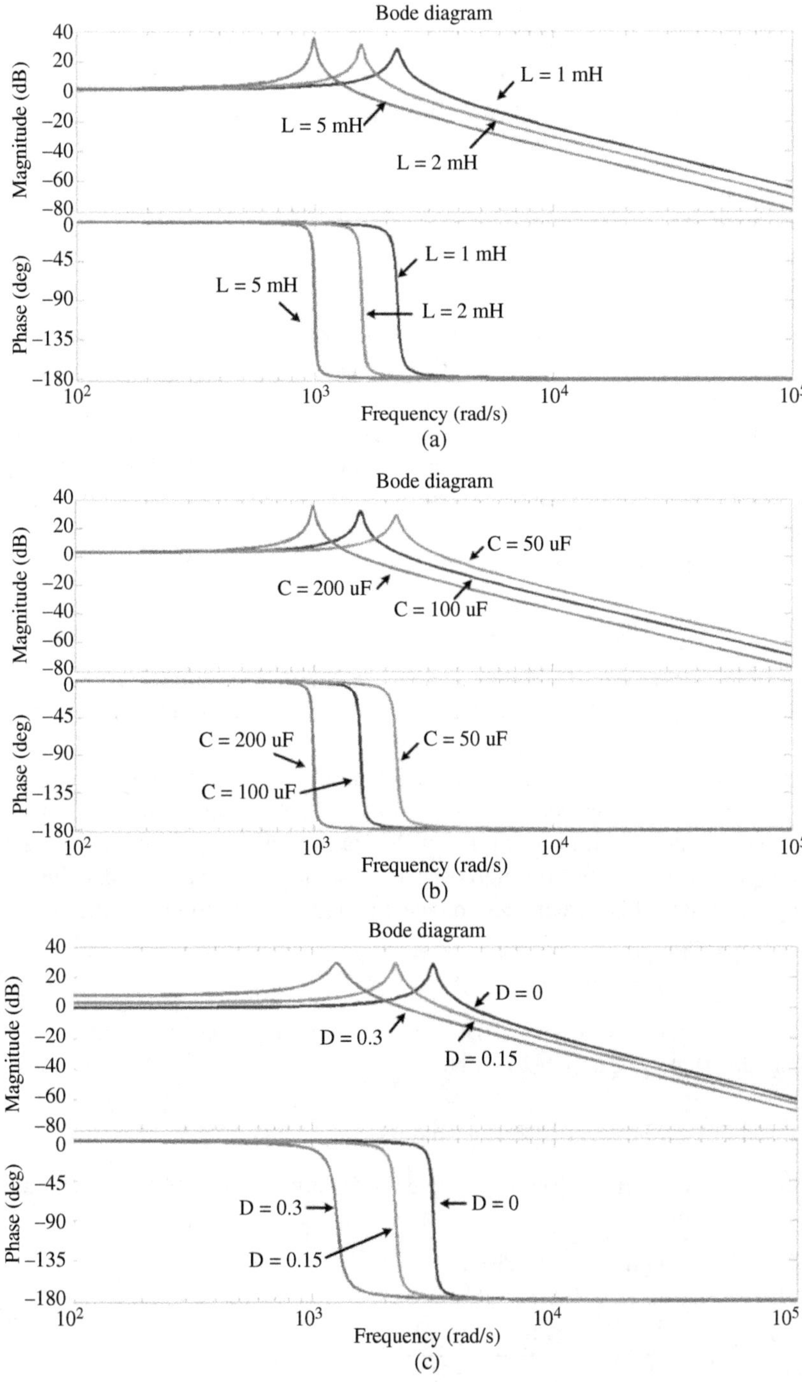

Figure 5.9 Bode diagrams of qZS network transfer function $G_{Lii}(s)$. (a) At inductance changes; (b) at capacitance changes; (c) at changes of shoot-through duty cycle.

for capacitor-C_2. To elaborate, $V_{C,\text{rated}}$ can be expressed as $DV_s/(1-2D)$, where V_s stands for the grid voltage's root mean square value. A notable observation is that the coefficients k_1 and k_2 have dependencies on the ripple associated with the switching frequency of the grid current $\boldsymbol{I}_{L1}$ as well as the qZS network's output voltage, $\boldsymbol{u}'$. Consequently, the required inductance for qZS should be:

$$L \geq \frac{U_m D(1-D)}{(1-2D) f_s k_1 I_{\text{rated}}} \tag{5.42}$$

The capacitance associated with qZS is defined as:

$$C \geq \frac{I_{\text{rated}}(1-2D)}{f_s k_2 V_s} \tag{5.43}$$

In the quest for optimal k_1 and k_2 values, various considerations become pivotal. One should note that a more substantial k_1 choice is instrumental in shrinking the requisite inductance. However, such a selection might inadvertently infuse pronounced harmonic elements into the grid structure. As a prudent measure, maintaining k_1 under a 5% threshold is advocated. Conversely, gravitating toward a magnified k_2 can be instrumental in curbing capacitance demands. Yet, this might pose risks to the consistency and quality of the output voltage inherent to the QZS-IMC. Shifting our focus to the QZS-IMC's virtual DC-link voltage (u_{DC}), emblematic of the qZS network's interline output voltage magnitude, we find its zenith value delineated by equation (5.31) as:

$$u_{DC,\max} = \sqrt{3}u'_{a,\max} = \frac{\sqrt{3}}{1-2D} u_{a,\max} = \frac{\sqrt{3}}{D} u_{Ca2,\max} \tag{5.44}$$

When we set the parameter D to a value of 1/3, the maximal possible u_{DC} is accurately represented as a product of 5.19 and $u_{Ca2,\max}$. From this, we can deduce a correlation between $k_2 V_{C,\text{rated}}$ and $u_{DC,\max}$; specifically, it's described by taking $u_{DC,\max}$, multiplying with k_2, and subsequently dividing the result by 5.19. An observation worth noting is that the qZS network's output voltage manifests a pronounced elevation in high-frequency voltage disturbances. This observation finds its roots in the synergistic influence of a pair of series capacitors, labeled as C_1 and C_2, operating within the qZS circuit during intervals devoid of shoot-through occurrences. This leads us to the conclusion that the proportionality factor of the high-frequency voltage fluctuation in the context of the virtual DC-link voltage is illustrated as:

$$\delta = 2k_2 V_{C,\text{rated}}/u_{DC,\max} = 2k_2/5.19 \tag{5.45}$$

Typically, the voltage ripple within the virtual DC-link remains negligible if δ is under 5%. Therefore, it's recommended that k_2 be kept below 13%.

5.6.2 Power Factor and Cut-off Frequency Requirements

As derived in reference [22], the input displacement angle for the innovative QZS-IMC is characterized as follows:

$$\begin{aligned}\theta_{ind} &= \arctan\frac{2\pi f_0 C u'}{I_{\text{rated}}} - \arctan\frac{2\pi f_0 L I_{\text{rated}}}{u'\left[1-4\pi^2 f_0^2 LC\right]} \\ &= \arctan\frac{2\pi f_0 C u_i}{(1-2D)I_{\text{rated}}} - \arctan\frac{2\pi f_0 (1-2D) L I_{\text{rated}}}{u_i\left[1-4\pi^2 f_0^2 LC\right]}\end{aligned} \tag{5.46}$$

Therefore, there will be

$$\cos\theta_{ind} \geq pf_{\min} \tag{5.47}$$

In our discourse, the parameter $pf_{\min}$ serves as an anchor, symbolizing the foundational power factor benchmark. Drawing upon the intricate signal evaluations detailed in Section 5.5 and taking into account the mathematical relations presented in equation (5.39), it becomes evident that the critical cut-off frequency for $G_{Lii}(s)$, identified as ω_n, must be strategically positioned. Specifically, it should reside within a spectrum that is distinctly below a magnitude of tenfold the primary switching frequency.

$$\omega_n = \frac{1-2D}{\sqrt{LC}} \leq \frac{2\pi f_s}{10} \tag{5.48}$$

Drawing insights from equations (5.42) to (5.48), we implement a structured approach for determining the qZS inductance and capacitance within the context of the advanced QZS-IMC. Specifically: Given a set grid voltage u_i (covering $i = a, b, c$), the stipulated grid current denoted as I_{rated}, and a fixed shoot-through duty ratio D, we define values for k_1 and k_2. Constraints for these are k_1 being at most 5% and k_2 capped at 13% based on prior discussions. Utilizing equations (5.42) and (5.43), we can then derive the L and C values, ensuring that the voltage and current ripples due to switching frequency are confined within acceptable thresholds. The L and C figures ascertained in the initial step are then cross-referenced to deduce the input power factor, termed $\cos\theta_{ind}$, and the cut-off frequency, ω_n, as guided by equations (5.46) through (5.48). To finalize the choice of L and C, two conditions are verified: the derived $\cos\theta_{ind}$ must exceed the baseline power factor $pf_{\min}$, and ω_n should be kept below a tenth of the prevailing switching frequency. If these criteria aren't met, the ripple constraints k_1 and k_2 are adjusted and the calculation process beginning from the first step is reiterated.

5.7 Simulation and Experimental Results

To ensure the reliability of the analytical outcomes of the LC-filter-integrated QZS-IMC shown in Figure 5.4, we carried out both simulation-based and hands-on evaluations. This also facilitated a comparison between this novel converter and the conventional QZS-IMC and ZS-IMC, as presented in Figure 5.3b,c. For these comparisons, a 1.5-kW system capacity was adopted for the ZS/QZS-IMC. Table 5.1 provides the comprehensive specifications.

The grid voltage is 50 V in magnitude. With a boosting factor, denoted by B, being 2, it gives rise to a duty ratio, symbolized as D, valued at 0.25. This results in an output voltage level of 100 V for the qZS system and a set switching frequency, f_s, at 10 kHz. Using the design approach detailed in Section 5.6, the qZS system's parameters, namely k_1 and k_2, are iteratively determined to be 5% and 10%, in that order. From these determinations, we obtain an inductance, L, for qZS of 1.9 mH and a capacitance, C, of 283 μF. The power factor is documented at $\cos\theta_{ind} = 0.957$, with a corresponding

Table 5.1 Parameters in simulation and experiment.

Description	Value
Power rating	1.5 kW
Capacitance of ZS/qZS network	300 μF
Inductance of ZS/qZS network	2 mH
Shoot-through duty cycle	0.15
Switching frequency, f_s	10 kHz
Input voltage frequency, f_0	50 Hz
Load resistance, R_LR$_L$	5 Ω
Load inductance, L_LL$_L$	3 mH

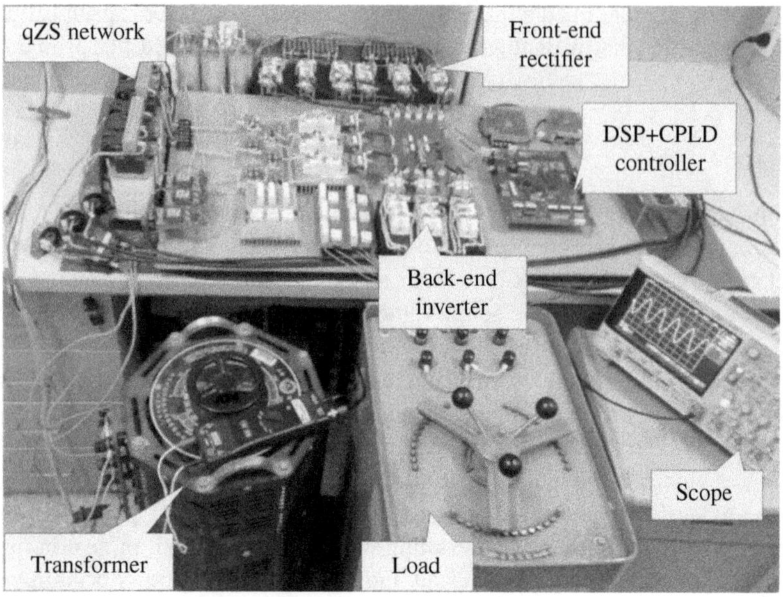

Figure 5.10 Experimental setup of the ZS/QZS-IMC.

cut-off frequency, ω_n, of 687 rad/s. Reflecting these outcomes, the constructed prototype incorporates a qZS inductance of 2 mH and a capacitance value of 300 μF.

The experimental configuration is depicted in Figure 5.10. Utilizing a digital signal processor (DSP), the controller oversees the primary algorithm, while the complex programmable logic device (CPLD) manages the gate signal extensions for the various switches. On the grid's end, a transformer is integrated to downscale the voltage magnitude prior to its introduction to the ZS/QZS-IMC. This testing setup offers flexibility, allowing for assessments of both the conventional QZS-IMC and ZS-IMC as displayed in Figure 5.3b,c. Moreover, by modifying specific connections and components, the LC-filter-integrated QZS-IMC from Figure 5.4 can also be evaluated.

5.7.1 Investigation of Modeling

Upon detailed evaluation of the small-signal model, spanning equations (5.11) through (5.21), we set out to juxtapose it with findings sourced from both circuit simulations and hands-on experiments, a necessary step to validate its efficacy and accuracy. These multidimensional insights are clearly showcased in Figure 5.11. Delving deeper, subsections (a), (b), and (c) of this figure offer a clear visualization of the qZS-network's nuanced metrics: input voltage (u_i), input current (i_L), and the pivotal output voltage (u'), respectively. A critical observation worth highlighting is that the pronounced peak evident in the qZS-network output voltage curve, seen in Figure 5.11a, has its roots firmly anchored in the derivations from the small-signal model. To fortify the credibility of our comparative assessment, all scenarios were consistently exposed to a grid voltage amplitude set at 50-V, harmonized with a 0.15 shoot-through duty cycle.

Upon examination of Figure 5.11, it's evident that there's a notable coherence among the outcomes derived from the small-signal model, the outputs from circuit simulations, and the empirical results obtained experimentally. Notwithstanding this consistency, a nuanced deviation is evident in the qZS-network's output voltage denoted as u'. This minor diminution in voltage, when juxtaposed with simulated data, likely stems from intrinsic voltage drops typical of inductors and operational power components.

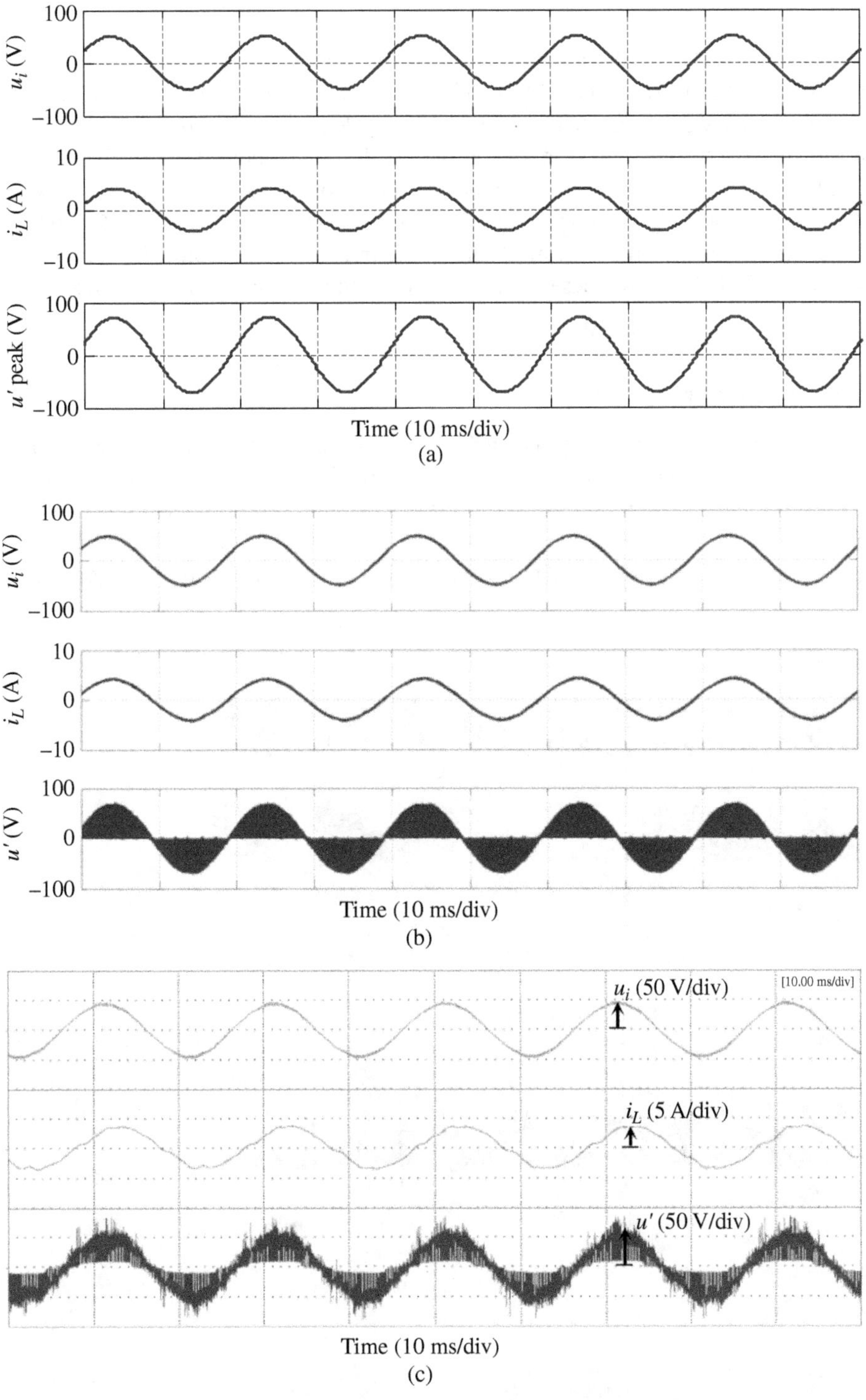

Figure 5.11 Analysis of the novel QZS-IMC's parameters: input voltage (u_i), current (i_L), and resulting output voltage (u') of the qZS-network. (a) Derived from small-signal model insights; (b) based on circuit simulation data; and (c) observations from practical experiments.

5.7.2 Voltage Gain Verification

In Figure 5.12, both parts (a) and (b) present outcomes from circuit simulations and experimental observations of the LC-filter-integrated QZS-IMC. These results consider a grid voltage amplitude of 50 V, an output frequency of 30 Hz, a shoot-through duty cycle set at 0.15, and simple boost control. When one disregards the stray resistance associated with the inductors, the calculated boost factor from (5.31) is determined to be 1.43. The simulated peak voltage for the DC-link showcased in Figure 5.12a measures at 71 V, yielding a boost factor of $B = 71/50 = 1.42$ (71/50). Meanwhile, the experimental peak voltage for the DC-link, as illustrated in Figure 5.12b, comes in at 69 V, with the resulting boost factor being 1.38. These findings exhibit a notable consistency with theoretical expectations.

5.7.3 Filtering Function Verification

To thoroughly understand the filtering proficiencies of various network configurations, namely the ZS, QZS, and LC-filter-integrated-QZS systems, we draw insights from Figures 5.3 and 5.4. The distinct configurations—ZS-IMC, QZS-IMC, and LC-filter-integrated-QZS-IMC—are precisely delineated in Figures 5.3b,c and 5.4, respectively. Complementing these, the real-time input current measurements for each converter system under experimental conditions are vividly portrayed in Figures 5.13–5.15.

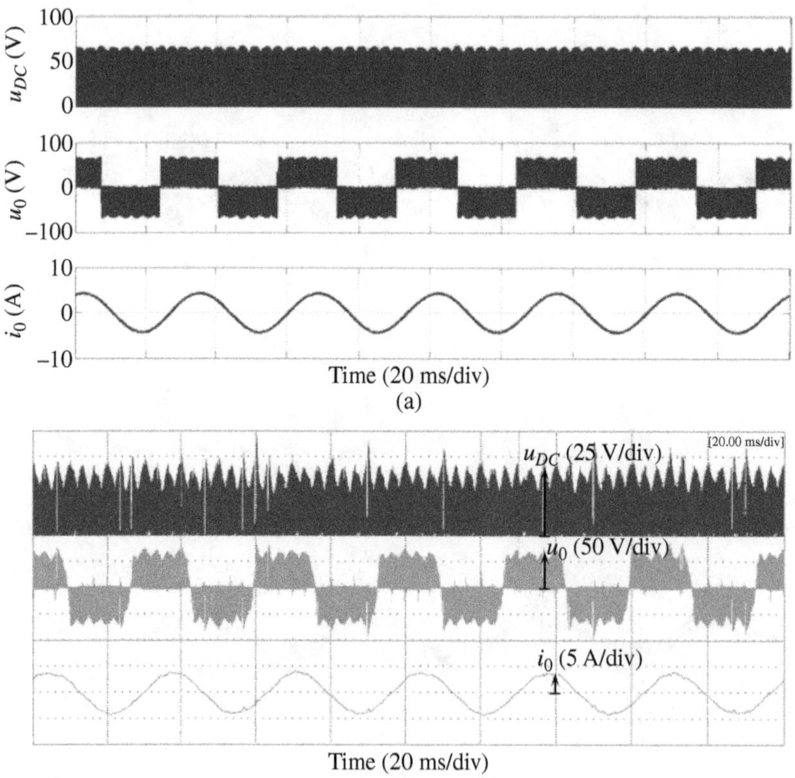

Figure 5.12 Depicting the attributes of the novel QZS-IMC—namely, the DC-link voltage (u_{DC}), the subsequent output pulse voltage (u_o), and the accompanying load current (i_o). (a) Presents the outcomes from simulation studies; (b) illustrates findings from hands-on experimentation.

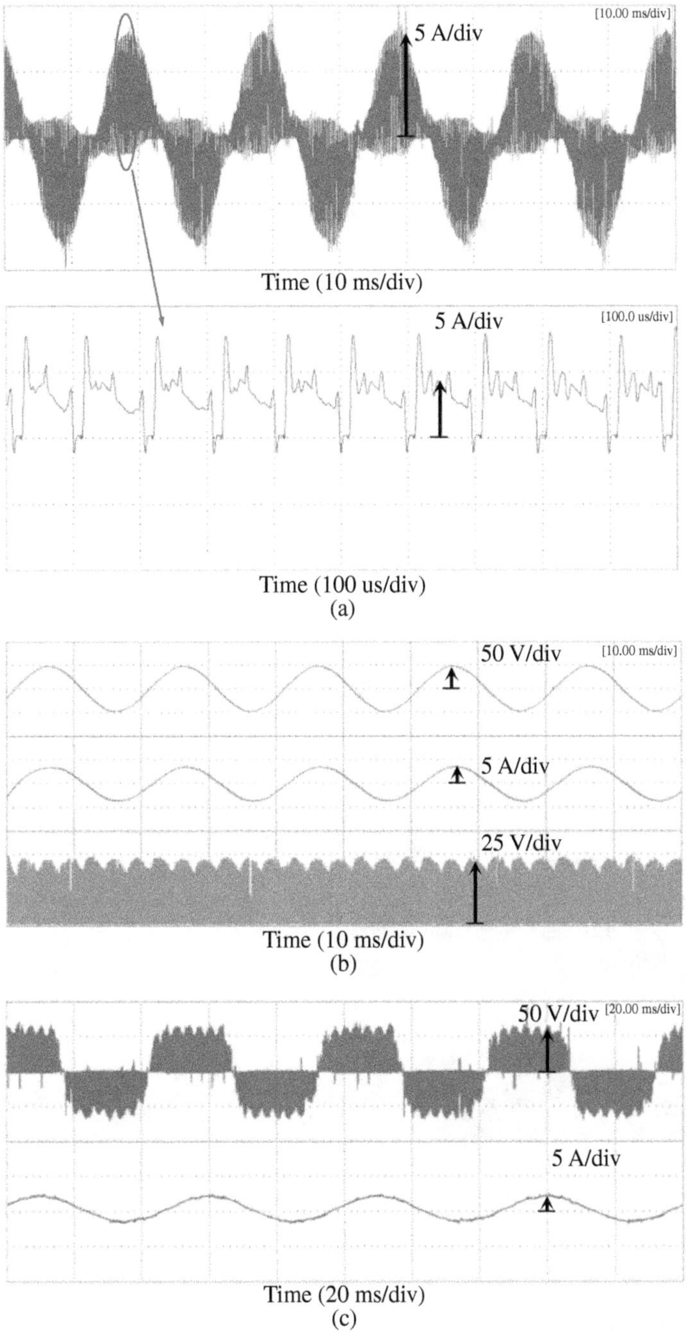

Figure 5.13 Assessment outcomes derived from evaluating the ZS-IMC depicted in Figure 5.3c. (a) Demonstrates the captured current from the ZS configuration; (b) portrays the grid's prevalent voltage alongside its current, emphasizing the value of DC-link voltage known as u_{DC}; (c) highlights the emitted voltage pulse together with the corresponding load current.

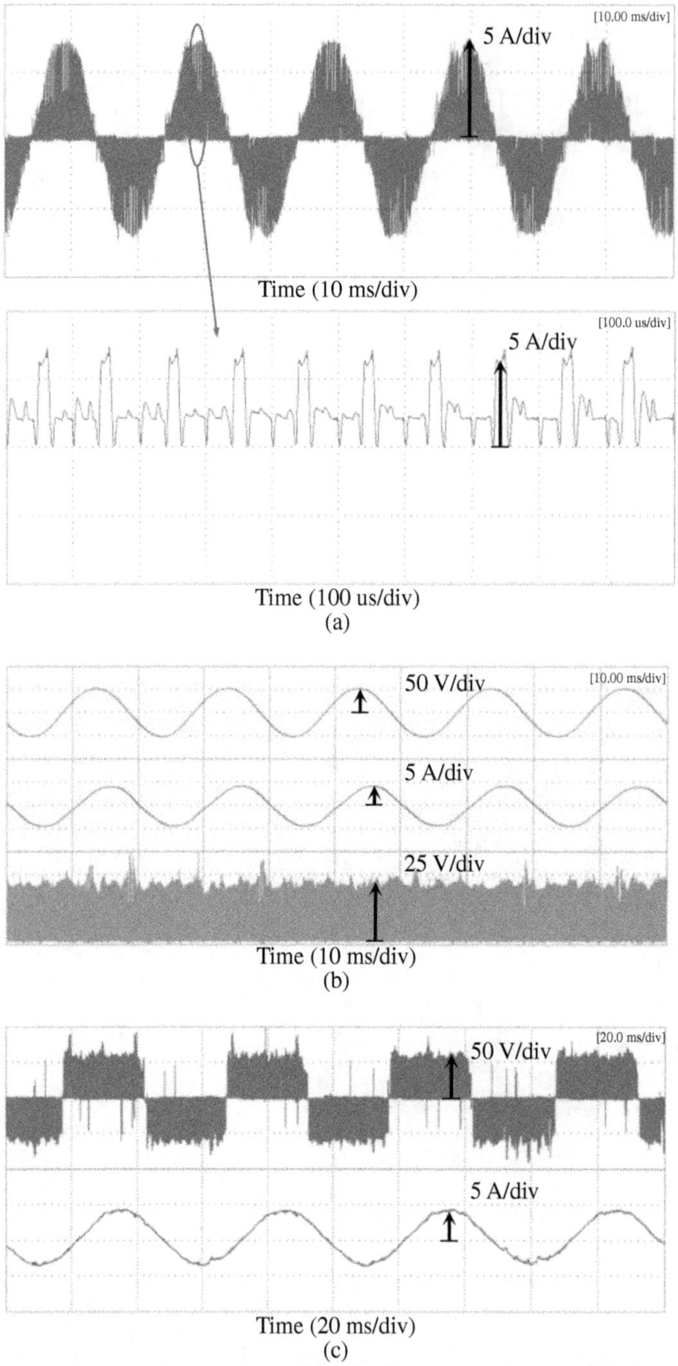

Figure 5.14 Outcome of the experimental assessment for the QZS-IMC as represented in Figure 5.3b. (a) Displays the input stream for the QZS configuration; (b) showcases both the grid's voltage and its subsequent current with an emphasis on the DC-link voltage, termed u_{DC}; (c) illustrates the emitted pulse voltage in conjunction with the prevailing load current.

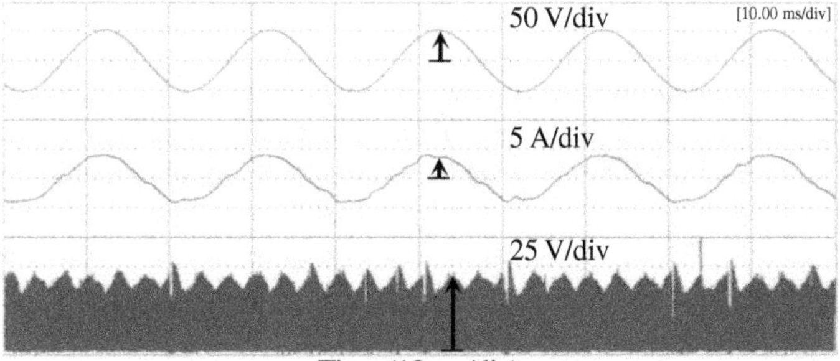

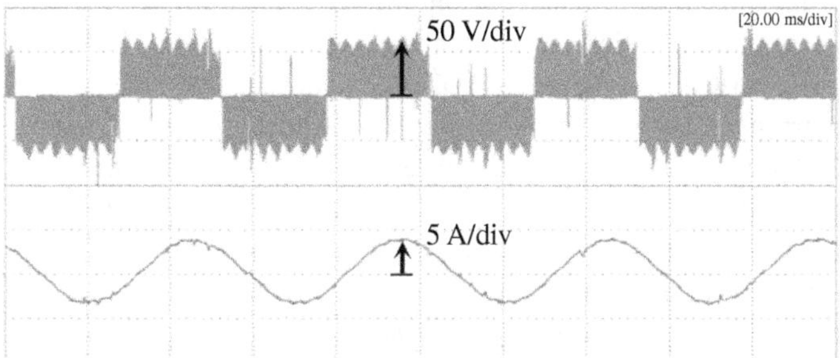

Figure 5.15 Results from an experimental assessment of the LC-filter-integrated QZS-IMC as depicted in Figure 5.4. (a) Depictions of both the grid voltage and associated current alongside the DC-link voltage, labeled as u_{DC}. Note: The grid current specifically represents the input stream for the LC-filter-integrated QZS system. (b) Demonstrations of the output's pulse voltage paired with the consequent load current.

The ZS-IMC, represented in Figure 5.3c, when assessed, unveils an intermittent waveform in Figure 5.13a, registering a THD value of 55.61%. In parallel, referencing the QZS-IMC detailed in Figure 5.3b, the non-continuous nature of the input current in Figure 5.14a culminates in a THD of 58.06%. Offering a contrasting view, the LC-filter-integrated QZS-IMC illustrated in Figure 5.4, upon examination, emanates a smooth input current waveform, as observed in Figure 5.15a, with a dramatically diminished THD of 5.22%.

A meticulous perusal of Figures 5.13–5.15 accentuates that the LC-filter-integrated QZS-IMC, as presented in Figure 5.4, is poised for direct grid integration, obviating the need for auxiliary input filters. The hallmark of this adaptation lies in the intrinsic filtering capability of the LC-filter-integrated QZS framework. Contrarily, the QZS and ZS models from Figure 5.3 mandate supplementary filtration measures prior to any grid assimilation. Striving to align with standard THD benchmarks (under 5%), these models integrate a supplementary input filter, endowed with a 3-mH inductance and a 50-μF capacitance. When placed side-by-side with the avant-garde QZS-IMC (Figure 5.4), it becomes evident that the QZS and ZS iterations become more resource-exhaustive, cumbersome, and prone to incremental energy dissipations.

Furthermore, a deep dive into Figures 5.13b and 5.14b offers grid current analytical data, revealing harmonically enriched waveforms with THDs of 2.08% and 2.94%, respectively. However, when evaluating energy efficiency, the data indicates the following: the LC-filter-integrated QZS-IMC achieves 89.67%; the ZS-IMC, 88.83%; and the QZS-IMC, 83.61%. This comparison clearly demonstrates the heightened energy wastage in ZS and QZS-IMCs, particularly when ancillary filters are considered.

Figures 5.13c, 5.14c, and 5.15b illustrate the 20-Hz load currents and their corresponding pulse voltages for the ZS-IMC, QZS-IMC, and the newly developed QZS-IMC, in that order. Upon examining Figure 5.13c, it becomes clear that the load current is comparatively lower than what's showcased in the subsequent diagrams. This decreased magnitude of load current can be attributed to the voltage gain of the ZS-IMC depicted in Figure 5.3c. Specifically, this gain doesn't match the elevated levels observed in Figures 5.3b and 5.4.

5.8 Conclusion

In this segment, we embarked on a thorough investigation into the LC-filter-integrated QZS-IMC, focusing on aspects like small-signal modeling, voltage gain appraisal, and its unique filtering characteristics, along with parameterization intricacies. By embedding the qZS network at the junction between the grid and the IMC bridge using the novel QZS-IMC methodology, we realized a two-tiered filtering process. This arrangement effectively eliminated the need for supplementary input filters while bestowing the IMC with a voltage gain that exceeds 0.866. Additionally, we delved into small-signal modeling pertaining to the qZS network. Building on this groundwork, we extracted the voltage gain, gauged its filtration capability, and pinpointed the most fitting values for qZS inductance and capacitance within this avant-garde IMC structure. As a testament to the theoretical propositions and parameter layouts, a tangible prototype was assembled. Minor variations arose between the small-signal representation, circuit emulation, and hands-on experimental data, attesting to the model's credibility. Notably, voltage gains derived from simulation efforts closely matched those from hands-on experiments, aligning with our calculated anticipations. Concluding our analysis, an impedance parameter design highlighted the commendable input current profiles and THD of the LC-filter-integrated QZS-IMC. This stands in sharp contrast to the conventional ZS-IMC and QZS-IMC, underscoring its superior filtration proficiency.

References

1. S. Safari, A. Castellazz, and P. W. Wheeler, "Experimental and analytical performanc evaluation of SiC power devices in the matrix converter," *IEEE Trans. Power Electron.*, vol. 29, no. 5, pp. 2584–2596, May 2014.
2. C. Xia, J. Zhao, Y. Yan, and T. Shi, "A novel direct torque control of matrix converter fed PMSM drives using duty cycle control for torque ripple reduction," *IEEE Trans. Ind. Electron.*, vol. 61, no. 6, pp. 2700–2713, Jun. 2014.
3. T. D. Nguyen and L. Hong-Hee, "Dual three-phase indirect matrix converter with carrier-based PWM method," *IEEE Trans. Power Electron.*, vol. 29, no. 2, pp. 569–581, Feb. 2014.
4. L. Empringham, J. W. Kolar, J. Rodriguez, P. W. Wheeler, and J. C. Clare, "Technological issues and industrial application of matrix converters: A review," *IEEE Trans. Ind. Electron.*, vol. 60, no. 10, pp. 4260–4271, Oct. 2013.
5. Y. D. Yoon and S. K. Sul, "Carrier-based modulation technique for matrix converter," *IEEE Trans. Power Electron.*, vol. 21, no. 6, pp. 1691–1703, Nov. 2006.
6. G. T. Chiang and J. Itoh, "Comparison of two over modulation strategies in an indirect matrix converter," *IEEE Trans. Ind. Electron.*, vol. 60, no. 1, pp. 43–53, Jan. 2013.
7. O. Ellabban, H. Abu-Rub, and S. Bayhan, "Z-source matrix converter: an overview," *IEEE Trans. Power Electron.*, vol. 31, no. 11, pp. 7436–7450. https://doi.org/10.1109/TPEL.2015.2471799.

8 P. Szczesniak, *Three-phase AC-AC Power Converters based on Matrix Converter Topology, Matrix-Reactance Frequency Converters Concept*. London: Springer-Verlag, 2013, ISBN: 978-1-4471-4895-1.

9 F. Z. Peng, "Z-source inverter," *IEEE Trans. Ind. Appl.*, vol. 39, no. 2, pp. 504–510, Mar/Apr., 2003.

10 J. Anderson and F. Z. Peng, "Four quasi-Z-Source inverters," in *Proc. PESC '08 -39th IEEE Annu. Power Electronics Specialists Conf.*, June 15, 2008 – June 19, 2008, Rhodes, Greece, 2008, pp. 2743–2749.

11 Y. P. Siwakoti, F. Z. Peng, F. Blaabjerg, P. C. Loh, and G. E. Town, "Impedance-source networks for electric power conversion part I: A topological review," *IEEE Trans. Power Electron.*, vol. 30, no. 2, pp. 699–716, Feb. 2015.

12 W. Song, Y. Zhong, H. Zhang, X. Sun, Q. Zhang, and W. Wang, "A study of Z-source dual-bridge matrix converter immune to abnormal input voltage disturbance and with high voltage transfer ratio," *IEEE Trans. Ind. Informat.*, vol. 9, no. 2, pp. 828–838, May 2013.

13 X. Liu, P. C. Loh, P. Wang, and X. Han, "Improved modulation schemes for indirect Z-source matrix converter with sinusoidal input and output waveforms," *IEEE Trans. Power Electron.*, vol. 27, no. 9, pp. 4039–4050, Sep. 2012.

14 E. Karaman, M. Farasat, and A. M. Trzynadlowski, "Permanent-magnet synchronous-generator wind-energy systems with boost matrix converters," in *Proc. IEEE Energy Conversion Congr. Expo. (ECCE)*, pp. 2977-2981, Sept. 15–19, 2013.

15 T. D. Nguyen, H. M. Nguyen, H. Lee, and T. Chun, "An approach of sparse matrix converter using Z-source network," in *Int. Power Electronics Conf. (IPEC)*, pp. 2774–2780, Jun. 21–24, 2010.

16 K. Park, K. Lee, and F. Blaabjerg, "Improving output performance of a Z-source sparse matrix converter under unbalanced input-voltage conditions," *IEEE Trans. Power Electron.*, vol. 27, no. 4, pp. 2043–2054, Apr. 2012.

17 B. Ge, Q. Lei, W. Qian, and F. Z. Peng, "A family of Z-source matrix converters," *IEEE Trans. Ind. Electron.*, vol. 59, no. 1, pp. 35–45. Jan. 2012.

18 S. Liu, B. Ge, H. Abu-Rub, and F. Z. Peng. "Modeling, analysis, and motor drive application of quasi-Z-source indirect matrix converter." *Int. J. Comput. Math. Electr. Electron. Eng.*, vol. 33 no. 1/2, pp. 28–28, 2013.

19 S. Liu, B. Ge, X. Jiang, H. Abu-Rub, and F. Z. Peng, "Comparative evaluation of three Z-source/quasi-Z-source indirect matrix converters," *IEEE Trans. Ind. Electron.*, vol. 62, no. 2, pp. 692–701, 2015.

20 S. Liu, B. Ge, Y. Liu, H. Abu-Rub, R. S. Balog and H. Sun, "Modeling, analysis, and parameters design of LC-filter-integrated quasi-Z-source indirect matrix converter," *IEEE Trans. Power Electron.*, vol. 31, no. 11, pp. 7544–7555, Nov. 2016.

21 IEEE, "*IEEE Recommended Practice and Requirements for Harmonic Control in Electric Power Systems*," in IEEE Std 519-2014 (Revision of IEEE Std 519-1992), vol., no., pp. 1–29, 11 June 2014, doi: 10.1109/IEEESTD.2014.6826459.

22 A.K. Sahoo, K. Basu, and N. Mohan, "Systematic input filter design of matrix converter by analytical estimation of RMS current ripple," *IEEE Trans. Ind. Electron.*, vol. 62, no. 1, pp. 132–143, Jan. 2015.

6

Z-Source/Quasi-Z-Source 3-1-Phase Matrix Converters

6.1 Introduction

Recently, the Matrix Converter (MC) has been attracting substantial attention in both academic circles and the industrial sector. As a groundbreaking apparatus designed specifically for direct power conversion, the MC is proficient at generating an output with variable magnitude and frequency directly from an AC source [1–12]. Intriguingly, this is achieved without the reliance on a DC-link capacitor. When setting it side by side with traditional back-to-back converters, the MC stands out, boasting of superior efficiency, diminished footprint, and an impressive longevity. Presently, the spotlight shines on the three-to-three-phase (3-3-phase) MC. This focus is hardly surprising given the widespread adoption of three-phase electrical systems across various industries. Several avenues are currently being explored concerning the MC. These include its tangible real-world applications [2–5], the development of novel control methodologies [6–8], the introduction of efficient fault identification mechanisms [9], state-of-the-art modeling approaches [10], and the implementation of strategies to minimize common-mode voltage disturbances [11, 12]. Additionally, the three-to-single-phase (3-1-phase) MC is making waves due to its multifaceted capabilities. Not only can it adjust phase, but it also offers direct frequency modifications. This versatile converter finds relevance in multiple domains such as electric motor drives [13], ensuring consistent single-phase AC power supply in both large-scale centralized grids and smaller distributed networks [14–16]. It's also making strides in the arena of one-stage wireless energy transfer [17] and is poised to revolutionize high-frequency applications [18, 19], with induction heating being a prime example.

In the research documented in [20], a variety of 3-1-phase MC layouts have been highlighted. These designs include both configurations that incorporate and lack a neutral line. Built with bidirectional switches, these converters leverage three-phase LC filters to minimize disturbances from high-frequency harmonic interferences in the input currents. Distinctively, they can generate single-phase voltage from a three-phase source sans the dependency on sizable capacitors. Their versatility extends to enabling bidirectional power flow and endorsing an all-silicon construction, optimizing spatial efficiency. However, they aren't without hurdles: a reduced input–output voltage transfer ratio, a pervasive challenge in many MC variants. To address the obstacle, insights from the 3-3-phase MC landscape have been considered. In works [21–23], integrating a ZS network into the MC's intermediary DC link proved effective in enhancing voltage gain, albeit at the sacrifice of the all-silicon theme. Following research, as depicted in [24, 25], introduced ZS and qZS impedance circuits on the MC's AC input side to meet similar goals, necessitating supplementary input filters for harmonics control.

In this chapter, we present an innovative design approach for a three-phase-to-single-phase quasi-Z-source matrix converter (3-1-phase qZS-MC) [26]. What sets this design apart is the

Impedance Source Matrix Converters and Control, First Edition.
Yushan Liu, Xiao Li, Baoming Ge, Haitham Abu-Rub, and Frede Blaabjerg.

strategic incorporation of a continuous qZS network, positioned between the three-phase AC source and the 3-1-phase MC. This inclusion significantly augments the voltage transfer ratio. Notably, this configuration eliminates the requirement for extra input filters, as it leverages the inherent filtering ability of the qZS network. In addition, we have developed an analytical framework that demystifies the interrelationships among harmonic components, output frequency, qZS impedance, and the shoot-through duty ratio. By harnessing this model, we can efficiently tailor the inductance and capacitance of the qZS network to ensure an input current with a sinusoidal profile by filtering out the double-line frequency power pulsations. We have conducted both simulation and experimental studies to corroborate the effectiveness of the proposed circuit topology and the accuracy of the established model.

6.2 Topology and Modulation of the 3-1-Phase QZS-MC

6.2.1 Topology

Figure 6.1 provides a comprehensive overview of the evolved 3-1-phase qZS-MC structure. This design harmoniously integrates the qZS impedance network, an advanced element, into the well-established three-phase/single-phase MC framework. Within this qZS assembly, one can observe three distinct sets of intricately connected inductors: firstly, L_{a1} aligning with L_{a2}; secondly, L_{b1} coordinating with L_{b2}; and thirdly, L_{c1} seamlessly integrating with L_{c2}. Additionally, this assembly showcases six meticulously placed capacitors, categorized as C_{a1} through C_{c2}, covering C_{a1}, C_{a2}, C_{b1}, C_{b2}, C_{c1}, and finally, C_{c2}. To enhance the flexibility and control of this design, an auxiliary set of bidirectional switches has been integrated. These are labeled as S_a, S_b, and S_c, and have been optimized to function in a synchronized manner, ensuring optimal power transition. The input voltages for each phase, represented as u_a, u_b, and u_c, are pivotal for this system's operation. Conversely, the outputs from the qZS segment are distinctly marked as u'_a, u'_b, and u'_c. Monitoring the network, currents labeled as i'_a, i'_b, and i'_c emerge, indicating the efficient flow from the qZS setup. To encapsulate the system's final output, u_o and i_o serve as indicators of the qZS-MC's respective output voltage and current.

6.2.2 Equivalent Circuits

The 3-1-phase qZS-MC transitions between two primary operational states: the non-shoot-through state and its counterpart, the shoot-through state. In the former state, the switches denoted as S_a, S_b, and S_c are actively engaged, a scenario depicted in Figure 6.2a. On the other hand, in the shoot-through state, these designated switches—namely S_a, S_b, and S_c—disengage, leading the MC

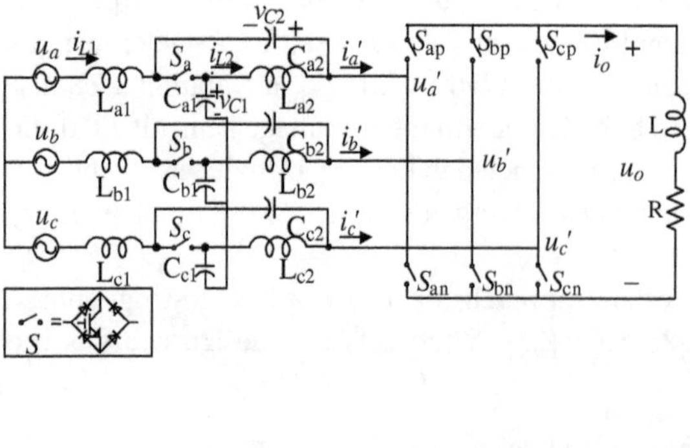

Figure 6.1 Topology of 3-1-phase qZS-MC.

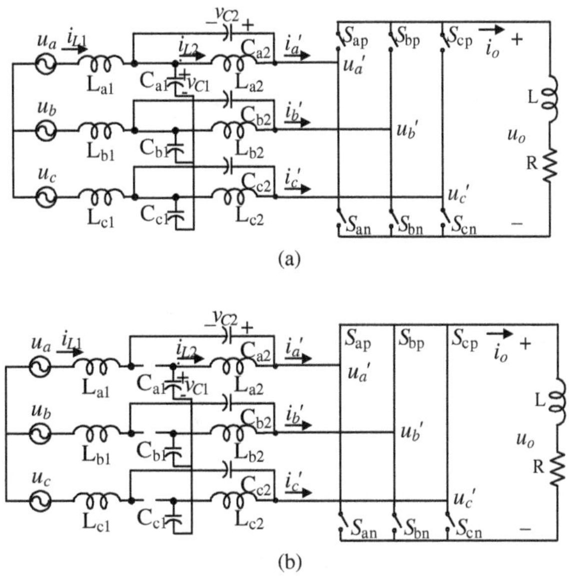

Figure 6.2 Circuit representations for three-phase-to-single-phase qZS-MC configurations. (a) State without shoot-through; (b) State exhibiting shoot-through.

into a short-circuit configuration, a situation captured in Figure 6.2b. For an enhanced comprehension of the operational dynamics, we'll focus our attention on phase "a." A closer examination of Figure 6.2a offers insights into the nuances of the non-shoot-through state as follows:

$$\begin{cases} L_1 \dfrac{di_{L1}}{dt} = u_a - v_{C1} \\ L_2 \dfrac{di_{L2}}{dt} = -v_{C2} \\ C_1 \dfrac{dv_{C1}}{dt} = i_{L1} - i'_a \\ C_2 \dfrac{dv_{C2}}{dt} = i_{L2} - i'_a \end{cases} \tag{6.1}$$

While in the shoot-through state, insights can be drawn from Figure 6.2b as follows:

$$\begin{cases} L_1 \dfrac{di_{L1}}{dt} = u_a + v_{C2} \\ L_2 \dfrac{di_{L2}}{dt} = v_{C1} \\ C_1 \dfrac{dv_{C1}}{dt} = -i_{L2} \\ C_2 \dfrac{dv_{C2}}{dt} = -i_{L1} \end{cases} \tag{6.2}$$

Given that the currents i_{L1} and i_{L2} and the voltages v_{C1} and v_{C2} oscillate, their mean values within one switching cycle don't equate to zero. Under the premise that $L_1 = L_2$ is represented as L and $C_1 = C_2$ is characterized as C, equations (6.1) and (6.2) reveal: Considering the oscillatory behavior of the currents i_{L1} and i_{L2} as well as the voltages v_{C1} and v_{C2}, it's evident that their average values over a single switching cycle aren't zero. Working under the assumption where L_1 is equivalent to L_2 (denoted as L) and C_1 matches C_2 (represented by C), the relationships can be drawn from equations (6.1) and (6.2):

$$\begin{cases} L\dfrac{di_{L1}}{dt} = u_a - (1-D)v_{C1} + Dv_{C2} \\ L\dfrac{di_{L2}}{dt} = -(1-D)v_{C2} + Dv_{C1} \\ C\dfrac{dv_{C1}}{dt} = (1-D)\left(i_{L1} - i'_a\right) - Di_{L2} \\ C\dfrac{dv_{C2}}{dt} = (1-D)\left(i_{L2} - i'_a\right) - Di_{L1} \end{cases} \tag{6.3}$$

Within this framework, D is defined as the shoot-through duty ratio. Its formulation can be expressed as $D = T_0/T_s$. Here, T_0 denotes the time period for the shoot-through duration, while T_s encompasses the total duration of the switching cycle.

6.2.3 Modulation Method

The dual voltage modulation strategy has undergone refinements, building upon the foundations laid out in [27, 28], for the effective control of the 3-1-phase qZS-MC. This advanced method computes the weighted average of the tri-phase input voltage to generate the target output voltage, which is flexible across varying amplitude and frequency spectra.

A visual representation of the sectors determining the MC's operational domain is presented in Figure 6.3. Diving deeper into the specifics, the input current reference that possesses the utmost absolute magnitude is labeled as "i_{base}." Definitions for "$i_{\max}$," "i_{mid}," and "$i_{\min}$" are elaborated in equation (6.4). Phase voltages from the source that correlate with "i_{base}," "$i_{\max}$," "i_{mid}," and "$i_{\min}$" are designated as "u_{base}," "$u_{\max}$," "u_{mid}," and "$u_{\min}$" in sequence. In the scenario where the MC functions within the boundary of sector 1, and there's an affirmative directive for the output voltage, we relate "$u_{\max}$" with "u_c," "u_{mid}" with "u_a," and both "$u_{\min}$" and "u_{base}" with "u_b." As an outcome, the "S_{bn}" switch is triggered to the ON position, and contrarily, the "S_{an}" and "S_{cn}" switches are rendered inactive. Modulation is directed toward the superior switches, namely "S_{ap}," "S_{bp},"

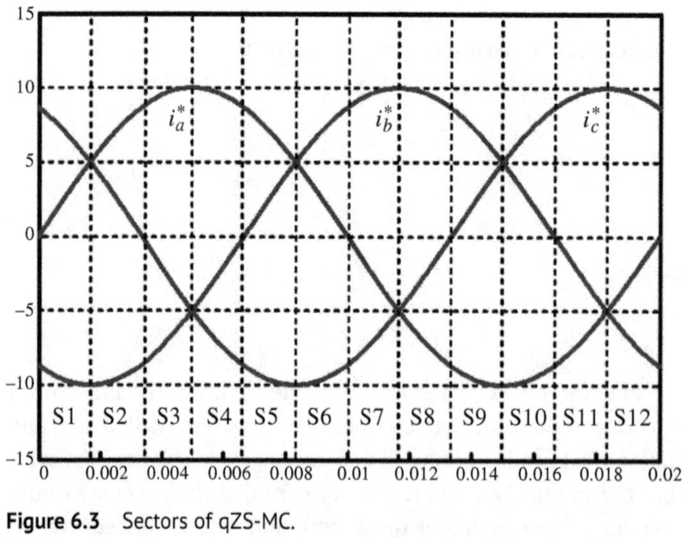

Figure 6.3 Sectors of qZS-MC.

and "S_{cp}." There are two distinct vector states framing the MC's switching dynamics: the active and the zero vector states. To amplify the voltage, the integration of a shoot-through zero vector state into the zero states is pivotal. Vector-specific duty cycles are extrapolated from equation (6.7); within which "d_3" represents the duty cycle corresponding to $\Delta u_{\max}$, "d_2" stands for Δu_{mid}, "d_4" is associated with shoot-through vector, and "d_0" is linked with the zero vector.

$$i_{\max} = MAX\left[i_a^*, i_b^*, i_c^*\right], i_{mid} = MID\left[i_a^*, i_b^*, i_c^*\right], i_{\min} = MIN\left[i_a^*, i_b^*, i_c^*\right] \tag{6.4}$$

$$\Delta u_{\max} = u_{\max} - u_{\min}, \quad \Delta u_{mid} = \begin{cases} u_{\max} - u_{mid}, u_{base} = u_{\max} \\ u_{mid} - u_{\min}, u_{base} = u_{\min} \end{cases} \tag{6.5}$$

$$\alpha = \begin{cases} i_{mid}/i_{\min}, u_{base} = u_{\max} \\ i_{mid}/i_{\max}, u_{base} = u_{\min} \end{cases} \tag{6.6}$$

$$\begin{cases} d_3 = \dfrac{|u_o|}{\alpha \Delta u_{mid} + \Delta u_{\max}} \\ d_2 = \alpha d_3 = \dfrac{\alpha |u_o|}{\alpha \Delta u_{mid} + \Delta u_{\max}} \\ d_4 = D \\ d_0 = 1 - d_2 - d_3 - d_4 \end{cases} \tag{6.7}$$

6.3 Modeling and Analysis of Three-Phase-to-Single-Phase qZS-MC

6.3.1 Model of Three-Phase-to-Single-Phase qZS-MC

To construct a robust analytical framework, we delineate specific parameters pertinent to the MC, as described in equations (6.8)–(6.11). In this schema, the output voltage and its concurrent current are encapsulated using the symbols u_o and i_o. In a contrasting manner, the denotations u_i and i^* are employed to characterize the input voltage and its corresponding current reference. Progressing to a nuanced examination of angular frequency, we allocate distinct notations: ω_o for output-associated frequencies and ω_i for those affiliated with the input. The intricacies of the phase angles linked with the AC output voltage and current are explicitly defined using the labels φ_{ou} and φ_{oi}. Furthermore, when dissecting the magnitudes associated with these entities, we utilize U_o for the voltage amplitude and I_o for the current's amplitude. In addition, it's essential to highlight that the divergence in current phase angle is denoted as φ_i. To conclude our parameterization, we emphasize the amplitude nuances for the input voltage and its current reference, expressed through U_i and i, respectively.

$$u_o = U_o \cos(\omega_o t + \varphi_{ou}) \tag{6.8}$$

$$i_o = I_o \cos(\omega_o t + \varphi_{oi}) \tag{6.9}$$

$$\begin{bmatrix} u_a \\ u_b \\ u_c \end{bmatrix} = U_i \begin{bmatrix} \cos(\omega_i t) \\ \cos(\omega_i t - 2\pi/3) \\ \cos(\omega_i t + 2\pi/3) \end{bmatrix} \tag{6.10}$$

$$\begin{bmatrix} i_a^* \\ i_b^* \\ i_c^* \end{bmatrix} = i \begin{bmatrix} \cos(\omega_i t - \varphi_i) \\ \cos(\omega_i t - \varphi_i - 2\pi/3) \\ \cos(\omega_i t - \varphi_i + 2\pi/3) \end{bmatrix} \tag{6.11}$$

Upon a detailed inspection of the MC's operational dynamics in sector 1, as graphically delineated in Figure 6.3, one can discern that the command for the output voltage exhibits a positive orientation. This operational nuance indicates a specific alignment: u_{max} is congruent with u_c, u_{mid} finds its counterpart in u_a, while both u_{min} and u_{base} resonate with u_b. Taking cues from the modulation technique employed, the subsequent conclusions can be drawn:

$$\alpha = \frac{i_a^*}{i_c^*} = \frac{\cos(\omega_i t - \varphi_i)}{\cos(\omega_i t - \varphi_i + 2\pi/3)} \tag{6.12}$$

$$\begin{aligned} d_3 &= \frac{U_o \cos(\omega_o t + \varphi_{ou})}{\alpha u_{ab} + u_{cb}} = \frac{U_o \cos(\omega_o t + \varphi_{ou})}{x} \\ &= \frac{U_o \cos(\omega_o t + \varphi_{ou}) \cos(\omega_i t - \varphi_i + 2\pi/3)}{1.5 U_i \cos\varphi_i} \end{aligned} \tag{6.13}$$

$$\begin{aligned} d_2 &= \frac{\alpha U_o \cos(\omega_o t + \varphi_{ou})}{\alpha u_{ab} + u_{cb}} \\ &= \frac{\frac{\cos(\omega_i t - \varphi_i)}{\cos(\omega_i t - \varphi_i + 2\pi/3)} U_o \cos(\omega_o t + \varphi_{ou})}{x} \\ &= \frac{U_o \cos(\omega_o t + \varphi_{ou}) \cos(\omega_i t - \varphi_i)}{1.5 U_i \cos\varphi_i} \end{aligned} \tag{6.14}$$

with

$$x = \frac{\cos(\omega_i t - \varphi_i)}{\cos(\omega_i t - \varphi_i + 2\pi/3)} \sqrt{3} U_i \cos(\omega_i t + \pi/6) + \sqrt{3} U_i \cos(\omega_i t + \pi/2)$$

Therefore, if S_{bn} and S_{ap} are ON state, $i'_a = i_o$, $i'_b = -i_o$, $i'_c = 0$ during $d_2 T_s$; if S_{bn} and S_{cp} are ON state, $i'_a = 0$, $i'_b = -i_o$, $i'_c = i_o$ during $d_3 T_s$; if S_{bn} and S_{bp} are ON state, $i'_a = i'_b = i'_c = 0$ during $d_0 T_s$. So the average values of i'_a, i'_b, and i'_c are

$$i'_a = \frac{i_o \cdot d_2 T_s}{(1-D)T_s} = \frac{y_0}{(1-D)1.5 U_i \cos\varphi_i} = \frac{y_1}{3 U_i \cos\varphi_i (1-D)} \tag{6.15}$$

$$i'_c = \frac{i_o \cdot d_3 T_s}{(1-D)T_s} = \frac{y_2}{(1-D)1.5 U_i \cos\varphi_i} = \frac{y_3}{3 U_i \cos\varphi_i (1-D)} \tag{6.16}$$

$$i'_b = -\left(i'_a + i'_c\right) = \frac{y_4}{3 U_i \cos\varphi_i (1-D)} \tag{6.17}$$

with

$$y_0 = I_o \cos(\omega_o t + \varphi_{oi}) U_o \cos(\omega_o t + \varphi_{ou}) \cos(\omega_i t - \varphi_i),$$

$$\begin{aligned} y_1 = U_o I_o \big[&\cos(\varphi_{ou} - \varphi_{oi}) \cos(\omega_i t - \varphi_i) \\ &+ 0.5 \cos(2\omega_o t + \omega_i t + \varphi_{ou} + \varphi_{oi} - \varphi_i) \\ &+ 0.5 \cos(2\omega_o t - \omega_i t + \varphi_{ou} + \varphi_{oi} + \varphi_i)\big], \end{aligned}$$

$$y_2 = I_o \cos(\omega_o t + \varphi_{oi}) U_o \cos(\omega_o t + \varphi_{ou}) \cos(\omega_i t - \varphi_i + 2\pi/3),$$

$$\begin{aligned} y_3 = U_o I_o \big[&\cos(\varphi_{ou} - \varphi_{oi}) \cos(\omega_i t - \varphi_i + 2\pi/3) \\ &+ 0.5 \cos(2\omega_o t + \omega_i t + \varphi_{ou} + \varphi_{oi} - \varphi_i + 2\pi/3) \\ &+ 0.5 \cos(2\omega_o t - \omega_i t + \varphi_{ou} + \varphi_{oi} + \varphi_i - 2\pi/3)\big], \end{aligned}$$

$$y_4 = U_o I_o \left[\cos(\varphi_{ou} - \varphi_{oi})\cos(\omega_i t - \varphi_i - 2\pi/3) + 0.5\cos(2\omega_o t + \omega_i t + \varphi_{ou} + \varphi_{oi} - \varphi_i - 2\pi/3) + 0.5\cos(2\omega_o t - \omega_i t + \varphi_{ou} + \varphi_{oi} + \varphi_i + 2\pi/3)\right]$$

Using (6.3) and (6.15), we can get the complete model of phase a, as follows

$$\begin{aligned} i_{L1} = {} & \frac{-CU_i\omega_i}{2}\left[\frac{1}{(1-2D)^2 - LC\omega_i^2} + \frac{1}{1-LC\omega_i^2}\right]\sin\omega_i t \\ & + \frac{U_o I_o \cos(\varphi_{ou} - \varphi_{oi})(1-2D)}{3U_i\cos\varphi_i\left[(1-2D)^2 - LC\omega_i^2\right]}\cos(\omega_i t - \varphi_i) \\ & + k_1\cos(2\omega_o t + \omega_i t + \varphi_{ou} + \varphi_{oi} - \varphi_i) \\ & + k_2\cos(2\omega_o t - \omega_i t + \varphi_{ou} + \varphi_{oi} + \varphi_i) \end{aligned} \quad (6.18)$$

$$\begin{aligned} i_{L2} = {} & \frac{-CU_i\omega_i}{2}\left[\frac{1}{(1-2D)^2 - LC\omega_i^2} - \frac{1}{1-LC\omega_i^2}\right]\sin\omega_i t \\ & + \frac{U_o I_o \cos(\varphi_{ou} - \varphi_{oi})(1-2D)}{3U_i\cos\varphi_i\left[(1-2D)^2 - LC\omega_i^2\right]}\cos(\omega_i t - \varphi_i) \\ & + k_1\cos(2\omega_o t + \omega_i t + \varphi_{ou} + \varphi_{oi} - \varphi_i) \\ & + k_2\cos(2\omega_o t - \omega_i t + \varphi_{ou} + \varphi_{oi} + \varphi_i) \end{aligned} \quad (6.19)$$

$$\begin{aligned} v_{C1} = {} & \frac{U_i}{2}\left[\frac{1-2D}{(1-2D)^2 - LC\omega_i^2} + \frac{1}{1-LC\omega_i^2}\right]\cos\omega_i t \\ & + \frac{L\omega_i U_o I_o \cos(\varphi_{ou} - \varphi_{oi})}{3U_i\cos\varphi_i\left[(1-2D)^2 - LC\omega_i^2\right]}\sin(\omega_i t - \varphi_i) \\ & + k_3\sin(2\omega_o t + \omega_i t + \varphi_{ou} + \varphi_{oi} - \varphi_i) \\ & + k_4\sin(2\omega_o t - \omega_i t + \varphi_{ou} + \varphi_{oi} + \varphi_i) \end{aligned} \quad (6.20)$$

$$\begin{aligned} v_{C2} = {} & \frac{U_i}{2}\left[\frac{1-2D}{(1-2D)^2 - LC\omega_i^2} - \frac{1}{1-LC\omega_i^2}\right]\cos\omega_i t \\ & + \frac{L\omega_i U_o I_o \cos(\varphi_{ou} - \varphi_{oi})}{3U_i\cos\varphi_i\left[(1-2D)^2 - LC\omega_i^2\right]}\sin(\omega_i t - \varphi_i) \\ & + k_3\sin(2\omega_o t + \omega_i t + \varphi_{ou} + \varphi_{oi} - \varphi_i) \\ & + k_4\sin(2\omega_o t - \omega_i t + \varphi_{ou} + \varphi_{oi} + \varphi_i) \end{aligned} \quad (6.21)$$

with

$$k_1 = \frac{U_o I_o (1-2D)}{\left[(1-2D)^2 - LC(2\omega_o + \omega_i)^2\right] 6U_i\cos\varphi_i}, \quad k_2 = \frac{U_o I_o (1-2D)}{\left[(1-2D)^2 - LC(2\omega_o - \omega_i)^2\right] 6U_i\cos\varphi_i},$$

$$k_3 = \frac{U_o I_o L(2\omega_o + \omega_i)}{\left[(1-2D)^2 - LC(2\omega_o + \omega_i)^2\right] 6U_i\cos\varphi_i}, \quad k_4 = \frac{U_o I_o L(2\omega_o - \omega_i)}{\left[(1-2D)^2 - LC(2\omega_o - \omega_i)^2\right] 6U_i\cos\varphi_i}$$

6.3.2 Voltage Gain Analysis

As illustrated in Figure 6.2a, the qZS network of the 3-1-phase qZS-MC functions in a non-shoot-through state. Under these circumstances, the output voltage, denoted as u'_a, is established by aggregating v_{C1} and v_{C2}. In stark contrast, Figure 6.2b showcases the qZS-MC in its shoot-through state, where the output voltage u'_a plummets, resulting in a null value. As an inference, the envelope of u'_a mirrors the cumulative effects of v_{C1} and v_{C2}.

$$
\begin{aligned}
u'_a = v_{C1} + v_{C2} = & \frac{(1-2D)U_i}{(1-2D)^2 - LC\omega_i^2}\cos\omega_i t \\
& + \frac{2L\omega_i U_o I_o \cos(\varphi_{ou} - \varphi_{oi})}{3U_i \cos\varphi_i\left[(1-2D)^2 - LC\omega_i^2\right]}\sin(\omega_i t - \varphi_i) \\
& + \frac{U_o I_o L(2\omega_o + \omega_i)\sin(2\omega_o t + \omega_i t + \varphi_{ou} + \varphi_{oi} - \varphi_i)}{3U_i \cos\varphi_i\left[(1-2D)^2 - LC(2\omega_o + \omega_i)^2\right]} \\
& + \frac{U_o I_o L(2\omega_o - \omega_i)\sin(2\omega_o t - \omega_i t + \varphi_{ou} + \varphi_{oi} + \varphi_i)}{3U_i \cos\varphi_i\left[(1-2D)^2 - LC(2\omega_o - \omega_i)^2\right]}
\end{aligned}
\tag{6.22}
$$

In determining the voltage boost factor B as presented in (6.22), we consider the fundamental component's amplitude $\overline{u}'_a$. It can be expressed as follows:

$$
B = \frac{\overline{u}'_a}{U_i} = \frac{\sqrt{\frac{(1-2D)^2 U_i^2}{\left[(1-2D)^2 - LC\omega_i^2\right]^2} + \frac{\left[2L\omega_i U_o I_o \cos(\varphi_{ou} - \varphi_{oi})\right]^2}{\left\{3U_i\cos\varphi_i\left[(1-2D)^2 - LC\omega_i^2\right]\right\}^2}}}{U_i} \tag{6.23}
$$

The examination of the MC reveals an amplification in its input voltage, characterized by a specific multiplier, referred to as B. Notably, the magnitude of this factor is intricately linked to both the shoot-through duty ratio, symbolized as D, and the inherent properties of the load. Parameters governing the acceptable boundaries for D are as follows:

$$
d_2 + d_3 + D \leq 1 \tag{6.24}
$$

Setting aside the influence of the load for a moment, equation (6.23) simplifies to:

$$
B = \frac{\overline{u}'_a}{U_i} = \frac{1-2D}{(1-2D)^2 - LC\omega_i^2} \tag{6.25}
$$

Based on the insights derived from equation (6.7), we can deduce the value of the output voltage, U_o, as:

$$
U_o = f(d_2, d_3)\cdot\overline{u}'_a \tag{6.26}
$$

For the 3-1-phase qZS-MC, the calculation of the voltage gain, denoted as G, is outlined below:

$$
G = \frac{U_o}{U_i} = \frac{f(d_2, d_3)\cdot BU_i}{U_i} = f(d_2, d_3)\cdot B \tag{6.27}
$$

In consolidating insights derived from equations (6.7) and (6.27), Figure 6.4 visually depicts the association between the voltage amplification and the cumulative value of d_2 and d_3. Notably, when one carefully determines the sum of d_2 and d_3 and ensures the shoot-through duty cycle is restricted below 0.5, it becomes feasible for the voltage gain to surpass the 1.5 threshold.

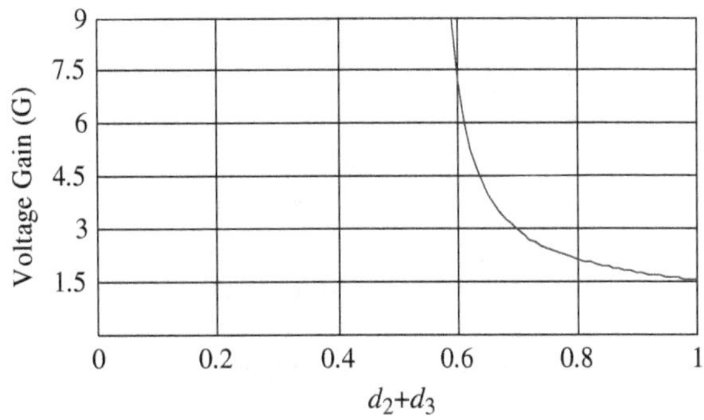

Figure 6.4 Voltage amplification in relation to $(d_2 + d_3)$ for single-phase qZS-MC.

6.4 Simulation and Experimental Tests

To validate the operational capabilities of the 3-1-phase qZS-MC and its accompanying computational model, a twofold strategy was employed involving both computer-based simulations and tangible experimental trials. The setup used for validation, as portrayed in Figure 6.5, was meticulously designed to achieve the desired precision. Our evaluation process was segmented into three pivotal stages: hands-on empirical assessments, computer-generated simulations, and the underpinning theoretical paradigm. Each of these methodologies adhered to a consistent set of parameters, encompassing a 1 mH inductance and 100 μF capacitance designated for the qZS, a predetermined switching frequency of 10 kHz, a resistive load benchmarked at 5 Ω, and an allocated load inductance measuring 5 mH. Further probing delved into assessing the AC outputs at two distinct frequencies, namely 50 Hz and 1000 Hz, exclusively for the 3-1-phase qZS-MC. The data collected from these meticulous investigations are visually represented in a series of diagrams, spanning Figures 6.6–6.8.

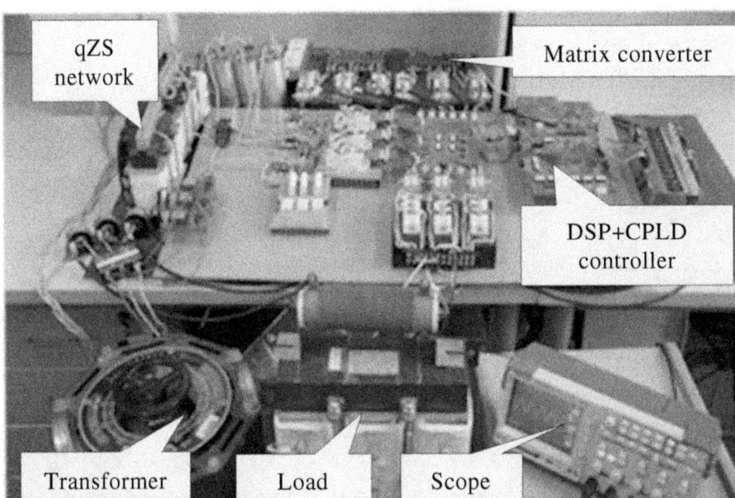

Figure 6.5 Experimental bench of the 3-1-phase qZS-MC.

6.4.1 Verification of Modeling

Diving into comparative analytics, Figure 6.6 serves as a crucible where outcomes sourced from the structured model, hands-on experiments, and intricate simulations for the 3-1-phase MC converge, all benchmarked against the parameters $D = 0.1$ and $f_o = 50$ Hz. The almost symphonic congruence across these diverse data sources vouches for the robustness and accuracy of the model, outlined by equations (6.18)–(6.21). A caveat, however, emerges when comparing practical on-ground measurements with theoretical postulations and digital simulations. This delta can be ascribed to inescapable voltage drop phenomena manifesting across the spectrum of inductors and the gamut of interconnected hardware.

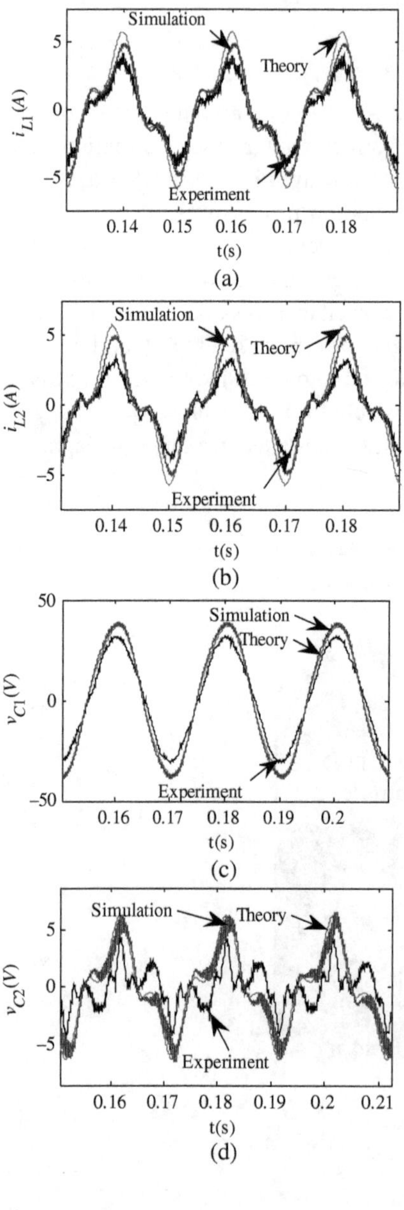

Figure 6.6 Analysis juxtaposing model-derived predictions, simulated data, and hands-on experimental outcomes under conditions of 50 Hz output frequency and $D = 0.1$. (a) Current reading from inductor L_{a1}; (b) flow metrics from inductor L_{a2}; (c) voltage assessment across capacitor C_{a1}; and (d) voltage metrics on capacitor C_{a2}.

6.4.2 Verification of Voltage Gain

The voltage amplification capability of the examined qZS-MC's qZS framework is elucidated by juxtaposing hands-on evaluations with theoretical predictions. As deduced from the prior dissection and showcased in Figure 6.4, the voltage enhancement can be ascertained using equation (6.27) and possesses the potential to exceed 1.5, contingent on the values chosen for $(d_2 + d_3)$ and the shoot-through duty cycle D, where D remains less than 0.5. To substantiate this assertion, a trio of scenarios were scrutinized on the experimental platform: firstly, $d_2 + d_3$ equaled 1; secondly, the sum was 0.9; and thirdly, it registered at 0.8. A synopsis of these observations is tabulated in Table 6.1, underscoring the congruence between theoretical postulations and actual findings, albeit with marginal deviations. The delineated qZS-MC trumps the voltage amplification confines inherent to the conventional 3-1-phase MC.

Figure 6.7(a) and (b) renders a depiction of empirical findings encompassing output voltage, the voltage and current of the load, as well as the input voltage and current, corresponding to conditions where the 3-1-phase qZS-MC operates at $D = 0.1$ and the sum of d_2 and d_3 equals 0.9. Conversely, Figure 6.8a,b encapsulates results observed upon the application of a traditional 3-1-phase MC.

For a setting where D equals 0.1 and the sum of d_2 and d_3 is 0.9, the anticipated voltage gain stands at 1.7. Analyzing the outcomes presented in Figures 6.7a and 6.8a, it's evident that the output voltage and current associated with the 3-1-phase qZS-MC under scrutiny surpass those of its

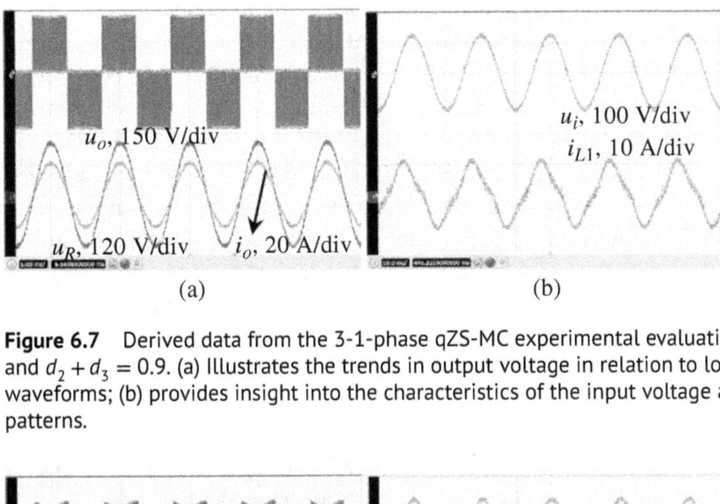

Figure 6.7 Derived data from the 3-1-phase qZS-MC experimental evaluations under conditions of $D = 0.1$ and $d_2 + d_3 = 0.9$. (a) Illustrates the trends in output voltage in relation to load voltage and current waveforms; (b) provides insight into the characteristics of the input voltage and its accompanying current patterns.

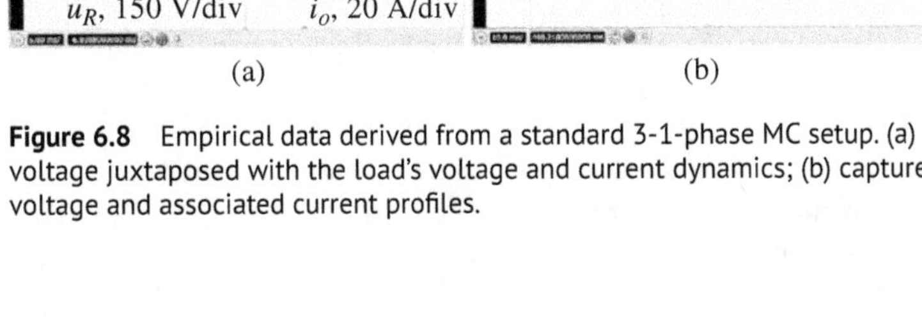

Figure 6.8 Empirical data derived from a standard 3-1-phase MC setup. (a) Depicts the variations in output voltage juxtaposed with the load's voltage and current dynamics; (b) captures the fluctuations in both input voltage and associated current profiles.

Table 6.1 Voltage gain comparison of theoretical and tested values.

$d_2 + d_3$	Theoretical value	Tested value
1	1.5	1.46
0.9	1.7	1.64
0.8	2.06	1.90

traditional counterpart. Such an observation underscores the fact that the qZS-MC in focus not only offers an enhanced voltage transfer ratio vis-a-vis the conventional MC but also betters the quality of the input current.

6.5 Conclusion

We introduced an innovative three-phase to single-phase qZS-MC aiming to augment the voltage gain typically seen in its traditional counterpart. One of the distinct advantages of integrating the qZS network is its inherent filtration capability, which obviates the need for separate input filters. A comprehensive analysis was performed to understand its topology and fundamental principles. An analytical framework was established for this qZS-MC variant, delineating the interplay among harmonic components, the qZS impedance, the output frequency, and the shoot-through duty ratio. Both simulation and empirical studies were conducted to validate the efficiency and reliability of the 3-1-phase qZS-MC and its associated model. The findings highlighted that the qZS impedance is influenced by the output voltage frequency. In scenarios with elevated output voltage frequencies, minimal qZS network inductance and capacitance suffice to generate sinusoidal input currents. Conversely, when dealing with lower output voltage frequencies, it necessitates a more substantial qZS network. However, it's noteworthy that even in the absence of supplementary input filters, the system could produce an input current exhibiting low THD and sinusoidal properties.

References

1 P. W. Wheeler, J. Rodriguez, J. C. Clare, L. Empringham, and A. Weinstein, "Matrix converter: A technology review," *IEEE Trans. Ind. Electron.*, vol. 49, no. 2, pp. 276–288, Apr. 2002.

2 L. Empringham, J. W. Kolar, J. Rodriguez, P. W. Wheeler, and J. C. Clare, "Technological issues and industrial application of matrix converters: A review," *IEEE Trans. Ind. Electron.*, vol. 60, no. 10, pp. 4260–4271, Oct. 2013.

3 C. Xia, J. Zhao, Y. Yan, and T. Shi, "A novel direct torque and flux control method of matrix converter-fed PMSM drives," *IEEE Trans. Power Electron.*, vol. 29, no. 10, pp. 5417–5430, Oct. 2014.

4 S. Safari, A. Castellazzi, and P. Wheeler, "Experimental and analytical performance evaluation of SiC power devices in the matrix converter," *IEEE Trans. Power Electron.*, vol. 29, no. 5, pp. 2584–2596, May 2014.

5 C. Xia, J. Zhao, Y. Yan, and T. Shi, "A novel direct torque control of matrix converter-fed PMSM drives using duty cycle control for torque ripple reduction," *IEEE Trans. Ind. Electron.*, vol. 61, no. 6, pp. 2700–2713, Jun. 2014.

6 M. Rivera, A. Wilson, C. A. Rojas, J. Rodriguez, J. R. Espinoza, P. W. Wheeler, and L. Empringham, "A comparative assessment of model predictive current control and space vector modulation in a direct matrix converter," *IEEE Trans. Ind. Electron.*, vol. 60, no. 2, pp. 578–588, Feb. 2013.

7 S. S. Sebtahmadi, H. Pirasteh, S. H. A. Kaboli, A. Radan, and S. Mekhilef, "A 12-sector space vector switching scheme for performance improvement of matrix-converter-based DTC of IM drive," *IEEE Trans. Power Electron.*, vol. 30, no. 7, pp. 3804–3817, Jul. 2015.

8 J. Rodrguez, M. Rivera, J. W. Kolar, and P. W. Wheeler, "A review of control and modulation methods for matrix converters," *IEEE Trans. Ind. Electron.*, vol. 59, no. 1, pp. 58–70, Jan. 2012.

9 C. Brunson, L. Empringham, L. de Lillo, P. Wheeler, and J. Clare, "Open-circuit fault detection and diagnosis in matrix converters," *IEEE Trans. Power Electron.*, vol. 30, no. 5, pp. 2840–2847, May 2015.

10 F. Villarroel, J. R. Espinoza, C. A. Rojas, J. Rodriguez, M. Rivera, and D. Sbarbaro, "Multi objective switching state selector for finite-states model predictive control based on fuzzy decision making in a matrix converter," *IEEE Trans. Ind. Electron.*, vol. 60, no. 2, pp. 589–599, Feb. 2013

11 H. Nguyen and H. Lee, "A DSVM method for matrix converters to suppress common-mode voltage with reduced switching losses," *IEEE Trans. Power Electron.*, vol. 31, no. 6, pp. 4020–4030. https://doi.org/10.1109/TPEL.2015.2471078.

12 R. Baranwal, K. Basu, and N. Mohan, "Carrier-based implementation of SVPWM for dual two-level VSI and dual matrix converter with zero common-mode voltage," *IEEE Trans. Power Electron.*, vol. 30, no. 3, pp. 1471–1487, Mar. 2015.

13 D. V. Nicolae, C. G. Richards, and P. Ehlers, "Small power, three to one phase matrix converte for wind generators," in *Int. Symp. Power Electronics Electrical Drives Automation and Motion (SPEEDAM)*, pp. 1339–1343, 2010.

14 C. Young, H. Chen, and M. Chen, "A cockcroft–walton voltage multiplier fed by a three-phase-to-single-phase matrix converter with PFC," *IEEE Trans. Ind. Appl.*, vol. 50, no. 3, pp. 1994–2004, May–Jun. 2014.

15 Y. Miura, T. Amano, and T. Ise, "Operating characteristics of a three-phase to single-phase matrix converter with hybrid control scheme of power compensation and modulation applied to gas engine cogeneration system," in *Proc. 2011-14th European Conf. Power Electronics and Applications (EPE 2011)*, pp. 1–10, Aug. 30, 2011–Sept. 1, 2011.

16 R. Yamada, N. Yamamura, and M. Ishida, "A study on load voltage control method for three-phase-to-single-phase matrix converter system with neutral line," in *Int. Conf. Electrical Machines and Systems (ICEMS)*, pp. 1836–1841, Oct. 26–29, 2013.

17 D. S. B. Weerasinghe, U. K. Madawala, D. J. Thrimawithana, and D. M. Vilathgamuwa, "A three-phase to single-phase matrix converter based bi-directional IPT system for charging electric vehicles," in *IEEE ECCE Asia Downunder (ECCE Asia)*, pp. 1240–1245, June 3–6, 2013.

18 N. Nguyen-Quang, D. A. Stone, C. M. Bingham, and M. P. Foster, "A three-to-one-phase matrix converter for high-frequency induction heating," in *Proc. 13th European Conf. Power Electronics and Applications (EPE'09)*, pp. 1–10, Sept. 8–10, 2009.

19 N. Nguyen-Quang, D. A. Stone, C. M. Bingham, and M. P. Foster, "Comparison of single-phase matrix converter and H-bridge converter for radio frequency induction heating," in *Proc. European Conf. Power Electronics and Applications (EPE'07)*, pp. 1–9, Sept. 2–5, 2007.

20 E. Babaei, A. Aghagolzadeh, and S. H. Hosseini, S. Khanmohammadi, "A new structure for three-phase to single-phase ac/ac matrix converters," in *Proc. 10th IEEE Electronics, Circuits and Systems (ICECS)*, pp. 36–39, Dec. 16–17, 2003.

21 E. Karaman, M. Farasat, and A. Trzynadlowski, "A comparative study of series and cascaded Z-source matrix converters," *IEEE Trans. Ind. Electron.*, vol. 61, no. 10, pp. 5164–5173, Oct. 2014.

22 W. Song, Y. Zhong, H. Zhang, X. Sun, Q. Zhang, and W. Wang, "A study of Z-source dual-bridge matrix converter immune to abnormal input voltage disturbance and with high voltage transfer ratio," *IEEE Trans. Industr. Inform.*, vol. 9, no. 2, pp. 828–838, May 2013.

23 X. Liu, P. C. Loh, P. Wang, and X. Han, "Improved modulation schemes for indirect Z-source matrix converter with sinusoidal input and output waveforms," *IEEE Trans. Power Electron.*, vol. 27, no. 9, pp. 4039–4050, Sept. 2012.

24 B. Ge, Q. Lei, W. Qian, and F. Z. Peng, "A family of Z-source matrix converters," *IEEE Trans. Ind. Electron.*, vol. 59, no. 1, pp. 35–46, Jan. 2012.

25 S. Liu, B. Ge, H. Abu-Rub, F. Z. Peng, and Y. Liu, "Quasi-Z-source matrix converter based induction motor drives," in *Proc. 38th Annu. Conf. IEEE Industrial Electronics Society (IECON)*, pp. 5303–5307, Oct. 25–28, 2012.

26 Y. Liu, W. Liang, B. Ge, and H. Abu-Rub, N. Nie, "Quasi-Z-source three-to-single-phase matrix converter and ripple power compensation based on model predictive control," *IEEE Trans. Ind. Electron.*, vol. 65, no. 6, pp. 5146–5156, Jun. 2018.

27 J. Oyama, T. Higuchi, E. Yamada, T. Koga, and T. Lipo, "New control strategy for matrix converter," in *Proc. 20th Annu. IEEE Power Electronics Specialists Conf.*, vol. 1, pp. 360–367, Jun. 26–29, 1989.

28 J. Oyama, T. Higuchi, and E. Yamada, "Displacement angle control for matrix converter," in *Proc. 28th Annu. IEEE Power Electronics Specialists Conf.*, vol. 2, pp. 1033–1039, Jun. 22–27, 1997.

7

Z-Source/Quasi-Z-Source 3-1-Phase Matrix Converters With Low-Frequency Power Compensation

7.1 Introduction

The 3-1-phase Matrix Converters (MCs) have ushered in a transformative capability, enabling the direct generation of single-phase voltage from a conventional three-phase source. This technological advance has rendered the need for the traditionally hefty DC-link capacitors obsolete. Their cutting-edge design not only champions bi-directional energy flow but also pivots on an innovative all-silicon conversion methodology, leading to a more streamlined and efficient configuration. Yet, as with many technological advancements, they aren't without challenges. The primary hurdle encountered by these converters is the diminished voltage transfer ratio between input and output. This isn't unique to the 3-1-phase MCs but is a prevalent challenge for many MC types. The second notable concern is the input current distortion. Especially evident during operations at lower output frequencies, this distortion stems from the double-line-frequency (2ω) power oscillations, a byproduct of the single-phase output. Solutions rooted in the use of impedance networks have been tabled to address the voltage transfer challenge. These networks can be seamlessly integrated either at the MC's intermediary DC link or directly at the AC source end and has been discussed in Chapter 6. This strategic exploration, especially with reference to the 3-1-phase MCs, has seen significant traction in recent times [1–8]. The distortion challenge, on the other hand, has inspired the development of compensation circuits. These circuits, often deploying auxiliary components like capacitors or inductors, aim to curtail the 2ω power fluctuations [9–12]. The end game is twofold: to preserve a constant power input and to guarantee a sinusoidal current input. An intriguing hybrid control method, which amalgamates the principles of power modulation with compensation, emerged in [13]. However, it's worth noting that minor distortions in three-phase currents remained a residual issue. A common thread across existing compensatory strategies is the utilization of a proportional-integral (PI) controller. This controller ensures that the compensatory capacitor voltage follows a preconceived waveform—a waveform which is extrapolated from the ideal conditions of sinusoidal load. However, such strategies primarily focus on mitigating the 2ω ripple power, often sidelining other potential harmonic interferences. In practical, real-world conditions, where load anomalies or the presence of nonlinear loads might distort the scenario, these traditional methods sometimes show their limitations, primarily due to the intricate task of pinning down the precise capacitor voltage.

In this chapter, we present a redefined 3-1-phase MC, specifically designed with a quasi-Z-source construct termed as qZS-MC. This structure harmoniously incorporates a continuous qZS network,

Impedance Source Matrix Converters and Control, First Edition.
Yushan Liu, Xiao Li, Baoming Ge, Haitham Abu-Rub, and Frede Blaabjerg.

further complemented by a compensation system anchored on model predictive control (MPC) [8]. Delving into its unique aspects:

The newly conceived 3-1-phase MC, built on a quasi-Z-source framework, is meticulously designed to counteract low-frequency ripple power anomalies. This innovative approach seamlessly integrates the qZS network strategically located between the primary AC source and the converter. This configuration not only enhances the voltage gain in a purely silicon-based system but also doubles as an effective current filter, obviating the necessity for auxiliary input filters. Furthermore, with an emphasis on low-frequency harmonic neutralization, this system promises a significant impedance parameter reduction when juxtaposed against its conventional 3-1-phase MC counterpart. This refined model draws from the intrinsic advantages of matrix converters, especially when benchmarked against dual converters dependent on voluminous DC capacitors. A notable feature to underscore is the AC-oriented operational mode of the converter's compensatory capacitor. This modality offers a succinct and reliable alternative to the prevalent DC-link capacitors found in typical dual configurations. Embracing a comprehensive silicon-based design, amalgamated with the multifaceted roles of the qZS network—namely current filtration and voltage magnification—and the AC-centric compensatory capacitor, results in a compact and robust converter. It's also imperative to differentiate this design from the T-bridge three-level inverter elaborated in [14], which fundamentally leans on two DC bus capacitors to produce zero-voltage output and incorporates three bi-directional switches connected at the capacitors' midpoint.

By capitalizing on a meticulously designed MPC-oriented framework, we target 2ω power disruptions with a greater sense of precision. Our approach proficiently neutralizes not just the specific 2ω ripple interferences but also an array of divergent harmonic perturbations. The model we present fosters instantaneous tracking of any ripple power fluctuations, independent of the intricate nuances of prevailing load voltage and current dynamics. The inherent rapid-response nature of MPC, often touted for its efficiency, promises enhanced ripple power stabilization throughout both fluctuating and consistent operational phases, distinctly outshining the somewhat delayed dynamics commonly found in PI-driven controls. A significant hallmark of our MPC methodology lies in its commendable simplicity. By design, it eliminates the need for superfluous modulators or overly intricate control mechanisms. Importantly, introducing the concept of using MPC for ripple rectification in this specific MC configuration heralds a new era in power compensation techniques.

For ease of understanding and systematic progression, this chapter is laid out in a structured manner. In Section 7.2, we embark on a comprehensive exploration of the 3-1-phase qZS-MC. Following this, Section 7.3 is devoted to illuminating existed harmonic components and required impedance parameters without ripple compensation. A deep dive into the ripple rectification branch that is adeptly steered by MPC directives is presented in Section 7.4. Transitioning to Section 7.5, we juxtapose meticulously gathered simulated data with empirical test results, thereby lending weight to the validation of our proposed structural innovations, regulatory strategies, and core theoretical concepts. Wrapping up our discourse, Section 7.6 encapsulates our holistic insights and reflective conclusions on the topic.

7.2 The 3-1-Phase QZS-MC with Input Low-Frequency Harmonic Elimination

Figure 7.1 meticulously delineates the intricate structure of the ripple power-compensated qZS-MC, providing a comprehensive visual representation of its design. At the heart of this schematic is the distinctive qZS configuration. This configuration comprises three pairs of

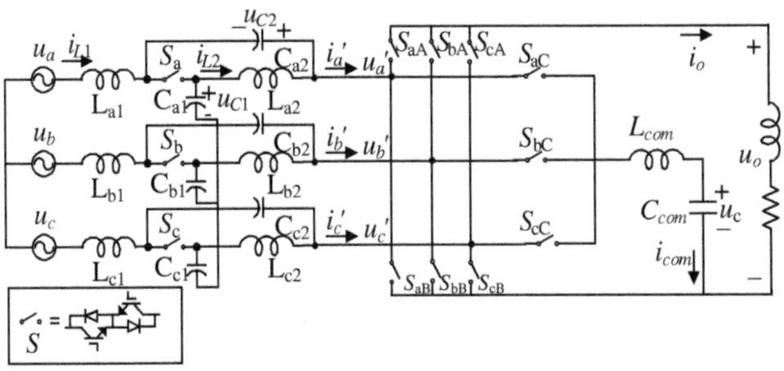

Figure 7.1 Configuration of 3-1-phase qZS-MC, incorporating ripple power adjustment.

meticulously aligned inductors, specifically L_{a1}–L_{a2}, L_{b1}–L_{b2}, and L_{c1}–L_{c2}. Furthermore, a precisely organized ensemble of capacitors, ranging from C_{a1} to C_{c2}, joins forces with a set of three bidirectional switches, labeled as S_a, S_b, and S_c. It's particularly noteworthy to emphasize that these switches, S_a through S_c, adhere to a uniform, meticulously planned switching paradigm. The input voltages, designated as u_a, u_b, and u_c, undergo a transition process, resulting in the respective qZS structure output voltages: u'_a, u'_b, and u'_c. Complementing these are the outputs currents, termed i'_a through i'_c. These currents, when combined with the final qZS-MC's output voltage and current, represented as u_o and i_o, respectively, underscore the effectiveness of the design. This ingeniously designed qZS interconnection, strategically positioned between the electrical grid and the matrix converter, plays a pivotal role. Its main function is to ensure a steady stream of input currents while adeptly acting as both an input current modulator and a voltage booster.

Moreover, an insightful inspection of Figure 7.1 reveals the nuanced inclusion of a specific capacitor, aptly termed C_{com}. This capacitor, in conjunction with an associated inductor labeled L_{com}, establishes a connection to the comprehensive tri-phasic system via an array of switches: S_{aC}, S_{bC}, and S_{cC}. Their core function revolves around efficiently channeling the 2ω power flux. This power, emanating from the single-phase extremity, gets stored into C_{com}. Strategic and well-coordinated operations of S_{aC}, S_{bC}, and S_{cC} in the compensatory framework ensure that this power is rerouted away from the input domain, maximizing the efficiency of the entire system.

7.3 Existed Harmonic Components and Required Impedance Parameters Without Ripple Compensation

Equations (A7.7)–(A7.10) indicate that qZS inductor currents and qZS capacitor voltages exhibit harmonic components at frequencies of $2\omega_o+\omega_i$ and $2\omega_o-\omega_i$. These harmonic amplitudes have an inherent relation to the output frequency, especially in the absence of a ripple compensation branch that buffers the 2ω power. By examining (A7.7) and (A7.8), we introduce $a1$ and $a2$ as ratios of harmonic amplitudes at frequencies $2\omega_o+\omega_i$ and $2\omega_o-\omega_i$ in relation to the fundamental component. These values, $a1$ and $a2$, are detailed in (7.1) and (7.2), respectively.

$$a1 = \frac{\left|\frac{U_o I_o (1-2D)}{6U_i \cos\varphi_i\left[(1-2D)^2 - LC(2\omega_o+\omega_i)^2\right]}\right|}{\sqrt{a3}} \tag{7.1}$$

$$a2 = \frac{\left|\frac{U_o I_o (1-2D)}{6U_i \cos\varphi_i\left[(1-2D)^2 - LC(2\omega_o-\omega_i)^2\right]}\right|}{\sqrt{a3}} \tag{7.2}$$

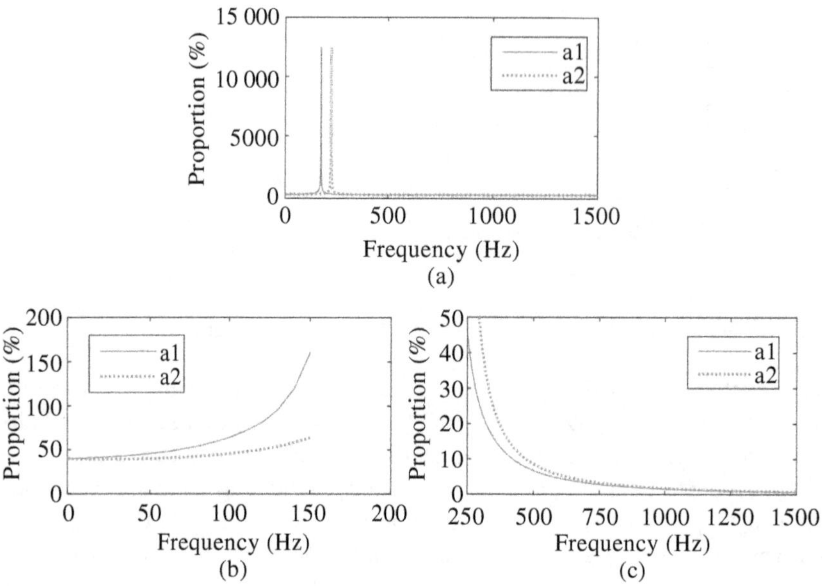

Figure 7.2 Harmonic amplitude correlation with output frequency. (a) Full spectrum; (b) lower frequencies; (c) upper frequencies.

with

$$a3 = \frac{C^2 U_i^2 \omega_i^2}{4}\left[\frac{1}{(1-2D)^2 - LC\omega_i^2} + \frac{1}{1-LC\omega_i^2}\right]^2 + \frac{[U_o I_o \cos(\varphi_{ou} - \varphi_{oi})(1-2D)]^2}{9U_i^2 \cos^2\varphi_i\left[(1-2D)^2 - LC\omega_i^2\right]^2}.$$

As illustrated in Figure 7.2, there's a notable interplay between $a1$ and $a2$ in relation to the output frequency. For this examination, we established parameters with D at 0.1, U_i set at 122.5 V, L equating to 1 mH, C valued at 100 μF, and ω_i marked at 314 rad/s. The choice of qZS impedance values—1 mH and 100 μF—intentionally surpasses the 150 μH and 90 μF used in our prior experimental arrangement. This adjustment was made to accommodate the absence of a compensation branch and to effectively counteract the ripple at lower frequencies. Delving into Figure 7.2a, it becomes evident that the output frequency wields a profound influence on the harmonic structures. Notably, the dual apex values showcased in Figure 7.2a are synonymous with two distinct resonance points, each relating to harmonic components. The mathematical underpinnings for these resonance points can be captured by subsequent formulas:

$$(1-2D)^2 - LC(2\omega_o + \omega_i)^2 = 0 \text{ and } (1-2D)^2 - LC(2\omega_o - \omega_i)^2 = 0 \tag{7.3}$$

The resonance determinants are inherently tied to the qZS impedance and the frequency from the input source. Analyzing Figure 7.2b,c reveals that as the output frequency rises in a high-frequency range, there's a diminishing trend in harmonic amplitudes. Contrarily, in the lower frequency domains, there's an uptick in these amplitudes with ascending output frequencies. Referring to equations (7.1) and (7.2), a correlation emerges between $a1$, $a2$, and the qZS impedance network's inductance and capacitance, as visualized in Figure 7.3, with set parameters: $D = 0.1$, $U_i = 122.5$ V, $\omega_i = 314$ rad/s, and $\omega_o = 6280$ rad/s. Observations from Figure 7.3 indicate that for L values above 0.1 mH and C values exceeding 100 μF, both $a1$ and $a2$ stay below a 5% threshold. Notably, when

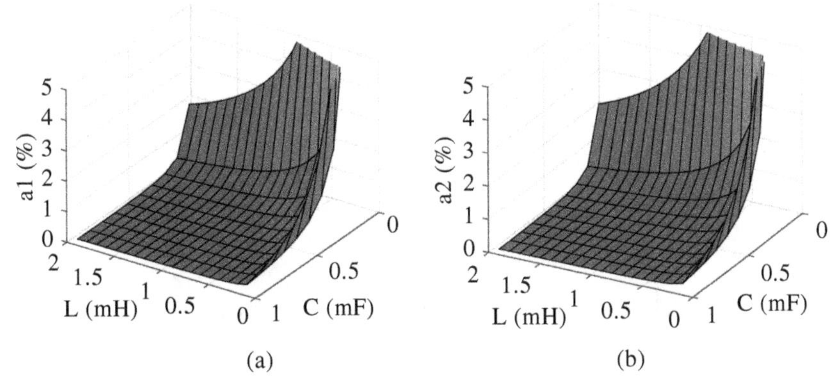

Figure 7.3 Relationship of harmonic amplitudes versus qZS impedance. (a) a1 versus qZS impedance; (b) a2 versus qZS impedance.

the 3-1-phase qZS-MC operates at elevated output frequencies, its input current exhibits reduced harmonic characteristics, irrespective of compensations for ripple power.

It is imperative for the 3-1-phase qZS-MC to maintain an output frequency that's not overly diminished. Such a precaution ensures that for minor qZS impedance magnitudes, the associated input currents, along with the currents of qZS inductors and the voltages of qZS capacitors, exhibit a sinusoidal pattern. In scenarios where the output frequency hovers on the lower end and there's an absence of ripple compensation, there arises a necessity for elevated qZS impedance levels, particularly for L and C.

7.4 Predictive Control of Ripple Compensation Branch

To effectively nullify the low-frequency ripple power at the input, the MPC takes charge of the switches S_{aC}, S_{bC}, and S_{cC}, operating them independently of the legs associated with the load. This control process encompasses four critical stages: Predicting the forthcoming sampling time's capacitor current, labeled as i_{com}. Estimating the anticipated power associated with capacitor C_{com}. Determining the ripple power on the single-phase end. Evaluating the cost function, which then guides the selection of the most favorable switching state for S_{aC}, S_{bC}, and S_{cC}. A more detailed elaboration is provided in the subsequent sections.

7.4.1 Current Model of Compensation Branch

The current methodology focuses on forecasting the i_{com} current in the LC branch for the upcoming sampling interval. As illustrated in Figure 7.1 and grounded in Kirchhoff's voltage law (KVL), the dynamics of the capacitor current, termed i_{com}, can be elucidated as:

$$L_{com}\frac{di_{com}(t)}{dt} = u_{com}(t) - u_c(t) \tag{7.4}$$

In this context, u_{c} represents the voltage across the capacitor C_{com}, whereas u_{com} denotes the voltage pertaining to the LC branch. Following this, the relationship can be articulated as:

$$\frac{di_{com}}{dt} = \frac{i_{com}(k+1) - i_{com}(k)}{T_s} \tag{7.5}$$

Let T_s represent the specific sampling duration. By integrating the relation from equation (7.5) into (7.4), the forecasted current for the succeeding (k+1) interval can be derived:

$$i_{com}(k+1) = i_{com}(k) + \frac{T_s}{L_{com}}[u_{com}(k) - u_c(k)] \tag{7.6}$$

At the specific kth moment, $i_{com}(k)$, $u_{com}(k)$, and $u_c(k)$ represent the current and voltage across the LC branch, as well as the voltage related to the capacitor C_{com}.

7.4.2 Power Model of Compensation Capacitor

As illustrated in Figure 7.1, the majority of the ripple power finds its buffer within the capacitor C_{com}. Meanwhile, the inductor L_{com}, possessing a relatively low value, serves primarily to mitigate the i_{com}'s switching-frequency ripple. Given these factors, it's plausible to equate the power of the compensation branch closely with that of the capacitor C_{com}. Considering the notably higher sampling frequency compared to the line frequency of the compensation branch, it can be inferred that the capacitor C_{com}'s voltage at the (k+1)th moment is nearly identical to $u_c(k)$. Consequently, the power for the compensation branch can be deduced as follows:

$$p_{com}(k+1) = i_{com}(k+1) \cdot u_c(k) \tag{7.7}$$

7.4.3 2ω Power of Single-Phase Side

For the single-phase side, the power corresponding to 2ω is encapsulated within the multiplication of the output voltage, denoted as $u_o(k)$, and the current, symbolized as $i_o(k)$. To put it explicitly, this relationship can be represented as P_{DC}, signifying the active power at that specific moment

$$\tilde{p}_o(k) = i_o(k) \cdot u_o(k) - P_{DC} \tag{7.8}$$

7.4.4 Cost Function

A distinctive cost function, denoted by g, is formulated in equation (7.9). This function serves to assess the variance between the ripple power and the subsequent sampling time's p_{com} power from the compensation branch. When subject to varying switching states, this function yields diverse outcomes. It's the optimal state that seeks to minimize this function's value. In alignment with the directional reference portrayed in Figure 7.1, we characterize the cost function in the following manner:

$$g = |p_{com}(k+1) + \tilde{p}_o(k)| \tag{7.9}$$

To conclude, the essence of predictive control lies in pinpointing the best switching state of S_{aC}, S_{bC}, and S_{cC} such that the cost function, g, is minimized. It's worth highlighting that the u_{com} voltage, as indicated in equation (7.8), is influenced by the S_{aB}, S_{bB}, S_{cB} switching state and the immediate voltages, u'_a, u'_b, and u'_c. These switching states are characterized by either an ON or OFF state, symbolized as "1" or "0," respectively. The value of u_{com} is thus determined by $u_{com} \in \{(S_{aC} \text{ AND } \overline{S}_{aB}) \times u'_a, (S_{bC} \text{ AND } \overline{S}_{bB}) \times u'_b, \text{ OR } (S_{cC} \text{ AND } \overline{S}_{cB}) \times u'_c\}$, where the "–" symbol signifies the Boolean "NOT."

7.5 Simulation and Experimental Tests

Through a combination of simulations and hands-on experiments, we put the 3-1-phase qZS-MC to the test, probing into its voltage gain and evaluating the efficiency of ripple power compensation.

A depiction of our experimental setup can be viewed in Figure 6.5. For the qZS structure, we chose parameters with an inductance value of $L = 150\ \mu$H and capacitance at $C = 90\ \mu$F. We maintained an RMS value for the inter-phase input voltage at 150 V, equivalent to a U_i value of 122.5 V. Essential operating conditions encompassed an input frequency, f_i, set at 50 Hz, and a f_s switching frequency pegged at 20 kHz. The connected load was characterized by a resistance of 10 Ω and an associated inductance of 5 mH. To streamline the current commutation process, we adopted a quad-tier strategy, drawing inspiration from the techniques outlined in [15]. The strategic placement of the qZS configuration between the primary AC source and the MC ensures the converter's heightened resistance to dead time. This not only amplifies its robustness but also enhances reliability, particularly when comparing the turn-on and turn-off time discrepancies in relation to a conventional MC.

Using both simulation and hands-on tests, the efficacy of MPC-based ripple power compensation was scrutinized. Key parameters include a shoot-through duty cycle D of 0.05 and modulation indices m_i and m_v of input current and output voltage, respectively, both set at 0.85. We fixed system parameters to delve into output frequencies at 50 Hz and 100 Hz. Notably, as output frequency ascends, there's a decrease in compensation capacitance and inductance. In this regard, for these frequencies, computations for compensation elements prioritize the 50 Hz metric, resulting in L_{com} values at 400 μH and C_{com} at 800 μF. Insight into the simulation outcomes for 3-1-phase qZS-MC at these frequencies can be gleaned from Figures 7.4 and 7.5. Conversely, Figure 7.6 offers a perspective on the 50-Hz scenario devoid of the compensation mechanism.

A scrutiny of Figures 7.4a, 7.5a, and 7.6a confirms the sinusoidal nature of output currents. Yet, a disparity emerges when comparing the input currents of qZS-MC across Figures 7.4b, 7.5b,

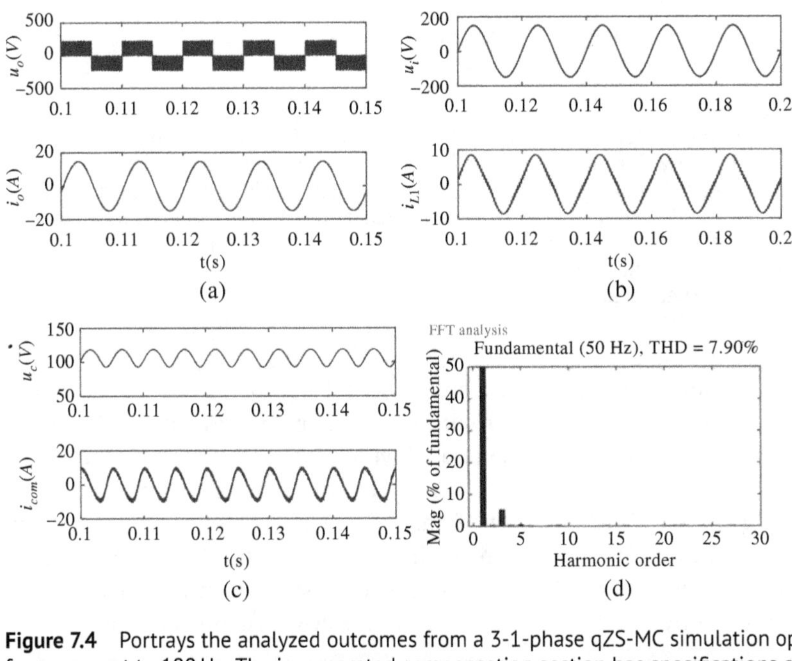

Figure 7.4 Portrays the analyzed outcomes from a 3-1-phase qZS-MC simulation operating at an output frequency set to 100 Hz. The incorporated compensation section has specifications of $L_{com} = 400\ \mu$H and $C_{com} = 800\ \mu$F. Details presented are as follows: (a) metrics associated with voltage and current at the output stage; (b) electrical values recorded for both voltage and current at the input stage; (c) voltage and current readings specific to the capacitor within the compensation unit; and (d) FFT graphical interpretation illuminating the spectrum attributes of the input current.

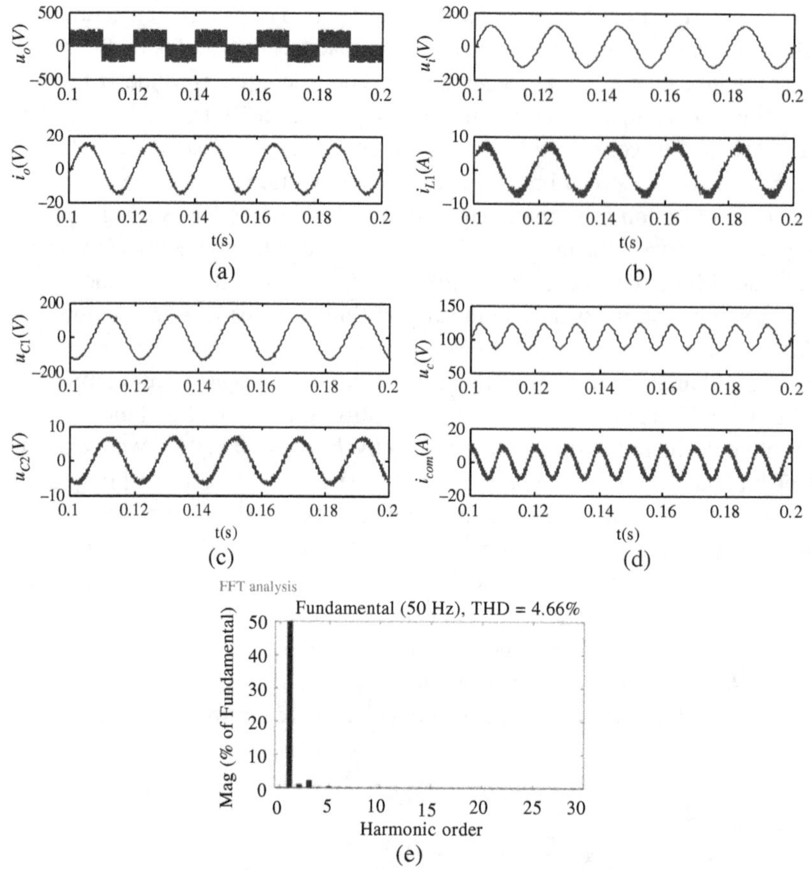

Figure 7.5 Offers insights into the modeling outcomes of the 3-1-phase qZS-MC functioning at an established output frequency of 50 Hz. For this specific configuration, the compensation section is characterized by L_{com} values at 400 μH and C_{com} at 800 μF. The insights detailed include: (a) ascertained values for output-driven voltage and current; (b) electrical parameters at the input encompassing voltage and current readings; (c) voltage measurements related to the Quasi-Z source capacitive elements; (d) specific readings regarding voltage and current for the compensating capacitor section; and (e) an FFT visual breakdown of the current's spectral attributes at the input.

and 7.6b. The latter, absent compensation, reveals a pronounced distortion. In stark contrast, the presence of compensation harmonizes the input current, as observed in Figures 7.4b and 7.5b, even under low-frequency operations as evidenced in Figure 7.2 with minimal impedance parameters. Figures 7.4c and 7.5d spotlight the voltage u_c and i_{com} current dynamics of the compensatory mechanism, underscoring the mitigated 2ω power from the singular phase. Key observations include u_c harboring DC and 2ω facets and i_{com} predominantly echoing the 2ω trait. Embracing the compensation pathway, Figures 7.4b, 7.4d, 7.5b, and 7.5e illustrate harmonized wave patterns and diminished input current THDs. Yet, the uncompensated scenarios in Figure 7.6b–e

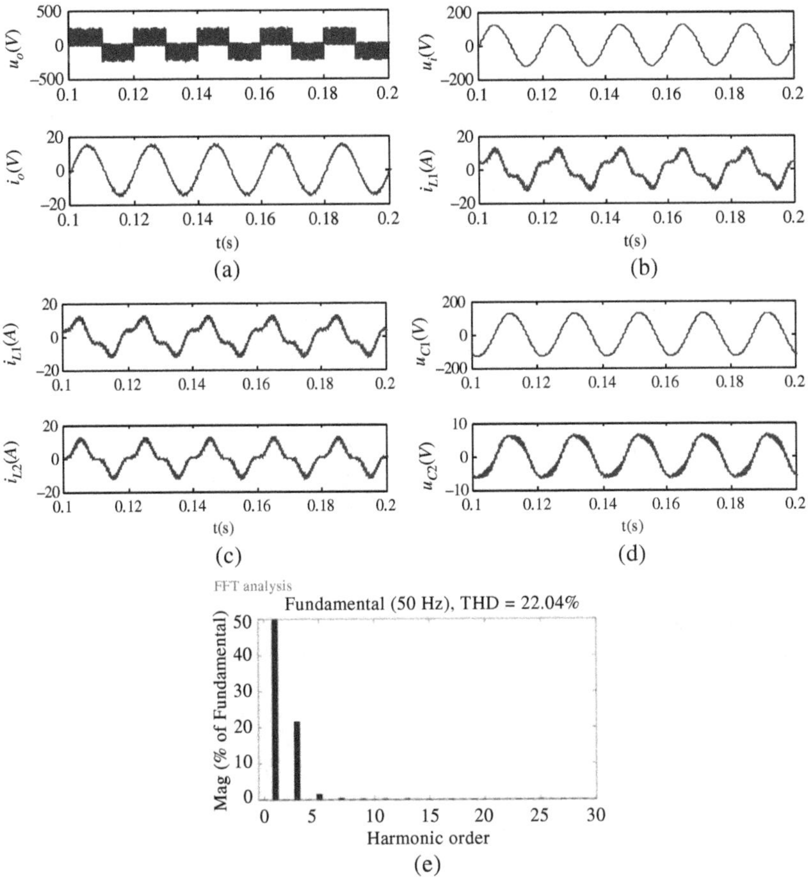

Figure 7.6 Depicts the analysis findings for the 3-1-phase qZS-MC, which operates at an output frequency of 50 Hz and lacks a compensation mechanism. The data captured includes: (a) the voltage and current at the system's output; (b) electrical properties at the input, both voltage and current; (c) current measurements for the Quasi-Z source inductors; (d) voltage values pertaining to the Quasi-Z source capacitors; and (e) FFT graphical representation of the input current's frequency distribution.

paint a picture of distorted input currents and escalated THD, stemming from unbuffered 2ω power.

Given a consistent load frequency of 100 Hz, there's a noteworthy reduction in compensation capacitance and inductance, registering values of $C_{com} = 300\,\text{F}$ and $L_{com} = 200\,\mu\text{H}$. Analyzing Figures 7.7–7.10 offers a lens into the simulated and hands-on outcomes when integrating or sidelining the MPC-based compensation mechanism at the stated 100-Hz load frequency. Drawing parallels between Figures 7.7a and 7.9a & 7.8a and 7.10a, especially with parameters set at $L_{com} = 200\,\mu\text{H}$ and $C_{com} = 300\,\mu\text{F}$, the benefits of the compensatory strategy in refining input current waveforms and diminishing THDs become evident. In scenarios devoid of this compensatory intervention, the 3rd and 5th harmonic domains dominate the input current's

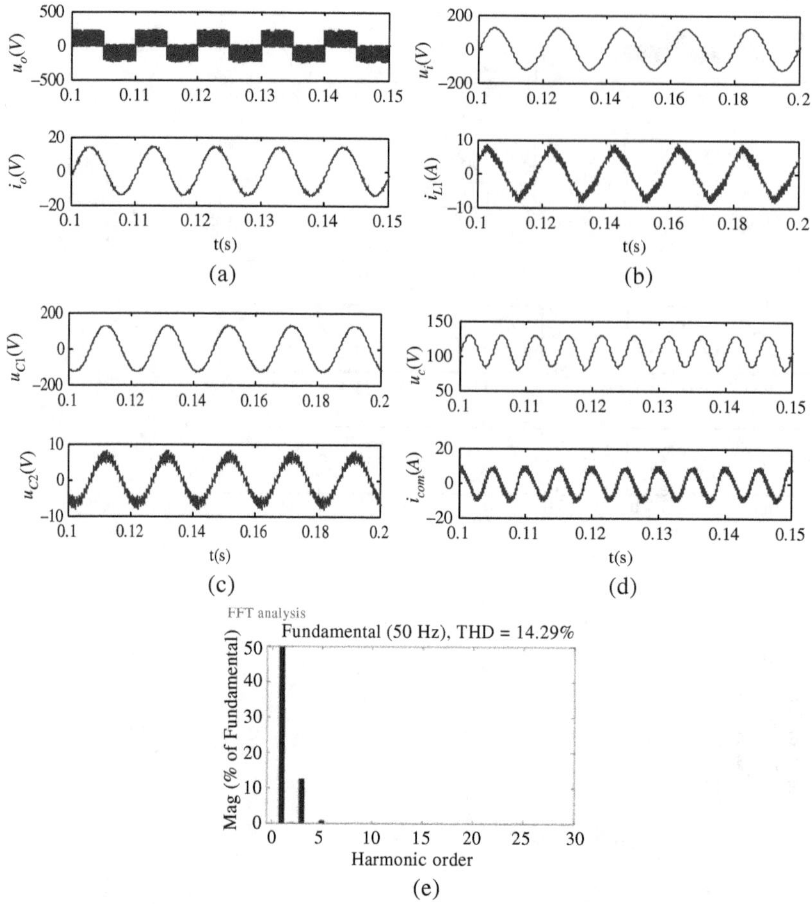

Figure 7.7 Showcases the outcomes of a simulation conducted on the 3-1-phase qZS-MC operating at an output frequency of 100 Hz, where the compensation segment is characterized by $L_{com} = 200\,\mu$H and $C_{com} = 300\,\mu$F. The visualizations include: (a) metrics for output voltage and associated current; (b) input electrical parameters—voltage and current; (c) voltage attributes of the Quasi-Z source capacitors; (d) measurements of voltage and current in the compensation segment's capacitor; and (e) FFT representation depicting the spectrum of the input current.

distorted signature, as evident in the FFT portrayals in Figures 7.8e and 7.10c. Simulated and empirical evaluations reported THDs of 52.61 and 54.20% under these conditions. However, once the compensatory system is operational, Figures 7.7e and 7.9d show the 3rd and 5th harmonics of input current plunging dramatically, yielding THDs of 14.29% and 17.94%. A deep dive into 7.9c shows u_c encompassing a mix of DC voltage and a 200 Hz component, aligning well with theoretical insights and simulated observations from Figures 7.4c and 7.7d.

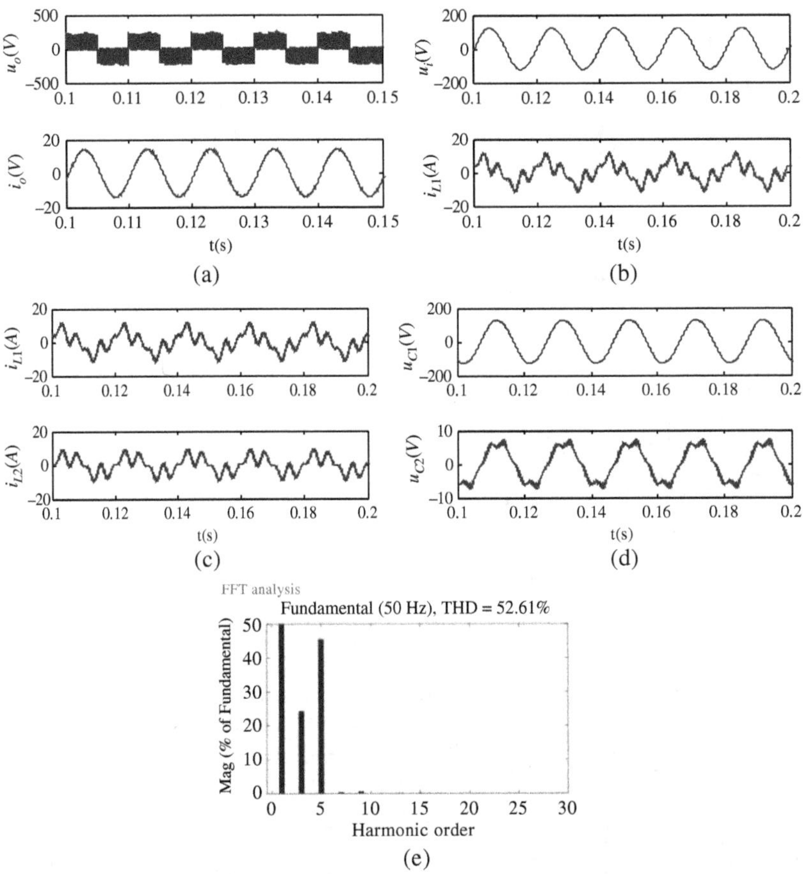

Figure 7.8 Presents the simulation outcomes for a 3-1-phase qZS-MC operating at a 100 Hz output frequency without the activation of the MPC compensation system. The following are illustrated: (a) voltage and current at the output; (b) voltage and current at the input; (c) flow of current through the Quasi-Z source inductors; (d) voltage readings across Quasi-Z source capacitors; and (e) FFT analysis showcasing the input current spectrum.

7.6 Conclusion

A redesigned 3-1-phase qZS-MC has been introduced, aiming to elevate the voltage gain beyond what the conventional 3-1-phase MC offers and simultaneously counteract the three-phase side's low-frequency ripple power. By integrating the qZS network, it takes on both the filtering and boosting responsibilities, negating the need for an extra input filter or step-up transformer. This integration paves the way for a streamlined design with fewer components. A detailed exploration

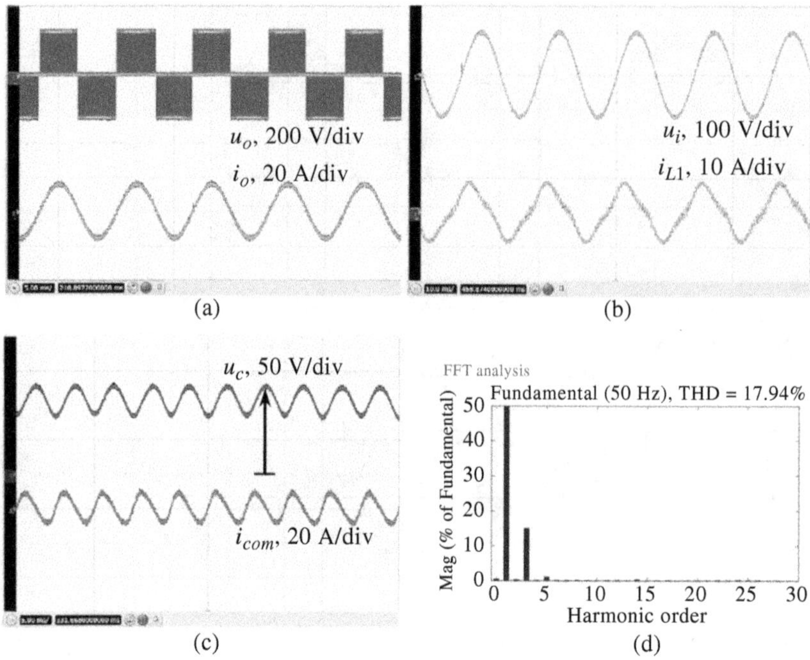

Figure 7.9 Data for the 3-1-phase qZS-MC at 100 Hz output frequency with compensation components L_{com} = 200 μH and C_{com} = 300 μF. (a) Output voltage and associated current profiles; (b) input voltage and associated current dynamics; (c) voltage and current characteristics of the compensation capacitor branch; (d) FFT spectral analysis highlighting harmonic components in the input current.

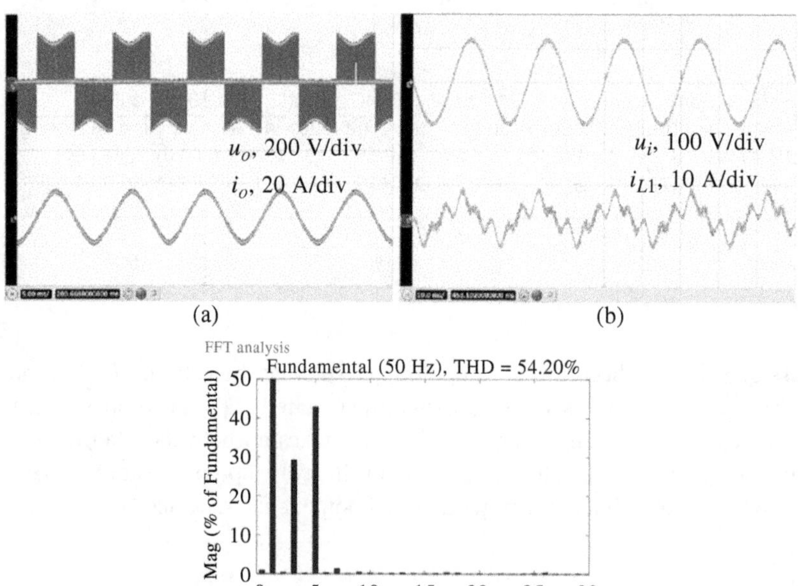

Figure 7.10 Data derived from the 3-1-phase qZS-MC tests, operating at an output frequency of 100 Hz with the MPC-based compensation mechanism turned off. (a) Provides a detailed view of the output voltage along with its associated current waveform; (b) captures the variations in input voltage and the consequent current behavior; while (c) depicts the FFT analysis of the input current for harmonic content.

into its fundamental structure, its functionality, and its analytical framework was undertaken to shed light on the qZS-MC's attributes. Moreover, a compensation technique, anchored on MPC, was developed to temper the 2ω power on the single-phase side, curbing its transfer to the three-phase domain. This tactic eradicated input current distortions at low frequencies and aided in optimizing qZS network parameters. Through simulations and hands-on tests, the efficacy of the revamped 3-1-phase qZS-MC, its conceptual model, and the ripple compensation strategy were corroborated.

A Appendix

For illustrative purposes, we're utilizing the MC functioning in sector 1 as a reference for deriving the model. As depicted in Figure 7.2, the commanded output voltage exhibits a positive value. Consequently, the values align as follows: $u_{\max}$ corresponds to u_c, u_{mid} is matched with u_a, and both $u_{\min}$ and u_{base} are synonymous with u_b. Based on the modulation techniques detailed in Section II.B, we deduce:

$$\alpha = \frac{i_a^*}{i_c^*} = \frac{\cos(\omega_i t - \varphi_i)}{\cos(\omega_i t - \varphi_i + 2\pi/3)} \tag{A7.1}$$

$$\begin{aligned} d_3 &= \frac{U_o \cos(\omega_o t + \varphi_{ou})}{\alpha u_{ab} + u_{cb}} = \frac{U_o \cos(\omega_o t + \varphi_{ou})}{x} \\ &= \frac{U_o \cos(\omega_o t + \varphi_{ou}) \cos(\omega_i t - \varphi_i + 2\pi/3)}{1.5 U_i \cos\varphi_i} \end{aligned} \tag{A7.2}$$

$$\begin{aligned} d_2 &= \frac{\alpha U_o \cos(\omega_o t + \varphi_{ou})}{\alpha u_{ab} + u_{cb}} = \frac{\frac{\cos(\omega_i t - \varphi_i)}{\cos(\omega_i t - \varphi_i + 2\pi/3)} U_o \cos(\omega_o t + \varphi_{ou})}{x} \\ &= \frac{U_o \cos(\omega_o t + \varphi_{ou}) \cos(\omega_i t - \varphi_i)}{1.5 U_i \cos\varphi_i} \end{aligned} \tag{A7.3}$$

with

$$x = \frac{\cos(\omega_i t - \varphi_i)}{\cos(\omega_i t - \varphi_i + 2\pi/3)} \sqrt{3} U_i \cos(\omega_i t + \pi/6) + \sqrt{3} U_i \cos(\omega_i t + \pi/2)$$

Based on the given conditions, we can deduce the following scenarios:

1) With S_{bn} and S_{ap} in the active state, the conditions stand as i'_a equals i_o, i'_b becomes the negative of i_o, and i'_c remains zero throughout the $d_2 T_s$ duration.
2) Activating both S_{bn} and S_{cp} results in i'_a being zero, i'_b as the negative counterpart of i_o, and i'_c equivalent to io over the $d_3 T_s$ span.
3) In the instance where S_{bn} and S_{bp} are engaged, all of i'_a, i'_b, and i'_c are zero for the duration of $d_0 T_s$.

Given these observations, the mean values for i'_a, i'_b, and i'_c are deduced as:

$$i'_a = \frac{i_o \cdot d_2 T_s}{(1-D)T_s} = \frac{y_0}{(1-D)1.5 U_i \cos\varphi_i} = \frac{y_1}{3U_i \cos\varphi_i (1-D)} \tag{A7.4}$$

$$i'_c = \frac{i_o \cdot d_3 T_s}{(1-D)T_s} = \frac{y_2}{(1-D)1.5 U_i \cos\varphi_i} = \frac{y_3}{3U_i \cos\varphi_i (1-D)} \tag{A7.5}$$

$$i'_b = -\left(i'_a + i'_c\right) = \frac{y_4}{3U_i \cos\varphi_i (1-D)} \tag{A7.6}$$

with

$$
\begin{aligned}
y_0 &= I_o \cos(\omega_o t + \varphi_{oi}) U_o \cos(\omega_o t + \varphi_{ou}) \cos(\omega_i t - \varphi_i), \\
y_1 &= U_o I_o \left[\cos(\varphi_{ou} - \varphi_{oi}) \cos(\omega_i t - \varphi_i) \right. \\
&\quad + 0.5 \cos(2\omega_o t + \omega_i t + \varphi_{ou} + \varphi_{oi} - \varphi_i) \\
&\quad \left. + 0.5 \cos(2\omega_o t - \omega_i t + \varphi_{ou} + \varphi_{oi} + \varphi_i)\right], \\
y_2 &= I_o \cos(\omega_o t + \varphi_{oi}) U_o \cos(\omega_o t + \varphi_{ou}) \cos(\omega_i t - \varphi_i + 2\pi/3), \\
y_3 &= U_o I_o \left[\cos(\varphi_{ou} - \varphi_{oi}) \cos(\omega_i t - \varphi_i + 2\pi/3) \right. \\
&\quad + 0.5 \cos(2\omega_o t + \omega_i t + \varphi_{ou} + \varphi_{oi} - \varphi_i + 2\pi/3) \\
&\quad \left. + 0.5 \cos(2\omega_o t - \omega_i t + \varphi_{ou} + \varphi_{oi} + \varphi_i - 2\pi/3)\right], \\
y_4 &= U_o I_o \left[\cos(\varphi_{ou} - \varphi_{oi}) \cos(\omega_i t - \varphi_i - 2\pi/3) \right. \\
&\quad + 0.5 \cos(2\omega_o t + \omega_i t + \varphi_{ou} + \varphi_{oi} - \varphi_i - 2\pi/3) \\
&\quad \left. + 0.5 \cos(2\omega_o t - \omega_i t + \varphi_{ou} + \varphi_{oi} + \varphi_i + 2\pi/3)\right].
\end{aligned}
$$

From (6.3) and the derived (6.15)–(6.17), we can get the complete model of phase a, as follows

$$
\begin{aligned}
i_{L1} &= \frac{-CU_i\omega_i}{2}\left[\frac{1}{(1-2D)^2 - LC\omega_i^2} + \frac{1}{1 - LC\omega_i^2}\right]\sin\omega_i t \\
&\quad + \frac{U_o I_o \cos(\varphi_{ou} - \varphi_{oi})(1-2D)}{3U_i \cos\varphi_i\left[(1-2D)^2 - LC\omega_i^2\right]}\cos(\omega_i t - \varphi_i) \\
&\quad + k_1 \cos(2\omega_o t + \omega_i t + \varphi_{ou} + \varphi_{oi} - \varphi_i) \\
&\quad + k_2 \cos(2\omega_o t - \omega_i t + \varphi_{ou} + \varphi_{oi} + \varphi_i)
\end{aligned} \tag{A7.7}
$$

$$
\begin{aligned}
i_{L2} &= \frac{-CU_i\omega_i}{2}\left[\frac{1}{(1-2D)^2 - LC\omega_i^2} - \frac{1}{1 - LC\omega_i^2}\right]\sin\omega_i t \\
&\quad + \frac{U_o I_o \cos(\varphi_{ou} - \varphi_{oi})(1-2D)}{3U_i \cos\varphi_i\left[(1-2D)^2 - LC\omega_i^2\right]}\cos(\omega_i t - \varphi_i) \\
&\quad + k_1 \cos(2\omega_o t + \omega_i t + \varphi_{ou} + \varphi_{oi} - \varphi_i) \\
&\quad + k_2 \cos(2\omega_o t - \omega_i t + \varphi_{ou} + \varphi_{oi} + \varphi_i)
\end{aligned} \tag{A7.8}
$$

$$
\begin{aligned}
v_{C1} &= \frac{U_i}{2}\left[\frac{1-2D}{(1-2D)^2 - LC\omega_i^2} + \frac{1}{1 - LC\omega_i^2}\right]\cos\omega_i t \\
&\quad + \frac{L\omega_i U_o I_o \cos(\varphi_{ou} - \varphi_{oi})}{3U_i \cos\varphi_i\left[(1-2D)^2 - LC\omega_i^2\right]}\sin(\omega_i t - \varphi_i) \\
&\quad + k_3 \sin(2\omega_o t + \omega_i t + \varphi_{ou} + \varphi_{oi} - \varphi_i) \\
&\quad + k_4 \sin(2\omega_o t - \omega_i t + \varphi_{ou} + \varphi_{oi} + \varphi_i)
\end{aligned} \tag{A7.9}
$$

$$\begin{aligned} v_{C2} &= \frac{U_i}{2}\left[\frac{1-2D}{(1-2D)^2 - LC\omega_i^2} - \frac{1}{1-LC\omega_i^2}\right]\cos\omega_i t \\ &+ \frac{L\omega_i U_o I_o \cos(\varphi_{ou} - \varphi_{oi})}{3U_i\cos\varphi_i\left[(1-2D)^2 - LC\omega_i^2\right]}\sin(\omega_i t - \varphi_i) \\ &+ k_3\sin(2\omega_o t + \omega_i t + \varphi_{ou} + \varphi_{oi} - \varphi_i) \\ &+ k_4\sin(2\omega_o t - \omega_i t + \varphi_{ou} + \varphi_{oi} + \varphi_i) \end{aligned} \tag{A7.10}$$

with

$$k_1 = \frac{U_o I_o (1-2D)}{\left[(1-2D)^2 - LC(2\omega_o + \omega_i)^2\right] 6U_i\cos\varphi_i}, \quad k_2 = \frac{U_o I_o (1-2D)}{\left[(1-2D)^2 - LC(2\omega_o - \omega_i)^2\right] 6U_i\cos\varphi_i},$$

$$k_3 = \frac{U_o I_o L(2\omega_o + \omega_i)}{\left[(1-2D)^2 - LC(2\omega_o + \omega_i)^2\right] 6U_i\cos\varphi_i}, \quad k_4 = \frac{U_o I_o L(2\omega_o - \omega_i)}{\left[(1-2D)^2 - LC(2\omega_o - \omega_i)^2\right] 6U_i\cos\varphi_i}.$$

References

1 O. Ellabban, H. Abu-Rub, and B. Ge, "A quasi-Z-source direct matrix converter feeding a vector controlled induction motor drive," *IEEE J. Emerg. Select. Top. Power Electron.*, vol. 3, no. 2, pp. 339–348, Jun. 2015.

2 E. Karaman, M. Farasat, and A. Trzynadlowski, "A comparative study of series and cascaded Z-source matrix converters," *IEEE Trans. Ind. Electron.*, vol. 61, no. 10, pp. 5164–5173, Oct. 2014.

3 L. He, S. Duan, and F. Peng, "Safe-commutation strategy for the novel family of quasi-Z-source AC-AC converter," *IEEE Trans. Ind. Inform.*, vol. 9, no. 3, pp. 1538–1547, Aug. 2013.

4 W. Song *et al.*, "A study of Z-source dual-bridge matrix converter immune to abnormal input voltage disturbance and with high voltage transfer ratio," *IEEE Trans. Ind. Inform.*, vol. 9, no. 2, pp. 828–838, May 2013.

5 X. Liu, P. C. Loh, P. Wang, and X. Han, "Improved modulation schemes for indirect Z-source matrix converter with sinusoidal input and output waveforms," *IEEE Trans. Power Electron.*, vol. 27, no. 9, pp. 4039–4050, Sept. 2012.

6 B. Ge, Q. Lei, W. Qian, and F. Z. Peng, "A family of Z-source matrix converters," *IEEE Trans. Ind. Electron.*, vol. 59, no. 1, pp. 35–46, Jan. 2012.

7 S. Liu, B. Ge, X. Jiang, H. Abu-Rub, and F. Z. Peng, "Comparative evaluation of three Z-source/quasi-Z-source indirect matrix converters," *IEEE Trans. Ind. Electron.*, vol. 62, no. 2, pp. 692–701, Feb. 2015.

8 Y. Liu, W. Liang, B. Ge, H. Abu-Rub, and N. Nie, "Quasi-Z-source three-to-single-phase matrix converter and ripple power compensation based on model predictive control," *IEEE Trans. Ind. Electron.*, vol. 65, no. 6, pp. 5146–5156, Jun. 2018.

9 M. Saito, "A novel single/three-phase matrix converter for high power integration," in *Proc. 13th Power Electronics and Motion Control Conf.*, pp. 2439–2444, Sept. 1–3, 2008.

10 Y. Sun, Y. Liu, M. Su, W. Xiong, and J. Yang, "Review of active power decoupling topologies in single-phase systems," *IEEE Trans. Power Electron.*, vol. 31, no. 7, pp. 4778–4794, Jul. 2016.

11 H. Wu, H. Ge, Y. Xu, and W. Zhang, "Thepower factor correctionof three-phase to single-phase matrix converter with an active power decoupling capacity," in *Proc. 2014 IEEE ITEC Asia-Pacific*, pp. 1–5, 2014.

12 Y. Furuhashi and T. Takeshita, "Single-phase to three-phase matrix converter with compensation for instantaneous-power fluctuation," in *Proc. IEEE 37th IECON*, pp. 1572–1577, 2011.

13 Y. Miura, T. Amano, and T. Ise, "Operating characteristics of a three-phase to single-phase matrix converter with hybrid control scheme of power compensation and modulation applied to gas engine cogeneration system," in *Proc. 2011-14th EPE 2011*, pp. 1–10, 2011.

14 M. Schweizer and J. W. Kolar, "Design and implementation of a highly efficient three-level T-type converter for low-voltage applications," *IEEE Trans. Power Electron.*, vol. 28, no. 2, pp. 899–907, Feb. 2013.

15 P. W. Wheeler, J. Rodriguez, J. C. Clare, *et al*, "Matrix converters: a technology review," *IEEE Trans. Ind. Electron.*, vol. 49, no. 2, pp. 276–288, Apr 2002.

8

Model Predictive Control of LC Filter-Integrated Quasi-Z-Source Indirect Matrix Converter

8.1 Introduction

Matrix Converters (MCs) are gaining traction in motor drives due to advantages like compactness, absence of large electrolytic capacitors in DC links, and ease of maintenance. Two primary MC types exist: the direct matrix converter (DMC) and the indirect matrix converter (IMC) [1–5]. IMC offers improvements over DMC by utilizing zero-current commutation, making it more suitable for AC motor drives. However, conventional MCs have certain drawbacks, including limited voltage gain and susceptibility to interference. To address these issues, new modulation strategies and auxiliary circuits have been introduced. Notably, Z-source and quasi-Z source (ZS/QZS) MCs stand out as effective solutions [6–10].

By innovating upon traditional DMC and IMC frameworks, ZS/QZS DMC and ZS/QZS IMC were introduced to achieve: A voltage gain exceeding the 0.866 limitation of conventional models enhanced resilience against input–output interference. In the ZS/QZS DMC design, ZS/QZS networks are integrated into the AC source and DMC bridge, allowing the voltage to be boosted. Two ZS/QZS IMC designs have emerged: one integrates the ZS/QZS networks into the IMC's DC link, while the other substitutes the grid-side LC filter with ZS/QZS networks. Both designs support voltage augmentation, but the latter also rids the need for an extra LC filter.

One variant of the QZS-IMC, the LC filter-integrated QZS-IMC, interposes a QZS network—comprising two inductors (or a single coupled inductor), two capacitors, and a bidirectional switch—between the grid and rectifier for each phase. This design offers several merits: Combined filtering and voltage boost functions eliminate the need for additional input filters. Extended voltage gain is achieved. Improved resistance against interference, alongside low voltage ride-through capability. Preservation of full-silicon features.

A comprehensive comparison of the LC filter-integrated QZS-IMC with other QZS-IMCs is detailed in [11]. Its design and modeling specifics are found in [12, 13], with a focus on reducing common-mode voltage in [14] and an optimal boost control discussed in [15]. Moreover, the induction motor drive application is highlighted in [16, 17]. Traditional IMCs use the proportional-integral (PI) control method, leveraging PI regulators and pulse width modulation (PWM). This becomes intricate due to the complex computation and additional control loop design. As a remedy, the model predictive control (MPC) method, previously applied to conventional IMCs [18–24], is now being considered for the QZS-IMC.

In this chapter, we introduce an MPC approach tailored for the LC filter-integrated QZS-IMC [25, 26]. Utilizing current measurements and established discrete predictive models, this method

Impedance Source Matrix Converters and Control, First Edition.
Yushan Liu, Xiao Li, Baoming Ge, Haitham Abu-Rub, and Frede Blaabjerg.

predicts circuit variables, evaluates an established cost function, and determines the optimal switching state for both rectifier and inverter stages within a singular control loop, without needing additional modulators or regulators. The following sections will delve deeper: Section 8.2 introduces the LC filter-integrated QZS-IMC; the foundation of MPC and the proposed approach are discussed in Sections 8.3 and 8.4; Section 8.5 presents the simulation and experimental outcomes; and Section 8.6 offers concluding remarks.

8.2 LC Filter-Integrated QZS-IMC

The schematic in Figure 8.1 outlines the LC filter-integrated QZS-IMC structure. This network comprises three sets of paired inductors, specifically, L_{a1} with L_{a2}, L_{b1} with L_{b2}, and L_{c1} with L_{c2}. Additionally, there are six capacitors, identified as C_{a1}, C_{a2}, C_{b1}, C_{b2}, C_{c1}, and C_{c2}. The system also uses three bidirectional switches: S_a, S_b, and S_c, all governed by a singular gate signal labeled S_x. The design of both the rectifier and inverter stages mirrors the typical IMC, each incorporating six bidirectional switches.

The operational modality includes two primary states: the shoot through (ST) and non-ST states. In the non-ST scenario, S_x remains active, enabling the QZS network to regulate input voltage and current, bypassing the ST behavior observed in the rectifier. Contrarily, the ST state sees the QZS network functioning in a dual capacity—filtering while also amplifying the voltage. This is achieved by introducing the ST vector to the zero-current vector in the rectifier stage. Here, the rectifier's top and bottom triad of switches operate concurrently, with the QZS network's switches in the passive state, as detailed in [11, 12].

Upon engagement of the ST state, the voltage amplification ratio, represented as B within the QZS network, is discussed in [15].

$$B = \frac{U_{im}}{U_{in}} = \frac{1}{1-2D} \tag{8.1}$$

In this context, "D" signifies the duty cycle associated with ST. The amplitude of the input phase voltage is represented as U_{in}. Meanwhile, U_{im} corresponds to the amplitude of the output phase voltage of the QZS network. This latter value is also indicative of the input phase voltage amplitude directed to the IMC rectifier.

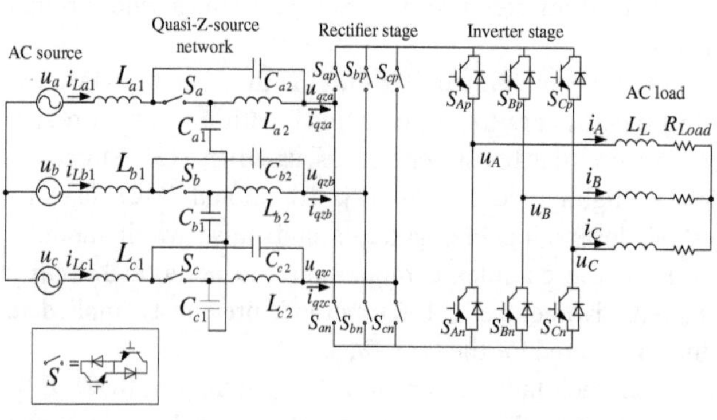

Figure 8.1 Topology of LC filter-integrated QZS-IMC.

Leveraging the dual space vector modulation approach for both the rectifier and inverter stages, the output phase voltage amplitude for the inverter stage can be deduced as elaborated in references [15, 16].

$$U_{om} = G \cdot U_{in} \tag{8.2}$$

$$G = \frac{U_{om}}{U_{in}} = Hm_o m_i B \cos \varphi_{qz} \tag{8.3}$$

Herein, G signifies the QZS-IMC's voltage gain, while m_i and m_o stand for the modulation indices of the rectifier and inverter stages, respectively. In the context of a traditional MC, the recognized voltage gain is 0.866, symbolized by H. Additionally, the rectifier's input power factor is represented by $\cos\varphi_{qz}$.

8.3 Principle of Model Predictive Control

The rapid state transitions and multiple control variables are notable traits of power electronic converters. Gradually, a finite control set (FCS)-MPC strategy has emerged addressing these aspects. Within this paradigm, power converters are viewed as discrete nonlinear entities. From a predetermined set of switching states, the FCS-MPC evaluates the system's variables using a discrete predictive framework, determines the optimal state through a cost function, and directly dispatches the control signal. In the context of this research, the FCS-MPC approach involves: (i) formulating system predictive models, identifying feasible switching states, and forecasting the forthcoming system state; (ii) designing a cost function to identify the most economical state; (iii) implementing the most compatible system state, or in other terms, the most cost-effective switching condition, to guide the inverter's switching mechanism.

As illustrated in Figure 8.2, the FCS-MPC's foundational concept is depicted. With a sampling period denoted as T_s, the system has n control actions, represented as S_i where $i \in [1, n]$. The state at any given time t is $x(t)$, while the reference state is $x^*(t)$. At instance t_k, the controller, leveraging current state values, prospective control actions, and forecast models, estimates the potential

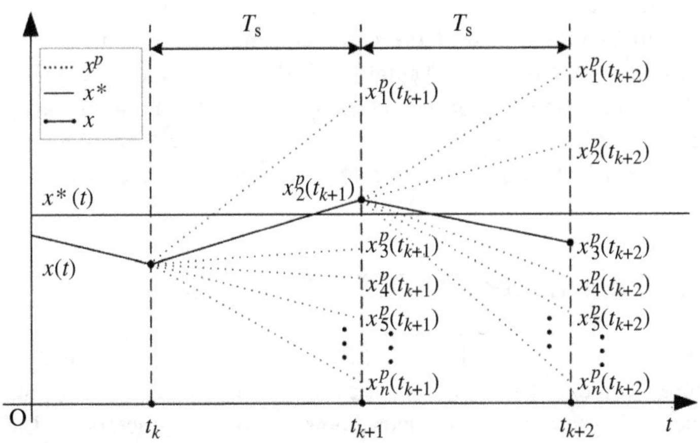

Figure 8.2 Principle of MPC.

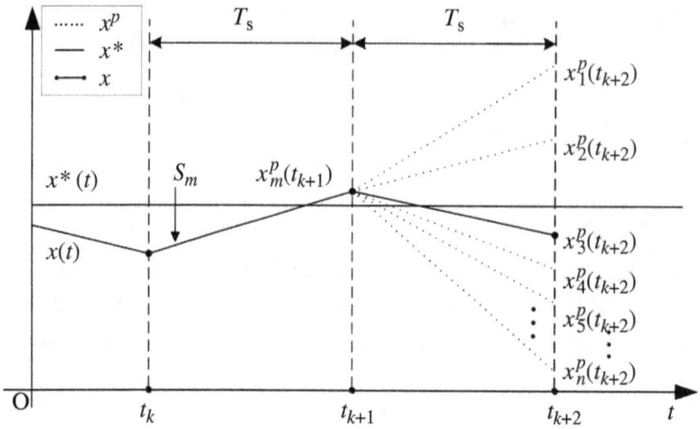

Figure 8.3 Principle of delay compensation.

state value for t_{k+1}: $\{x_1^p(t_{k+1}), x_2^p(t_{k+1}), x_3^p(t_{k+1}) \cdots x_n^p(t_{k+1})\}$. To pinpoint the optimal state value, the controller outputs the corresponding control action by adopting a cost minimization criterion. The procedure continues by estimating for t_{k+2}: $\{x_1^p(t_{k+2}), x_2^p(t_{k+2}), x_3^p(t_{k+2}) \cdots x_n^p(t_{k+2})\}$. In most instances, cost functions are sculpted to slash the deviation between forecasted and reference values. For instance, as Figure 8.2 showcases, the forecasted value $x_2^p(t_{k+1})$ trims down the cost function leading to the S_2 control signal being chosen. In a parallel fashion, for t_{k+2}, the $x_3^p(t_{k+2})$ prediction shrinks the cost function, making S_3 the preferred output.

In practical engineering scenarios, delays arising from the digital signal processor, sampling circuit, gate driver, and switching device cannot be sidestepped. Until the new switching state is updated and employed, the previous optimal state persists. Such latencies can impinge on system performance and warrant compensation. Figure 8.3 depicts the delay mitigation strategy adopted herein. The strategy encompasses: (i) At the onset of t_k, the optimal control action, S_m from the last cycle, is outputted. Subsequently, leveraging the present state $x(t_k)$, the control action S_m, and the predictive blueprint, the projected value $x_m^p(t_{k+1})$ for t_{k+1} is ascertained. (ii) Using $x_m^p(t_{k+1})$ as a pivot, the next prediction step unveils all feasible states for t_{k+2}: $\{x_1^p(t_{k+2}), x_2^p(t_{k+2}), x_3^p(t_{k+2}) \cdots x_n^p(t_{k+2})\}$. (iii) The cost function discerns the ideal control action tied to the most cost-effective state value. This is retained by the controller to be transmitted at the commencement of the subsequent cycle. This approach culminates in the compensation of a single cycle delay, T_s. As represented in Figure 8.3, at t_k, the previously determined optimal control action S_m is applied, and the two-step continuous prediction yields $x_3^p(t_{k+2})$ as the value that optimally reduces the cost function. Consequently, control action S_3 is earmarked for output initiation in the approaching cycle.

8.4 Proposed MPC for LC Filter-Integrated QZS-IMC

The schematic representation of the suggested MPC for an LC filter-integrated QZS-IMC can be viewed in Figure 8.4. The procedure commences by measuring the immediate voltages and currents of the circuit. Utilizing the predefined system predictive paradigms, various parameters for the upcoming $k+1^{\text{th}}$ control sample—like AC load current, the voltage at the entry of the rectifier stage, and the AC source's reactive power—are anticipated. Following this, the predictions are juxtaposed with their corresponding references to determine the cost function, culminating in the choice of

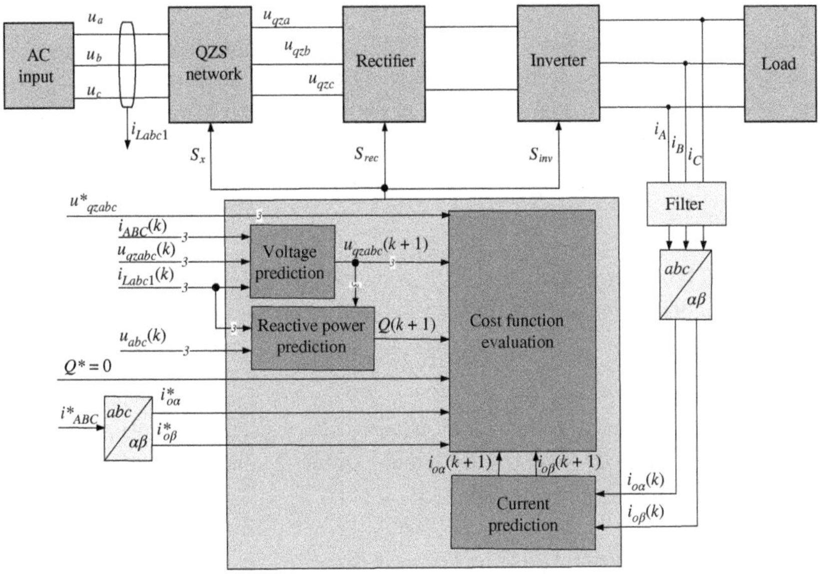

Figure 8.4 Proposed MPC for LC filter-integrated QZS-IMC.

the switching state. Consequently, the power components of both the rectifier and inverter stages are governed by the derived optimal switching control signals.

8.4.1 Modeling of IMC

Upon inspecting Figure 8.1, we can ascertain the representations for the three-phase grid voltage $\mathbf{U}_s$ and current $\mathbf{I}_s$. Analogously, the voltage $\mathbf{U}_i$ and current $\mathbf{I}_i$ at the input stage of the rectifier are delineated as:

$$\mathbf{U}_s = [u_a \ u_b \ u_c]^T,\ \mathbf{I}_s = \left[i_{La1} \ i_{Lb1} \ i_{Lc1}\right]^T,\ \mathbf{U}_i = \left[u_{qza} \ u_{qzb} \ u_{qzc}\right]^T,\ \mathbf{I}_i = \left[i_{qza} \ i_{qzb} \ i_{qzc}\right]^T \quad (8.4)$$

Vectors in this context are symbolized by bold characters. Moving forward, the voltages $\mathbf{U}_o$ at the inverter's output stage and the associated load currents $\mathbf{I}_o$ are defined by:

$$\mathbf{U}_o = \left[u_A \ u_B \ u_C\right]^T,\ \mathbf{I}_o = \left[i_A \ i_B \ i_C\right]^T \quad (8.5)$$

The connection between the input currents $\mathbf{I}_i$ of the rectifier stage and the output currents $\mathbf{I}_o$ of the inverter stage is demonstrated in [17]:

$$\mathbf{I}_i = \begin{bmatrix} S_{ap} - S_{an} \\ S_{bp} - S_{bn} \\ S_{cp} - S_{cn} \end{bmatrix} \cdot \left[S_{Ap} \ S_{Bp} \ S_{Cp}\right] \cdot \mathbf{I}_o \quad (8.6)$$

Furthermore, the relationship between the output voltages $\mathbf{U}_o$ of the inverter stage and the input voltages $\mathbf{U}_i$ of the rectifier stage can be described as:

$$\mathbf{U}_o = \begin{bmatrix} S_{Ap} - S_{An} \\ S_{Bp} - S_{Bn} \\ S_{Cp} - S_{Cn} \end{bmatrix} \cdot \left[S_{ap} - S_{an} \ S_{bp} - S_{bn} \ S_{cp} - S_{cn}\right] \cdot \mathbf{U}_i \quad (8.7)$$

For the purposes of clarity in equations (8.6) and (8.7), the term $S_{kj} \in \{0, 1\}$ identifies the switch's state as depicted in Figure 8.1. For instance, the suffix k can represent the phases of the rectifier stage when drawn from the set $\{a, b, c\}$ or the inverter stage phases from the set $\{A, B, C\}$. Meanwhile, the suffix j chosen from $\{p, n\}$ signifies either the upper or lower branch of the IMC's single-phase leg.

8.4.2 Predictive Models

8.4.2.1 Predictive Model of AC Load Current

Referencing Figure 8.1, we can describe the dynamics of the AC load current as follows:

$$\frac{d\mathbf{I}_o}{dt} = \frac{1}{L_L}\mathbf{U}_o - \frac{R_{Load}}{L_L}\mathbf{I}_o \tag{8.8}$$

In this context, both L_L and R_{Load} stand for the load inductance and resistance, respectively. Utilizing the principles of the Euler method, we can extrapolate a predictive framework for the load currents in the subsequent $(k+1)$th sampling interval, T_s:

$$\mathbf{I}_o(k+1) = \frac{T_s}{L_L}\mathbf{U}_o(k) + \left(1 - \frac{R_{Load}T_s}{L_L}\right)\mathbf{I}_o(k) \tag{8.9}$$

8.4.2.2 Predictive Model of QZS Network

As portrayed in Figure 8.5, the QZS-IMC operates in both non-ST and ST states, with their respective equivalent circuits. Assuming symmetrical components for simplification, the parameters are defined such that L_{a1} through L_{c2} equate to L and similarly, C_{a1} through C_{c2} are equivalent to C. For ease of representation, let's consider a singular phase, omitting the phase indication. This allows us to depict the dynamics of the QZS inductor currents in the non-ST state and the capacitor voltages in the ST state as follows:

$$\begin{pmatrix} L & 0 & 0 & 0 \\ 0 & L & 0 & 0 \\ 0 & 0 & C & 0 \\ 0 & 0 & 0 & C \end{pmatrix} p \begin{pmatrix} i_{L1} \\ i_{L2} \\ u_{C1} \\ u_{C2} \end{pmatrix} = \begin{pmatrix} -(R_C+R_L) & 0 & -1 & 0 \\ 0 & -(R_C+R_L) & 0 & -1 \\ 1 & 0 & 0 & 0 \\ 0 & 1 & 0 & 0 \end{pmatrix} \begin{pmatrix} i_{L1} \\ i_{L2} \\ u_{C1} \\ u_{C2} \end{pmatrix} + \begin{pmatrix} 1 & R_C \\ 0 & R_C \\ 0 & -1 \\ 0 & -1 \end{pmatrix} \begin{pmatrix} u \\ i_{qz} \end{pmatrix} \tag{8.10}$$

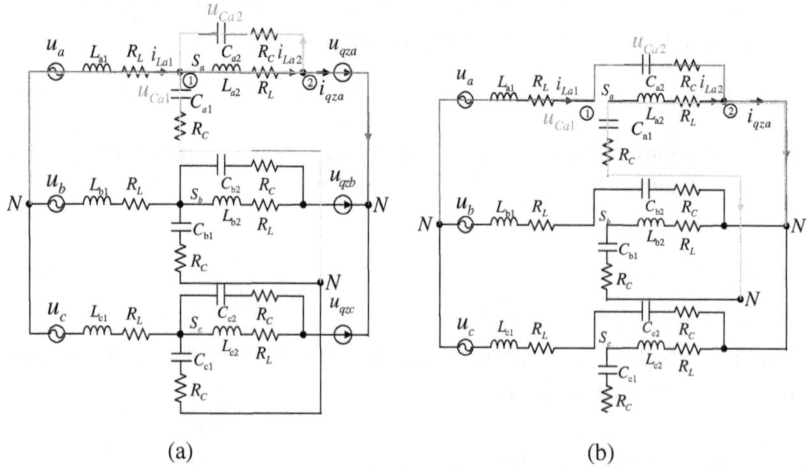

Figure 8.5 Equivalent circuits of LC filter-integrated QZS-IMC in (a) the non-ST state and (b) the ST state.

$$\begin{pmatrix} L & 0 & 0 & 0 \\ 0 & L & 0 & 0 \\ 0 & 0 & C & 0 \\ 0 & 0 & 0 & C \end{pmatrix} p \begin{pmatrix} i_{L1} \\ i_{L2} \\ u_{C1} \\ u_{C2} \end{pmatrix} = \begin{pmatrix} -(R_C + R_L) & 0 & 0 & 1 \\ 0 & -(R_C + R_L) & 1 & 0 \\ 0 & -1 & 0 & 0 \\ -1 & 0 & 0 & 0 \end{pmatrix} \begin{pmatrix} i_{L1} \\ i_{L2} \\ u_{C1} \\ u_{C2} \end{pmatrix} + \begin{pmatrix} 1 & 0 \\ 0 & 0 \\ 0 & 0 \\ 0 & 0 \end{pmatrix} \begin{pmatrix} u \\ i_{qz} \end{pmatrix} \tag{8.11}$$

In these equations, R_L symbolizes the QZS inductor's internal resistance, while R_C is representative of the capacitor's internal resistance. The term "p" stands for the derivative operator.

Utilizing the Euler method in conjunction with equation (8.10), we can establish predictive models for the QZS inductor- L_1 current, denoted as i_{L1}, as well as the rectifier-stage input voltage, symbolized as u_{qz}, when operating in the non-ST state. These are expressed as:

$$i_{L1}(k+1) = \left(1 - \frac{R_L T_s}{L}\right) i_{L1}(k) - \frac{T_s}{2L} u_{qz}(k) + \frac{T_s}{2L} u(k) \tag{8.12}$$

$$u_{qz}(k+1) = u_{qz}(k) + \frac{2T_s}{C} i_{L1}(k) - \frac{2T_s}{C} i_{qz}(k) \tag{8.13}$$

Here, the equation $u_{qz} = u_{C1} + u_{C2}$ represents the summation envelope of the rectifier-stage input voltage. Notably, the capacitor's internal resistance, R_C, is disregarded given its negligible magnitude. In a parallel vein, using equation (8.11), predictive frameworks for i_{L1} and u_{qz}, under the ST state conditions, are articulated as:

$$i_{L1}(k+1) = \left(1 - \frac{R_L T_s}{L}\right) i_{L1}(k) + \frac{T_s}{2L} u_{qz}(k) + \frac{T_s}{2L} u(k) \tag{8.14}$$

$$u_{qz}(k+1) = u_{qz}(k) - \frac{2T_s}{C} i_{L1}(k) \tag{8.15}$$

8.4.2.3 Cost Function Evaluation and Switching States Selection

The application of the cost function aids in pinpointing the optimal solution for the ensuing time step, evaluating the deviation between predictive values and reference standards. The primary aims of the QZS-IMC's control strategy encompass meeting the demands of load power and ensuring stability in the rectifier-stage input voltage. The first objective gets addressed by managing the load current given a specific load resistance, while the second involves stabilizing the output voltage of the inverter stage. Furthermore, curtailing the instantaneous reactive power on the QZS-IMC's input end is essential to enhance the circuit's reactive power efficiency, as supported by [18]. Given these factors, and accounting for the one-time delay compensation, we define our cost function as:

$$\begin{aligned} g = {} & |i^*_{o\alpha}(k) - i_{o\alpha}(k+2)|^2 + \left|i^*_{o\beta}(k) - i_{o\beta}(k+2)\right|^2 \\ & + \lambda_q (0 - |u_{s\alpha}(k) i_{L\beta 1}(k+2) - u_{s\beta}(k) i_{L\alpha 1}(k+2)|)^2 \\ & + \lambda_u \left(\left|u^*_{qz\alpha}(k) - u_{qz\alpha}(k+2)\right| + \left|u^*_{qz\beta}(k) - u_{qz\beta}(k+2)\right|\right) \end{aligned} \tag{8.16}$$

In the α-β stationary frame, the forecasted load current $\mathbf{I}_o$ is represented by $i_{o\alpha}$ and $i_{o\beta}$. Meanwhile, the reference current i^*_{ABC} within the α–β axis is depicted as $i^*_{o\alpha}$ and $i^*_{o\beta}$. When transposed to the α–β axis, the source grid voltage $\mathbf{U}_s$ corresponds to $u_{s\alpha}$ and $u_{s\beta}$, and the predicted grid current $\mathbf{I}_s$ aligns with $i_{L\alpha 1}$ and $i_{L\beta 1}$. Similarly, $u_{qz\alpha}$ and $u_{qz\beta}$ outline the anticipated rectifier-stage input voltage $\mathbf{U}_i$ in the same α–β stationary frame. The reference voltage u^*_{qzabc} within the α-β framework is denoted by $u^*_{qz\alpha}$ and $u^*_{qz\beta}$. Lastly, the reactive power and rectifier-stage input voltage control terms are weighted by factors λ_q and λ_u, respectively.

As depicted in Figure 8.6, an illustrated flowchart outlines the conceptual approach of the suggested MPC for the LC filter integrated QZS-IMC. Initial assessments are made to determine

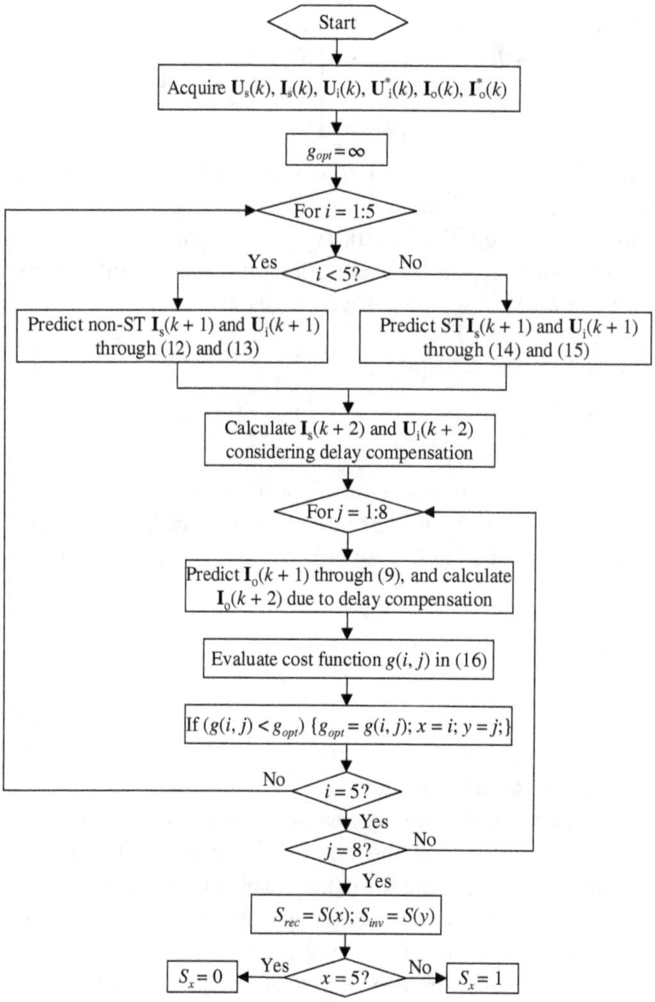

Figure 8.6 Flowchart of proposed MPC for the LC filter-integrated QZS-IMC.

between non-ST and ST states. When the converter exists in the ST state (specifically when $i = 5$), the projections for $\mathbf{I}_s$ and $\mathbf{U}_i$ follow the guidelines provided in equations (8.14) and (8.15). However, in alternative scenarios, the predictions abide by (8.12) and (8.13). It's crucial to highlight that the rectifier stage comprises 10 different switching states. This includes nine for the non-ST state and a singular one for the ST state, as detailed in Table 8.1. With the IMC's DC-link voltage remaining positive, one among the three non-ST zero states gets chosen for current commutation. Consequently, the rectifier stage comprises five unique active switching states, inclusive of the ST state.

When observing the inverter stage, it fully employs the eight available switching states of the conventional IMC, leading to a total of 40 active switching states. Each state is then evaluated against the cost function (8.16) to identify the optimal state with the least associated cost. Following this assessment, the prime switching states for both the rectifier and inverter stages, S_{rec} and S_{inv}, are chosen and executed in tandem.

Table 8.1 Switching states of rectifier stage.

	Switching states (S_{rec})						DC-Link voltage
	S_{a1}	S_{b1}	S_{c1}	S_{a2}	S_{b2}	S_{c2}	V_{dc}
1	**1**	0	0	0	0	**1**	u_{ac}
2	0	**1**	0	0	0	**1**	u_{bc}
3	0	**1**	0	**1**	0	0	$-u_{ab}$
4	0	0	**1**	**1**	0	0	$-u_{ac}$
5	0	0	**1**	0	**1**	0	$-u_{bc}$
6	**1**	0	0	0	**1**	0	u_{ab}
7	**1**	0	0	**1**	0	0	0
8	0	**1**	0	0	**1**	0	0
9	0	0	**1**	0	0	**1**	0
10	**1**	**1**	**1**	**1**	**1**	**1**	0

Furthermore, when the superior state of S_{rec} is identified as ST, an OFF signal is promptly dispatched to the S_x to engage the QZS network switches.

8.5 Simulation and Experimental Results

In an attempt to assess the efficacy of the MPC suggested for the QZS-IMC combined with the LC filter, simulations were carried out using the MATLAB/Simulink platform, which was then subsequently subjected to prototype testing. Delving into the system's detailed specifications, as delineated in Table 8.2, it's worth noting that the system boasts a rated power of 2 kW. The QZS framework's defining parameters include an inductance of 500 μH and a capacitance of 50 μF. Concurrently, the system's load inductance stands at 10 mH. With the power undergoing variations, the load resistance is subjected to alterations. Furthermore, a study was conducted to evaluate the load current's amplitude transition—moving from 10 A to a lower bracket of 5 A. This effectively signifies a shift in load power from the initially defined 2 kW to half, at 1 kW. A comparative assessment was executed, juxtaposing results acquired from the MPC methodology and the traditional PI control technique, illustrated vividly in Figures 8.7 through 8.10.

Table 8.2 System specifications.

Parameter	Value
Rated power of QZS-IMC, P_m	2 kW
Grid voltage amplitude, U_s	311 V
Grid voltage frequency, f_g	50 Hz
QZS network inductance, L	500 μH
QZS network capacitance, C	50 μF
Load inductance, L_L	10 mH

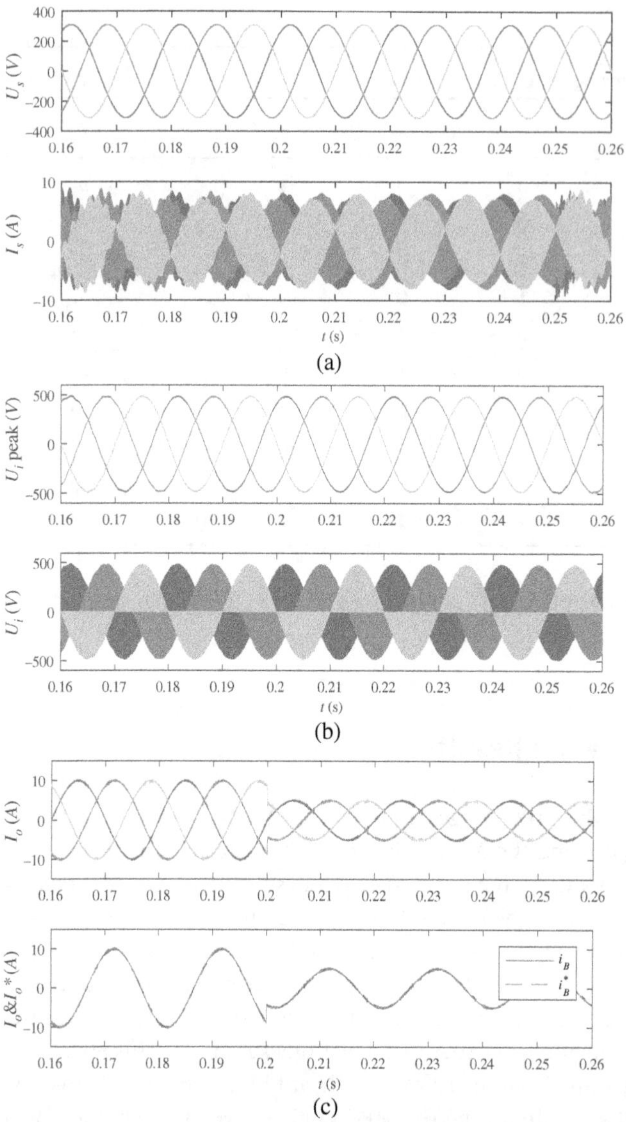

Figure 8.7 Simulation results for the QZS-IMC when integrated with an LC filter, utilizing the introduced MPC. Specifically: (a) voltage dynamics across a tri-phased grid alongside the associated current input; (b) captured voltage at the rectifier level complemented by its encompassing outline; and (c) current values over a tri-phase load, also highlighting the singular-phase current i_B juxtaposed with its benchmark i^*_B.

Illustrative results derived from simulations and experiments, portraying the three-phase grid voltage, grid input current, rectifier stage input voltage, and three-phase load currents of the LC filter paired with QZS-IMC, are exhibited in Figures 8.7 and 8.8. These results pertain to its management via the new MPC approach. Contrarily, results produced under the governance of the PI control are depicted in Figures 8.9 and 8.10. Significantly, Fast Fourier Transform (FFT) analyses of experimental load currents, based on the two aforementioned control techniques, are presented in Figures 8.8e and 8.10e.

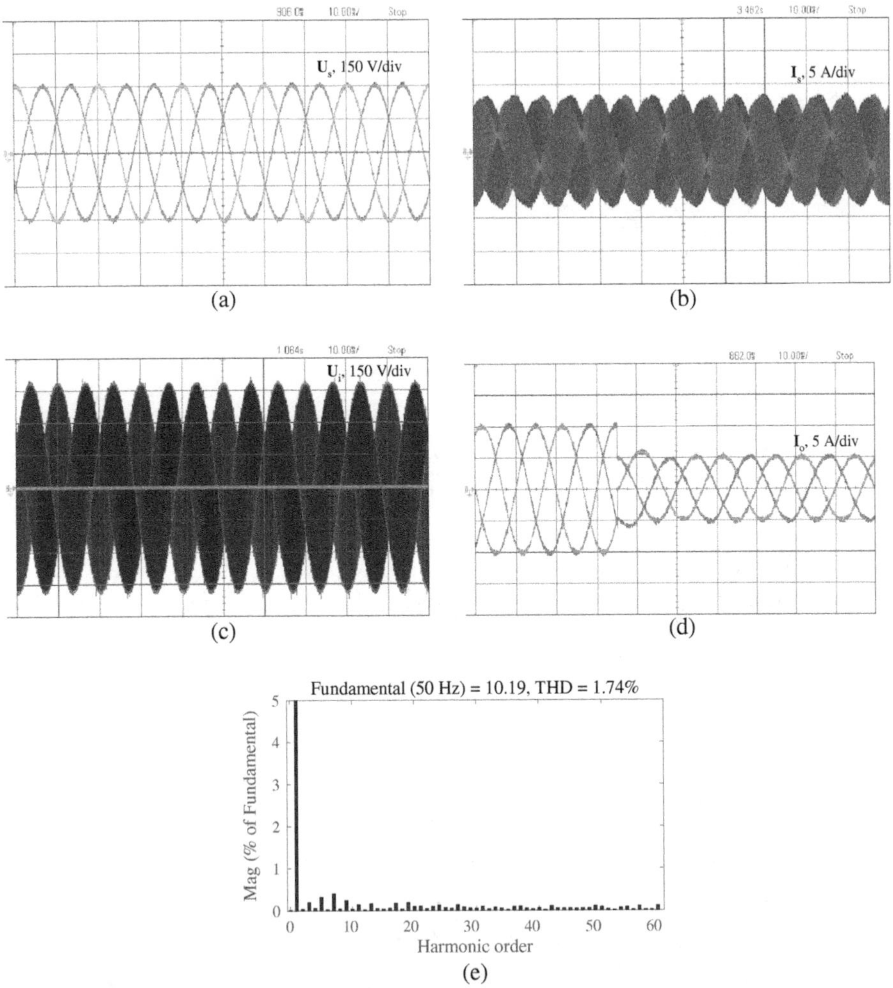

Figure 8.8 Experimental results for the QZS-IMC integrated with an LC filter when managed by the suggested MPC. As detailed: (a) voltage patterns across the three-phase grid; (b) grid's current influx; (c) voltage metrics at the rectifier phase; (d) current distribution over the three-phase load; and (e) load current's FFT under nominal conditions.

In surveying figures 8.7 to 8.10, it becomes evident that grid input current behaviors and rectifier stage input voltages remain unswayed by shifts in load power. Moreover, a discernible escalation in grid voltage amplitude transpires upon its introduction into the rectifier stage—a fact underscored by juxtaposing grid voltages in the Figures 8.7 and 8.9 series against rectifier stage input voltages in Figures 8.8 and 8.10. This phenomenon attests to the voltage augmentation capabilities inherent to the LC filter integrated QZS-IMC.

Comparative scrutiny of Figures 8.7 through 8.10 reveals a load current amplitude transition, from 10 A descending to 5 A, congruent with the prescribed values. Remarkable fidelity in dynamic tracking coupled with minuscule steady-state deviations are hallmarks of both the innovative MPC and its PI counterpart. Nonetheless, the MPC manifests superior tracking agility. This can be extrapolated from how swiftly the load current adheres to the reference amid fluctuating loads, in contrast to its PI-controlled counterpart.

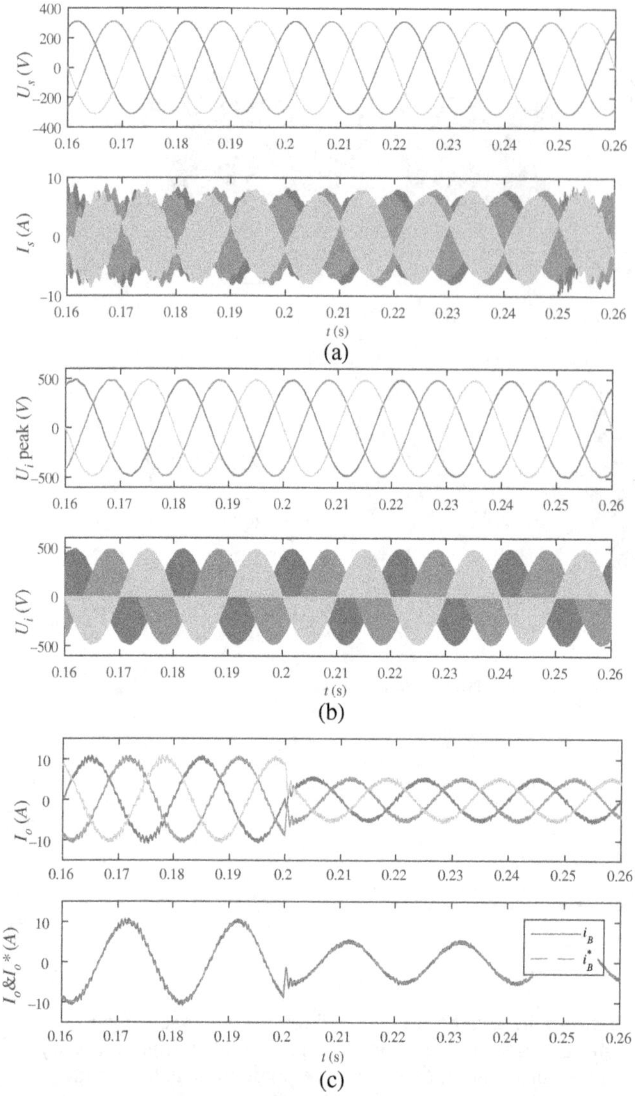

Figure 8.9 Simulation results for the QZS-IMC integrated with an LC filter operating under traditional PI control. Displayed are: (a) voltage profiles across the three-phase grid alongside current inputs; (b) voltage characteristics of the rectifier stage accompanied by its envelope representation; and (c) current values spanning the three-phase load, with a focus on the phase i_B current and its reference counterpart, i^*_B.

FFT outcomes in Figures 8.8e and 8.10e unveil a salient total harmonic distortion (THD) reduction, approaching half, when using the MPC, relative to the conventional PI approach. Quantitatively, the THD values were 1.74% and 3.25%, respectively. Yet, this performance comes at the trade-off of computational duration—with MPC's average sampling duration being approximately 30 μs and PI's stretching to 100 μs. When governed by the PI control, conspicuous harmonic magnitudes emerge predominantly at lower harmonic orders, as visualized in Figure 8.10e.

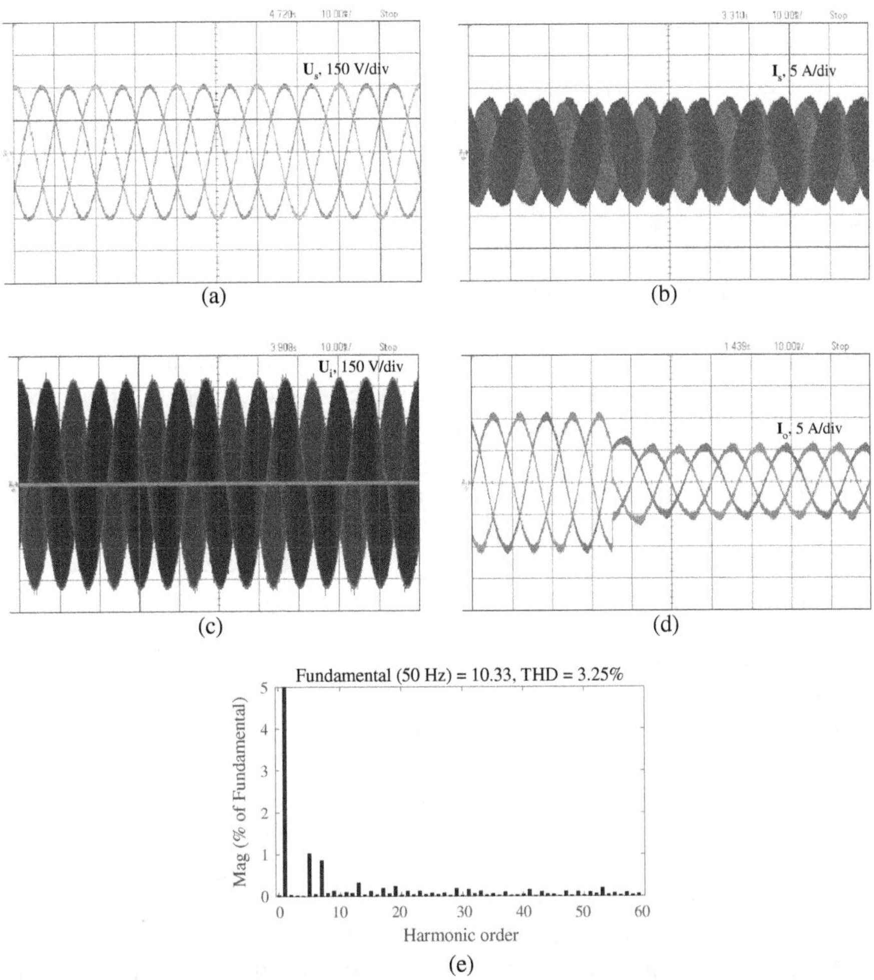

Figure 8.10 Empirical results showcasing the performance of the QZS-IMC integrated with an LC filter under traditional PI control. The panels illustrate: (a) the voltage across all three grid phases; (b) current entering the grid; (c) voltage at the rectifier stage input; (d) current distribution across the three-phase load; and (e) an FFT analysis of the load current under standard operational conditions.

8.6 Conclusion

In this study, we introduced an MPC tailored for the combination of an LC filter and QZS-IMC. By capturing data related to the three-phase grid's voltage and current, the envelope of the rectifier stage's input voltage, and the load current, we could anticipate control variables for the imminent sampling phase. This data, juxtaposed with reference points, facilitated the computation of a predefined cost function across various switching states. The switching state registering the minimal cost function was then prioritized for execution across the power components of both the rectifier and inverter phases at the onset of the subsequent sampling interval. The fruits of our simulation and tangible experiments underscore the adeptness of our MPC in streamlining the intricate processes typically associated with PI control in the QZS-IMC framework.

References

1 P. W. Wheeler, J. Rodriguez, J. C. Clare, L. Empringham, and A. Weinstein, "Matrix converters: A technology review," *IEEE Tran. Ind. Electron.*, vol. 49, no. 2, pp. 276–288, Aug. 2002.

2 L. Empringham, J. W. Kolar, J. Rodriguez, P. W. Wheeler, and J. C. Clare, "Technological issues and industrial application of matrix converters: A review," *IEEE Tran. Ind. Electron.*, vol. 60, no. 10, pp. 4260–4271, Sep. 2012.

3 M. Israyelu and S. Sashidhar, "Comparison of switching patterns of SVM for a novel high gain indirect matrix converter," in *IECON 2021 – 47th Annu. Conf. IEEE Industrial Electronics Society*, Toronto, ON, Canada, 2021, pp. 1–6.

4 A. Ammar, H. Y. Kanaan, M. Hamouda, and K. Al-Haddad, "Experimental application of double space vector pulse width modulation on three phase indirect matrix converters," in *IECON 2020 The 46th Annu. Conf. IEEE Industrial Electronics Society*, Singapore, 2020, pp. 2333–2337.

5 A. Ammar, H. Y. Kanaan, N. Moubayed, M. Hamouda, and K. Al-Haddad, "Comparative analysis attributed to DSVPWM-mode versus SPWM-mode indirect matrix converter," in *IECON 2019-45th Annu. Conf. IEEE Industrial Electronics Society*, Lisbon, Portugal, 2019, pp. 4694–4699.

6 B. Ge, Q. Lei, W. Qian, and F. Z. Peng, "A family of Z-source matrix converters," *IEEE Tran. Ind. Electron.*, vol. 59, no. 1, pp. 35–45. Jan. 2012.

7 O. Ellabban, H. Abu-Rub, and B. Ge, "A quasi-Z-source direct matrix converter feeding a vector controlled induction motor drive," *IEEE J. Emerg. Select. Top. Power Electron.*, vol. 3, no. 2, pp. 339–348, Jun. 2015.

8 O. Ellabban, H. Abu-Rub, and S. Bayhan, "Z-source matrix converter: An overview," *IEEE Trans. Ind. Electron.*, vol. 31, no. 11, pp. 7436–7450, 2016.

9 S. Liu, B. Ge, H. Abu-Rub, and F. Z. Peng, "Modeling, analysis, and motor drive application of quasi-Z-source indirect matrix converter," *Int. J. Comput. Math. Electric. Electron. Eng.*, vol. 33, no. 1/2, pp. 28–28, 2013.

10 S. Liu, B. Ge, X. Y. You, X. Jiang, H. Abu-Rub, and F. Z. Peng, "A novel quasi-Z-source indirect matrix converter," *Int. J. Circ. Theory Appl.*, vol. 43, no. 4, pp. 438–454, Apr. 2015.

11 S. Liu, B. Ge, X. Jiang, H. Abu-Rub, and F. Z. Peng, "Comparative evaluation of three Z-source/quasi-Z-Source indirect matrix converters," *IEEE Trans. Ind. Electron.*, vol. 62, no. 2, pp. 692–701, Feb. 2015.

12 S. Liu, B. Ge, Y. Liu, H. Abu-Rub, R. S. Balog, and H. Sun, "Modeling, analysis, and parameters design of LC-filter-integrated quasi-Z-source indirect matrix converter," *IEEE Trans. Power Electron.*, vol. 31, no. 11, pp. 7544–7555, Apr. 2016.

13 M. Guo *et al.*, "Modeling and analysis of LC filter integrated quasi-Z source indirect matrix converter," *Int. J. Circ. Theory Appl.*, vol. 48, no. 4, pp. 567–586, Apr. 2020.

14 X. You, B. Ge, S. Liu, N. Nie, X. Jiang, and H. Abu-Rub, "Common mode voltage reduction of quasi-Z source indirect matrix converter," *Int. J. Circ. Theory Appl.*, vol. 44, no. 1, pp. 162–184, Jan. 2016.

15 M. Guo, Y. Liu, B. Ge, and H. Abu-Rub, "Optimum boost control of quasi-Z source indirect matrix converter," *IEEE Trans. Ind. Electron.*, vol. 65, no. 10, pp. 8393–8404, Oct. 2018.

16 M. Guo *et al.*, "A quasi-Z source indirect matrix converter-fed induction motor drive," *IET Electr. Power Appl.*, vol. 14, no. 5, pp. 797–808, May 2020.

17 M. Guo, Y. Liu, B. Ge, X. Li, A. T. de Almeida, and F. J. T. E. Ferreira, "Dual, three-level, quasi-Z-source, indirect matrix converter for motors with open-ended windings," in *IEEE Trans. Energy Conv.*, vol. 38, no. 1, pp. 64–74, Mar. 2023.

18 M. Rivera, J. Rodriguez, B. Wu, J. R. Espinoza, and C. A. Rojas, "Current control for an indirect matrix converter with filter resonance mitigation," *IEEE Trans. Ind. Electron.*, vol. 59, no. 1, pp. 71–79, Jan. 2012.

19 M. Rivera, J. Rodriguez, J. R. Espinoza, and H. Abu-Rub, "Instantaneous reactive power minimization and current control for an indirect matrix converter under a distorted AC supply," *IEEE Trans. Industr. Inform.*, vol. 8, no. 3, pp. 482–490, Aug. 2012.

20 L. Tarisciotti *et al.*, "Modulated predictive control for indirect matrix converter," *IEEE Trans. Ind. Appl.*, vol. 53, no. 5, pp. 4644–4654, Sept.-Oct. 2017.

21 C. F. Garcia, M. E. Rivera, J. R. Rodríguez, P. W. Wheeler, and R. S. Peña, "Predictive current control with instantaneous reactive rower minimization for a four-leg indirect matrix converter," *IEEE Trans. Ind. Electron.*, vol. 64, no. 2, pp. 922–929, Feb. 2017.

22 W. Song, Y. Yang, W. Qin, and P. Wheeler, "Switching state selection for model predictive control based on genetic algorithm solution in an indirect matrix converter," *IEEE Trans. Transp. Electrif.*, vol. 8, no. 4, pp. 4496–4508, Dec. 2022.

23 H. M. Basri and S. Mekhilef, "Predictive current control for three-level four-leg indirect matrix converter under unbalanced input voltage," *IEEE J. Emerg. Select. Top. Power Electron.*, vol. 11, 3738–3750.

24 Z. Gong, J. Li, P. Dai, D. Su, and X. Wu, "Design and evaluation of a virtual vector based modulated model predictive control for the indirect matrix converters with improved performance," *IEEE Trans. Ind. Electron.*, vol. 69, no. 12, pp. 12019–12029, Dec. 2022.

25 Y. Liu and B. Ge, "A model predictive control method of quasi-Z-source indirect matrix converter," China Patent ZL201811604068.5, Jan. 7, 2020.

26 Y. Liu, X. Feng, X. Li, B. Ge, M. Guo, and Y. Zhang, "Model predictive control of LC filter integrated quasi-Z-source indirect matrix converter," in *IECON 2023-49th Annu. Conf. IEEE Industrial Electronics Society*, Singapore, Singapore, 2023, pp. 1–6.

9

Optimum Boost Control of LC Filter-Integrated Quasi-Z-Source Indirect Matrix Converter

9.1 Introduction

In the fast-paced evolution of power semiconductor materials and comprehensive integrated systems, Matrix Converters (MCs) are garnering a surge of both scholarly attention and applied endeavors, as is evidenced by a spate of recent publications [1–5]. These versatile power electronics converters are categorically subdivided into two predominant configurations, namely Direct Matrix Converters (DMCs) and Indirect Matrix Converters (IMCs) [1, 6]. Unlike a host of alternative AC-to-AC power conversion technologies currently under investigation, MCs exhibit the salient advantage of boasting a remarkably streamlined, silicon-based architectural framework. This characteristic feature singularly distinguishes them, elevating their prominence within the global spectrum of ongoing power electronics research. More pertinently, the architecture of IMCs incorporates and extends the functional benefits inherent in DMCs. One noteworthy enhancement is the incorporation of zero-current commutation capabilities, which significantly propels the IMC configuration as a pioneering force in AC power conversion technologies. This makes IMCs particularly well suited for deployment in intricate AC motor drive systems, as discussed in existing literature [2, 7–12].

In applications that deploy AC variable-frequency drives, the expansion of system voltage gain functions as a key enabler for extending the operational speed range of the motors involved. However, the inheritable limitation concerning the voltage gain in IMCs has been a factor that restricts their widespread applicability in diverse industrial settings [2]. The academic community is currently investigating multiple pathways to escalate the voltage gain of IMCs. These pathways are generally bifurcated into two main categories: modulation strategy refinement and auxiliary circuit incorporation [13–17]. Specific modulation enhancements under exploration include techniques such as over-modulation as well as composite modulation schemes, among other nuanced approaches [13–17]. While these modulation advancements indeed elevate the gain, they also inject additional layers of system complexity. Conversely, the introduction of auxiliary circuits generally involves the integration of buck–boost topologies or Z-source/Quasi-Z source (ZS/QZS) frameworks within the transitional DC bus [16–18]. Buck–boost approaches, while effective, invert the polarity of the DC bus voltage and necessitate the integration of more active components. This adds complexity to the overall control architecture [16, 17]. In contrast, ZS/QZS circuits, which are predominantly designed using passive elements or at most a single active device, afford both a balanced design architecture and a less complex control mechanism. This simplicity, combined

Impedance Source Matrix Converters and Control, First Edition.
Yushan Liu, Xiao Li, Baoming Ge, Haitham Abu-Rub, and Frede Blaabjerg.

with their potential for significantly increasing the voltage gain, has led to extensive scholarly focus on ZS/QZS configurations as a potent method for augmenting IMC voltage gain [18–23].

Several configurations have been explored to integrate QZS circuits into MCs, each with its own merits and shortcomings. Notable designs comprise QZS-Direct Matrix Converters (QZS-DMC) [21], DC-QZS-Indirect Matrix Converters (DC-QZS-IMC), where the QZS framework is situated within the intermediary DC bus [22, 23], and another variant is the QZS-IMC, in which the QZS components are interlaced between the AC source and individual phases of the IMC [24–29]. The QZS-DMC configuration, though innovative, doesn't offer the advantages intrinsic to the IMC design. On the flip side, the DC-QZS-IMC model carries over these IMC benefits but compromises the all-silicon architecture, which can be problematic for streamlined system integration and manufacturing processes [24–29]. Contrastingly, the third structural arrangement—the QZS-IMC—harmonizes the strengths inherent to the IMC model and further amplifies them through the integration of compact power modules. These enhancements are enabled by the voltage-boosting and current-filtering features of the QZS circuitry, all while maintaining the fully silicon-based topology [25, 26].

Since its introduction in 2015, the QZS-IMC has been the subject of multifaceted analysis, focusing on aspects like circuit topology, gain characteristics, common-mode voltage effects, and the filtering performance. Additionally, studies have looked into optimizing its parameters and evaluating its utilization in motor drive applications [24–33]. Generally, existing control strategies rely on a constant shoot-through duty cycle (denoted as "D") to manage a predetermined modulation index ("M") within the rectifier stage, aimed at securing the desired voltage gain. However, a specific research endeavor has proposed the innovative concept of a variable "D" to gradually tweak the DC-link voltage, while keeping "M" constant throughout this modification process [34].

By integrating the QZS network into the existing IMC configuration, the voltage gain is substantially enhanced, obviating the requirement for an additional input filter. While this optimization certainly serves to reduce both the system's cost and physical dimensions, it introduces a complex question. Specifically, when the shoot-through duty cycle, labeled as "D," is a non-zero value, a wide range of "D," along with the modulation index at the rectifier stage ("m_i") and at the inverter stage ("m_o"), can yield the same voltage gain. Importantly, these parameters play a crucial role in affecting both the switching and conduction losses within the power devices. Consequently, it poses a critical question: What specific set of values for "D", "m_i", and "m_o" can be identified as the most efficient in terms of minimizing power device losses?

Apart from standard considerations, the shoot-through duty cycle, often designated as "D," is further influenced by fluctuations in both the inductor current and the capacitor voltage, as described in equations (9.1) and (9.2) [26]. It is noteworthy that the values for capacitance and inductance in the QZS network are pre-set to function as filters for the grid current. An increment in "D" invariably causes an intensification of these oscillations in the inductor current and the capacitor voltage. Therefore, striking a balance by minimizing "D" to achieve the required voltage gain, while at the same time limiting these oscillations becomes an essential aspect to consider.

$$\Delta i_L = \frac{D(1-D)}{1-2D} k_L \tag{9.1}$$

$$\Delta u_C = \frac{D(1-D)}{1-2D} k_C \tag{9.2}$$

In equation (9.3), the terms k_L and k_C are thoroughly defined. Here, T_s represents the duration of the switching cycle, P_{in} is indicative of the power input, $\cos(\psi)$ serves as the marker for the input power factor, while u_i designates the voltage of the AC source.

$$k_L = \frac{T_s u_i}{L}, kc = \frac{T_s P_{in}}{3u_i \cos(\psi) C} \tag{9.3}$$

The objective of this chapter is to present a control methodology grounded in constrained optimization techniques. The method is formulated to identify the most effective set of parameters—comprising the shoot-through duty cycle D, modulation index at the rectifier stage m_i, and the corresponding modulation index at the inverter stage m_o—during the functioning of a QZS-IMC [27]. The ultimate goal is to achieve this without escalating power losses.

9.2 Gain Model and Modulation of QZS-IMC System

9.2.1 Derivation and Analysis of Gain Model

As depicted in Figure 9.1, the structural configuration of the QZS-IMC involves three primary elements. The QZS impedance circuit consists of inductors L_{x1} and L_{x2}, capacitors C_{x1} and C_{x2}, and a dedicated switch S_x for each phase (where x can be a, b, or c). The rectifier portion utilizes bidirectional switches, labeled as S_{xp} and S_{xn}, while the inverter segment incorporates switches S_{Xp} and S_{Xn} corresponding to each phase (where X can be A, B, or C). The system operates in two distinct states: shoot-through and non-shoot-through, as delineated in reference [26]. For the purpose of this discussion, let D be defined as the shoot-through duty ratio, calculated as $D = T_0/T_s$. In this equation, T_s represents the complete control cycle duration, whereas T_0 and T_1 denote the lengths of the shoot-through and non-shoot-through intervals in each total cycle, following the relationship $T_0 + T_1 = T_s$.

The voltage boost factor for the QZS network, referred to as B, is mathematically articulated as follows:

$$B = \frac{U_{im}}{U_{in}} = \frac{1}{1-2D} \tag{9.4}$$

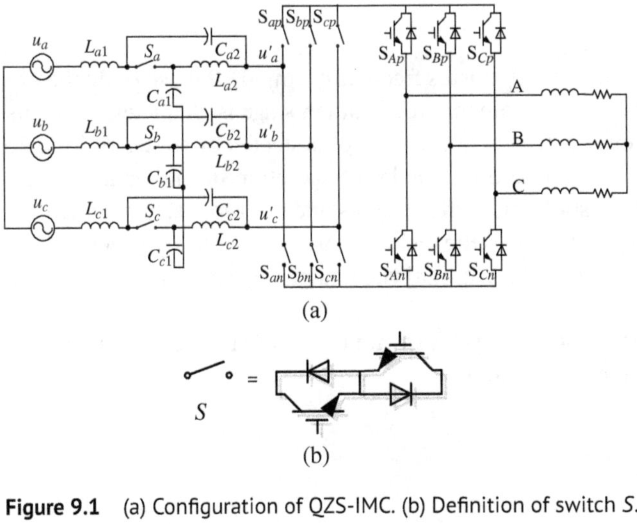

Figure 9.1 (a) Configuration of QZS-IMC. (b) Definition of switch S.

Here, U_{in} represents the input phase voltage amplitude originating from the power source, while U_{im} signifies the amplitude of the phase voltage at the QZS network's output, which is also the input phase voltage for the IMC system. The shoot-through duty ratio D ranges between 0 and 0.5 [26].

When a dual space vector modulation (SVM) approach incorporating a zero vector is utilized in the rectifier phase [23, 29], the average DC bus voltage can be mathematically described as follows:

$$\overline{U}_{dc} = \frac{3}{2} m_i U_{im} \cos \varphi_{qz} \tag{9.5}$$

In such a framework, the modulation index linked to the rectifier stage is signified by m_i, while $\cos \varphi_{qz}$ represents the power factor of the output from the QZS network. Here, φ_{qz} designates the angular displacement between the phase voltage and current within the Quasi-Z Source network. The amalgamation of equations (9.4) and (9.5) culminates in equation (9.6).

$$\overline{U}_{dc} = \frac{3}{2} B m_i\, U_{in} \cos \varphi_{qz} \tag{9.6}$$

Concerning the inverter stage, the modulation index, denoted by m_o, is mathematically expressed in the subsequent section of this chapter.

$$m_o = \frac{U_{ol}}{\overline{U}_{dc}} = \frac{\sqrt{3} U_{om}}{\overline{U}_{dc}} \tag{9.7}$$

In the realm of QZS-IMC systems, two primary parameters are often considered: U_{ol} and U_{om}. These parameters uniquely depict the magnitudes associated with the output line voltage and the phase voltage, respectively. A thorough analysis, which synthesizes the information from equations (9.6) and (9.7), facilitates our comprehension of the total voltage gain. This understanding aligns with the discussions presented in references [25–28].

$$G = \frac{U_{om}}{U_{in}} = H m_o m_i B \cos \varphi_{qz} \tag{9.8}$$

$H = \frac{\sqrt{3}}{2}$ represents the voltage gain of conventional MC. Factor G, which exhibits dependence on variables such as $\cos \varphi_{qz}$, D, m_o, and m_i, plays a pivotal role. The primary goal remains to ascertain the optimal combination of these variables, aiming for the targeted voltage gain and concurrently reducing energy wastage.

9.2.2 Modulation Method

In the design of QZS-IMC systems, control mechanisms frequently employ the use of dual-SVM methods which integrate zero vectors. To elaborate, the rectification stage is characterized by its use of input current-directed SVM. Conversely, the inversion stage is steered by strategies hinging on output voltage-centric SVM. When it comes to the modes of operation in the rectifier stage, they can be classified into distinct vector states: the active vectors and the zero vectors, which are detailed in Table 9.1. An interesting aspect to note is the inclusion of the shoot-through zero vector state in the rectifier's zero states; this action results in a short circuit across the tri-phase AC sources, thereby enhancing the voltage.

For a comprehensive understanding of the duty cycles within the rectifier phase, it is advisable to delve into the subsequent mathematical computations:

$$\begin{cases} d_\alpha = m_i \sin(60^\circ - \theta_i) \\ d_\beta = m_i \sin \theta_i \\ d_{0r} = 1 - d_\alpha - d_\beta - d_{st} \\ d_{st} = D = const \end{cases} \tag{9.9}$$

Table 9.1 Input current vectors, switching combinations, and DC-link voltage.

N	U_{dc}	S_{ap},S_{bp},S_{cp}	S_{an},S_{bn},S_{cn}	S_x	Vector	Description
1	u_{ac}	1, 0, 0	0, 0, 1	1	I_1	Active Vector
2	u_{bc}	0, 1, 0	0, 0, 1	1	I_2	Active Vector
3	$-u_{ab}$	0, 1, 0	1, 0, 0	1	I_3	Active Vector
4	$-u_{ac}$	0, 0, 1	1, 0, 0	1	I_4	Active Vector
5	$-u_{bc}$	0, 0, 1	0, 1, 0	1	I_5	Active Vector
6	u_{ab}	1, 0, 0	0, 1, 0	1	I_6	Active Vector
7	0	1, 0, 0	1, 0, 0	1	I_7	Zero Vector
8	0	0, 1, 0	0, 1, 0	1	I_8	Zero Vector
9	0	0, 0, 1	0, 0, 1	1	I_9	Zero Vector
10	0	1, 1, 1	1, 0, 0	0	I_{10}	Shoot-Through Vector
		1, 1, 1	0, 1, 0	0	I_{11}	
		1, 1, 1	0, 0, 1	0	I_{12}	
11	0	1, 0, 0	1, 1, 1	0	I_{13}	Shoot-Through Vector
		0, 1, 0	1, 1, 1	0	I_{14}	
		0, 0, 1	1, 1, 1	0	I_{15}	

Note: S_x represents the switch S_a, S_b, and S_c, as shown in Figure 9.1.

Within the context of this equation, the symbol θ_i characterizes the directional angle associated with the input current vector. The duty cycles, denoted as d_α, d_β, d_{st}, and d_{0r}, correspond to distinct current vectors present during an individual switching period. Moving to the inverter section, it features six proactive vectors complemented by two zero vectors. To discern the duty cycles for this particular phase, one would refer to the ensuing calculations:

$$\begin{cases} d_\mu = m_o \sin(60° - \theta_o) \\ d_v = m_o \sin\theta_o \\ d_{0i} = 1 - d_\mu - d_v \end{cases} \tag{9.10}$$

In the context of this representation, θ_o is an indicative symbol for the directional angle pertaining to the output voltage vector. On the other hand, the variables d_μ, d_v, and d_{0i} stand for the respective duty cycles associated with the proactive output voltage vectors and their corresponding zero vector.

The modulation techniques employed in each stage exhibit mutual dependence. To be more precise, during the transitional phases of commutation, the inverter stage mandates alignment with its rectifier counterpart. The intricacies of their interconnected duty cycle can be discerned via the following mathematical analysis:

$$\begin{cases} d_{\alpha\mu} = d_\alpha d_\mu \\ d_{\beta\mu} = d_\beta d_\mu \\ d_{\alpha v} = d_\alpha d_v \\ d_{\beta v} = d_\beta d_v \\ d_{\alpha 0i} = d_\alpha d_{0i} \\ d_{\beta 0i} = d_\beta d_{0i} \end{cases} \tag{9.11}$$

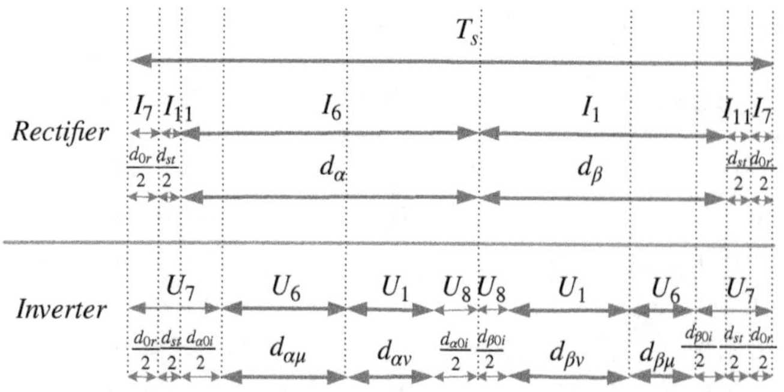

Figure 9.2 Sequential cooperation of two stages for the QZS-IMC.

To minimize potential risks linked to distortion, time intervals are distributed uniformly. A graphical representation of the ordered interaction between the two stages can be seen in Figure 9.2.

A close examination of Figure 9.2 reveals that when the power switches of the rectifier stage are in the midst of their commutation process, the modulation patterns in the inverter stage coincide exactly with periods marked by zero vectors. As a result, the intermediate DC link is free of any active current, and the power switches of the rectifier stage remain untouched by any current passage. Such an orchestrated pattern is pivotal in achieving a commutation devoid of current within the rectifier stage.

9.3 Multi-Constraints Optimization and Operation Control for QZS-IMC

9.3.1 Constrained Optimization Method

The optimal solution can be achieved by adeptly managing multiple constraints. Such a process involves the transformation of a constrained problem into an unconstrained one by the application of a specific penalty function, as referenced in [34, 35]. The subsequent sections outline the architecture of this approach and the techniques used for problem solving.

$$\left.\begin{array}{lll}\text{Objective Function} & \min & f(x) \\ \text{Constraint Condition} & s.t & g_i(x) \geq 0, i = 1, \ldots, m\end{array}\right\} \tag{9.12}$$

To tackle this issue, an auxiliary function is defined as:

$$F(x, \delta) = f(x) + \delta \sum_{i=1}^{m} [\max\{0, -g_i(x)\}]^2 \tag{9.13}$$

In this case, δ is a positive infinite number. At a feasible point "x", the following holds true:

$$\max\{0, -g_i(x)\} = 0 \tag{9.14}$$

Conversely, at a non-feasible point "x", the relationship is:

$$\max\{0, -g_i(x)\} = -g_i(x) \tag{9.15}$$

Consequently, (9.12) is altered into an optimization problem devoid of constraints, expressed as:

$$\min \quad F(x, \delta) = f(x) + \delta P(x) \tag{9.16}$$

Delving deeper, it is evident that boundary analysis is instrumental in elucidating the path toward determining the most optimal resolution. When considering a feasible point denoted as "x", one discerns that the value $P(x)$ seamlessly transitions to zero. This behavior further translates to $F(x,\delta)$ essentially being tantamount to f(x). In contrast, situations where "x" emerges as a non-feasible point present a different scenario. Here, the metric $\delta P(x)$ tends to soar toward the realm of positive infinity. Such a pronounced inclination effectively serves as a stringent penalty for those data points that venture away from the predetermined admissible domain. Intriguingly, the inherent design of the function proactively drives the iterative data point, ensuring its alignment with the acceptable zone, especially during the exhaustive optimization steps. A noteworthy observation is that as the value of δ escalates, there is an evident convergence between the derived solutions of (9.16) and (9.12). Elaborating on terminologies, $\delta P(x)$ is systematically acknowledged as the penalty term. Concurrently, δ is conceptualized as the penalty coefficient, and the entity $F(x,\delta)$ is squarely defined as the encompassing penalty function.

9.3.2 Optimal Function Curve of QZS-IMC

Drawing insights from equations (9.4) and (9.8), one can infer the following:

$$G = \frac{U_{om}}{U_{in}} = Hm_o m_i \cos\varphi_{qz} \frac{1}{1-2D} \tag{9.17}$$

9.3.2.1 Pre-constrained Condition

To streamline the problem-solving process, three constraints are taken into account.

a) Maintaining a system operation at $\cos\varphi_{qz} = 1$.
   In reality, the QZS-IMC system consistently operates at $\cos\varphi_{qz} = 1$, and hence from equation (9.17), we have

$$G = Hm_o m_i \frac{1}{1-2D} \tag{9.18}$$

b) Holding the inverter modulation index at unity.
   Assume $m_o = 1$, from (9.18), there is by setting $m_o = 1$, it follows from equation (9.18) that

$$G = Hm_i \frac{1}{1-2D} \tag{9.19}$$

   Two crucial elements are highlighted here: one is the simplification of the optimization procedure, and the other is the maximization of the inverter stage's voltage gain, coupled with the optimal utilization of the DC link voltage.

c) Ensuring the duty ratio, D, is zero when $G \leq H$.
   When $G \leq H$, the D is zero and the QZS network plays the role of a filter. From (9.19), there is a scenario where $G \leq H$, D becomes zero, thereby enabling the QZS network to function as a filter. Consequently, from equation (9.19), it is deduced that

$$G = Hm_i \tag{9.20}$$

9.3.2.2 Constrained Condition When D is Nonzero

Utilizing the SVM approach with a zero vector in the rectifier stage allows the circuit to act as a current source converter. This capability enables control over input currents, which in turn provides an avenue for input power factor modification. The representative input current vector, denoted as I_{ref}, is graphically illustrated in Figure 9.3, as referenced in prior studies [25, 29].

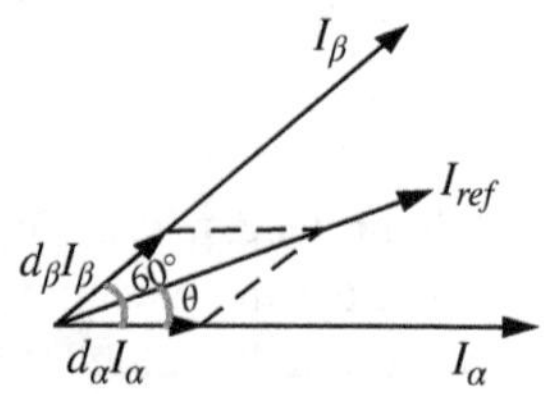

Figure 9.3 Diagram of the input current vector.

In the rectifier stage, the duty ratios of all vectors are represented by equation (9.21).

$$\begin{cases} d_\alpha = m_i \sin(60° - \theta_i) \\ d_\beta = m_i \sin \theta_i \\ d_{0r} = 1 - d_\alpha - d_\beta - d_{st} \\ d_{st} = D \end{cases} \tag{9.21}$$

These vectors converge at point $d_\alpha + d_\beta + d_{0r} + d_{st} = 1$, with d_α and d_β denoting the effective duty ratios for vectors I_α and I_β, respectively. d_{0r} signifies the duty ratio of the non-shoot-through zero vector I_{0r}, whereas d_{st} symbolizes the duty ratio of the shoot-through zero vector I_{st}.

Upon conducting an in-depth analysis, a constrained condition must be satisfied when G exceeds H, as depicted in equation (9.22).

$$d_\alpha + d_\beta + D \leq 1 \tag{9.22}$$

By incorporating equation (9.21) into equation (9.22), the resultant relationship can be mathematically defined by equation (9.23).

$$m_i \cos(30° - \theta_i) + D \leq 1 \tag{9.23}$$

For illustrative purposes, let's consider the first sector where the angle θ_i falls within the range of 0°–60°. Under these conditions, cos(30°–θ_i) varies between 0.866 and 1. This leads to a new constraint described by equation (9.24), which can be formulated as (0.866 to 1)$m_i + D \leq 1$.

$$(0.866 \sim 1)m_i + D \leq 1 \tag{9.24}$$

9.3.2.3 Optimal Function at $D \neq 0$

Drawing insights from equation (9.19), an objective function is articulated in equation (9.25).

$$\min\ D = \frac{1}{2} - \frac{Hm_i}{2G} \tag{9.25}$$

Subsequently, a constraint pertaining to the value of D under non-zero operation can be derived from equation (9.24), denoted as equation (9.26).

$$s.t\quad 1 - m_i - D \geq 0 \tag{9.26}$$

To solve this constrained optimization problem, we introduce an auxiliary function, formulated as equation (9.27).

$$F(x, \delta) = f(x) + \delta P(x) \tag{9.27}$$

Within this function, specific terms and variables are further explicated in equation (9.28).

$$P(x) = [\max\{0,\ D + m_i - 1\}]^2 = \begin{cases} 0 & m_i \leq 1 - D \\ (D + m_i - 1)^2 & m_i > 1 - D \end{cases} \tag{9.28}$$

Continuing with the analysis, the derivative of the auxiliary function $F(x,\delta)$ is obtained, as indicated by equation (9.29).

$$\frac{dF}{dm_i} = \begin{cases} -\frac{H}{2G} & m_i \le 1 - D \\ -\frac{H}{2G} + 2\delta(D + m_i - 1) & m_i > 1 - D \end{cases} \tag{9.29}$$

By setting this derivative to zero, a relationship is established, encapsulated in equation (9.30).

$$D + m_i - 1 = \frac{H}{4\delta G} \tag{9.30}$$

As δ approaches infinity, the corresponding behavior is outlined in equation (9.31).

$$D + m_i - 1 = \frac{H}{4\delta G} \sim 0 \tag{9.31}$$

As a result, when m_i serves as a feasible point, the optimal solution for minimizing the objective function defined in equation (9.25) is expressed as $m_i = 1{-}D$. This leads us to further deduce equations (9.32) and (9.33),

$$D_{\min} = \frac{H - G}{(H - 2G)} \tag{9.32}$$

$$m_i = \frac{-G}{H - 2G} \tag{9.33}$$

which are congruent with the initial set of equations and constraints.

$$m_i + D_{\min} = 1$$

9.3.2.4 Full-range Optimal Operation Curve

In the scenario where G is less than or equal to H, the governing operational function can be identified as equation (9.34).

$$G = HBm_i \tag{9.34}$$

In this case, the modulation index "m_i" remains confined within the range greater than zero and less than or equal to one. Simultaneously, the value of D is established at zero.

On the other hand, when G surpasses H, equation (9.33) lays the foundation for the operational function described in equation (9.35).

$$G = H\frac{m_i}{2m_i - 1} \tag{9.35}$$

Here, the parameter D varies from zero to just below 0.5, while "m_i" fluctuates between a lower bound of 0.5 and an upper limit below one.

Figure 9.4 graphically illustrates these operating curves, adding a visual dimension to our numerical findings.

9.3.3 Optimal Operation Control of QZS-IMC

As visually outlined in Figure 9.5, the control scheme for achieving optimal operation consists of three key components: the principal circuitry, the modulation algorithmic procedures, and the overarching control strategy. The modulation algorithm is subdivided into two parts, dedicated to the rectifier and inverter stages, respectively. Specifically, the QZS network modulation is

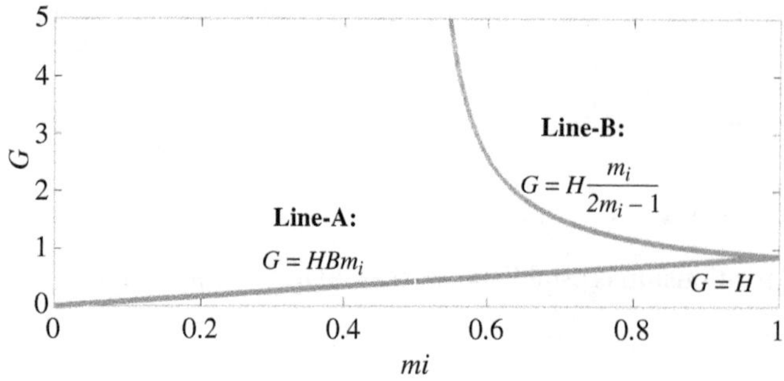

Figure 9.4 Full-range operation curve of the QZS-IMC.

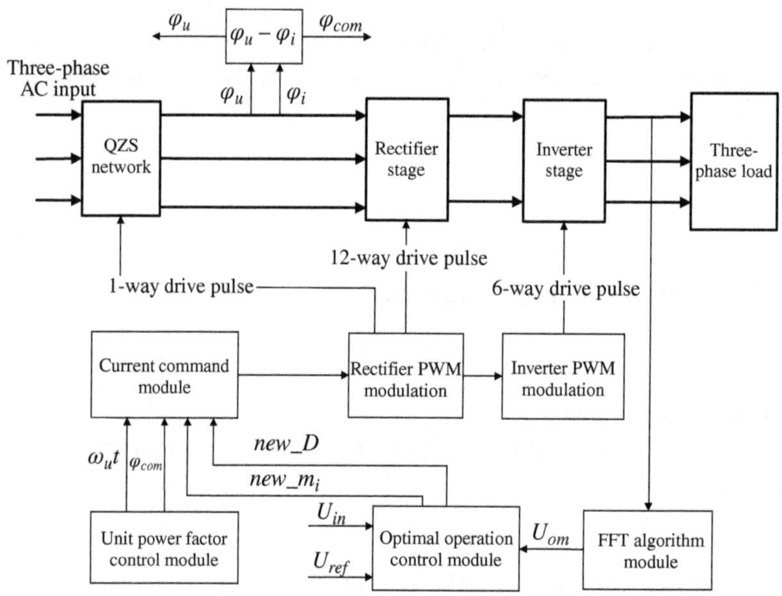

Figure 9.5 Schematic diagram of optimal operation control.

generated within the rectifier stage through selected pulse-width modulation (PWM) methods. These algorithms serve the essential role of generating the requisite driving pulses for both the QZS network and the IMC. Turning our attention to the control algorithm, it encompasses multiple facets, including the regulation of unity power factor, the optimization of parameter *D*, and the execution of Fast Fourier Transform (FFT) analyses on the output voltage. Operating at a top-tier level, this control algorithm is pivotal in capturing and processing the prescribed input signals.

In the realm of electrical engineering, the *D*-optimization method has emerged as a prominent strategy, primarily because it harnesses the power of real-time FFT strategies. This intricate process gauges the magnitude of the resultant voltage in dynamic scenarios. The voltage data, once harvested, is synergized with two pivotal parameters. The first is the magnitude of the entry voltage, symbolically represented by U_{in}. The second, conversely, relates to the reference voltage value, distinctively tagged as U_{ref}. Drawing insights from these coalesced parameters, the efficient operation control methodology is designed to make an inference. Specifically, it determines if the system necessitates a voltage elevation, often referred to as boosting, or whether a reduction, known in technical terms as bucking, would be more apt. This form of dynamic assessment is crucial for maintaining optimal operational standards. Once this evaluative stage is complete, the method takes another insightful leap. It references a pre-defined optimization theory trajectory to meticulously generate refined, more precise values for both the D and m_i parameters. The importance of this step cannot be overstated; it's here that the real-time adjustments truly come to life. These freshly adjusted values play a critical role. They not only inform but also guide the modulation technique used in the rectifier phase. This ensures instantaneous online modifications to the voltage output, aligning it more closely with real-time requirements. The overarching ambition of this intricate control system is threefold. Firstly, it seeks the synchronization of the system's output voltage with its predetermined reference counterpart. Secondly, the method aspires to achieve a unity power factor—a testament to its efficiency. Lastly, it endeavors to ensure that the system operates in adherence to the prime trajectory, maximizing its functional capability. To demystify this complex process and offer a comprehensive view, the entire approach is systematically illustrated via a series of meticulously crafted flow diagrams. These graphical representations span from the principal flow diagram to the more specific boost-centric blueprint and the buck-centric outline. For readers' convenience and clarity, they are showcased in a sequential manner from Figures 9.6 to 9.8.

9.3.3.1 Main Flow Chart

Figure 9.6 articulately lays out the workings of the primary flow chart, highlighting its meticulous reliance on distinct data points. Three major parameters anchor its logical functioning: firstly, U_{in}, which stands as a symbolic representation of the input phase voltage's amplitude. Secondly, there's U_{om-ref}, encapsulating the benchmark for the output phase voltage. Lastly, in the real-time scenario, the output phase voltage is capably captured by $U_{om-real}$. In its foundational operational phase, the system is equipped to juxtapose two pivotal factors. On one side is $G_{\_real}$, capturing the real-time amplification metrics, and on the other, we find the matrix converter's pre-established gain, aptly titled H. It's through this systematic contrasting that the operational pathway gets illuminated. This phase also sets the stage for the instantaneous calculation of the intertwined D and m_i parameters, crucial for the system's adaptability. Pivoting from this, a subsequent evaluation homes in on the correlation between $G_{\_real}$ and its reference counterpart, succinctly represented as $G_{\_ref}$. Grounded in this analytical comparison, a consequential decision emerges. It outlines the system's operational stance, ascertaining whether its functionality should lean toward boost mode, migrate to buck mode, or sustain its prevailing operational mode.

9.3.3.2 Flow Chart of Boost Mode

As shown in Figure 9.7, boost mode includes:

(a) $G = HBm_i$; (b) $G = H\dfrac{m_i}{2m_i - 1}$;

(c) From $G = HBm_i$ to $G = H\dfrac{m_i}{2m_i - 1}$.

Moreover, considering practical application, the minimum m_i is 0.55, and the D is 0.45.

9.3.3.3 Flow Chart of Buck Mode

As shown in Figure 9.8, buck mode includes:

(a) $G = HBm_i$; (b) $G = H\dfrac{m_i}{2m_i - 1}$;

(c) From $G = H\frac{m_i}{2m_i-1}$ to $G = HBm_i$.

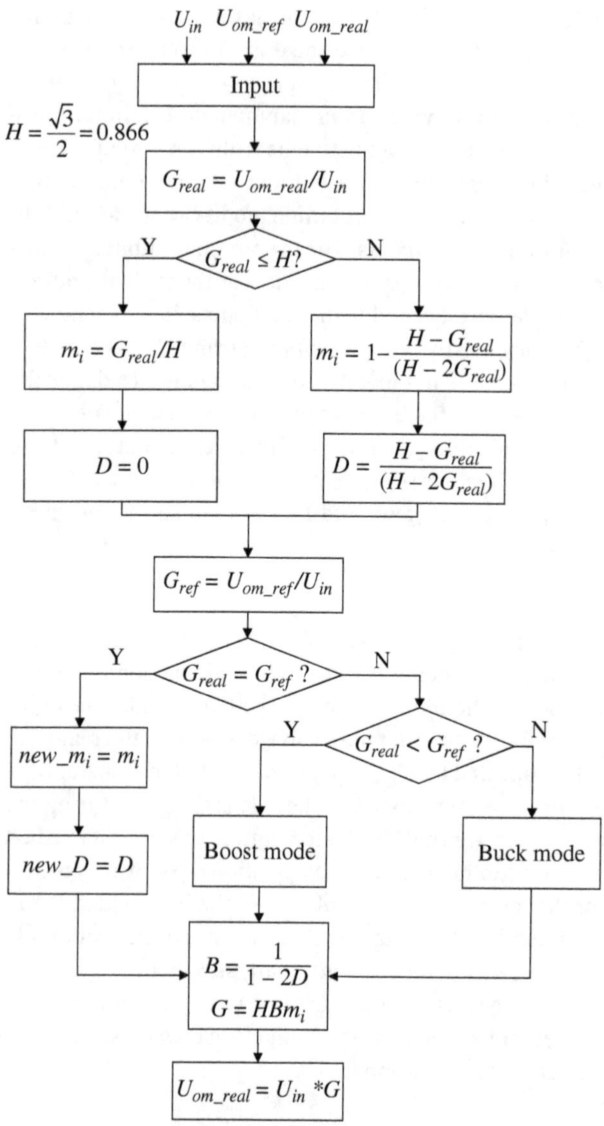

Figure 9.6 Main flow chart.

9.4 Simulation and Experimental Verifications

9.4.1 Verification of Optimal Operation Control

The outlined flow chart, dedicated to efficient operational control, places an emphasis on dual modes discerned both in simulation and real-world testing scenarios. Specifically, the analytical focus zeroes in on the upward trajectory from line-A toward line-B, termed "boost mode," as shown in Figure 9.4, and its reverse, from line-B descending to line-A, christened "buck mode." Within these modal contexts, certain frequency markers and associated load inductance benchmarks are placed under the microscope. In the realm of the boost mode, a harmonized union of a 200-Hz output frequency and a 500-μH load inductance is subjected to meticulous observation. On the flip

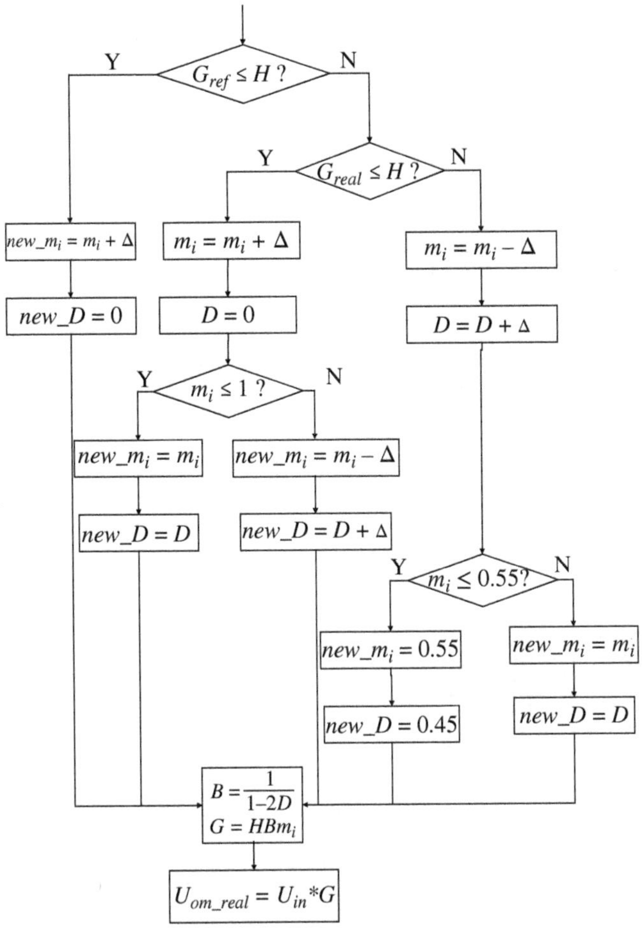

Figure 9.7 Flow chart of boost mode.

side, the buck mode sees a detailed exploration at the confluence of a 50-Hz output frequency and a 2-mH load inductance setting. For those interested in the finer details that govern these investigative paradigms, Table 9.2 provides an exhaustive breakdown. To further edify readers, a graphical representation of the universally adopted test configuration can be located in Figure 9.9.

9.4.1.1 Case 1: Boost Mode From Line-A to Line-B

Utilizing an initial output phase voltage of 100 V for line-A and a concluding value of 180 V for line-B, the system waveforms under optimal control settings are depicted in Figures 9.10–9.15 Specifically, Figures 9.10–9.12 provide both simulated and empirical data at an output frequency of 200 Hz, while Figures 9.13–9.15 reveal similar findings but at a lower output frequency of 50 Hz.

In the case of boost mode functioning at 200 Hz, Figure 9.10 offers a series of subfigures illustrating the evolution of D and m_i parameters. Subfigure 9.10a captures the gradual ascent of D from an initial zero level to a value of approximately 0.1688, maintaining m_o at unity. Subfigure 9.10b presents the m_i parameter experiencing a change from 0.697 to about 0.83 over the operation process. Meanwhile, Subfigure 9.10c visualizes how the sum of D and m_i initially elevates from zero, stabilizing at unity as the system operates optimally.

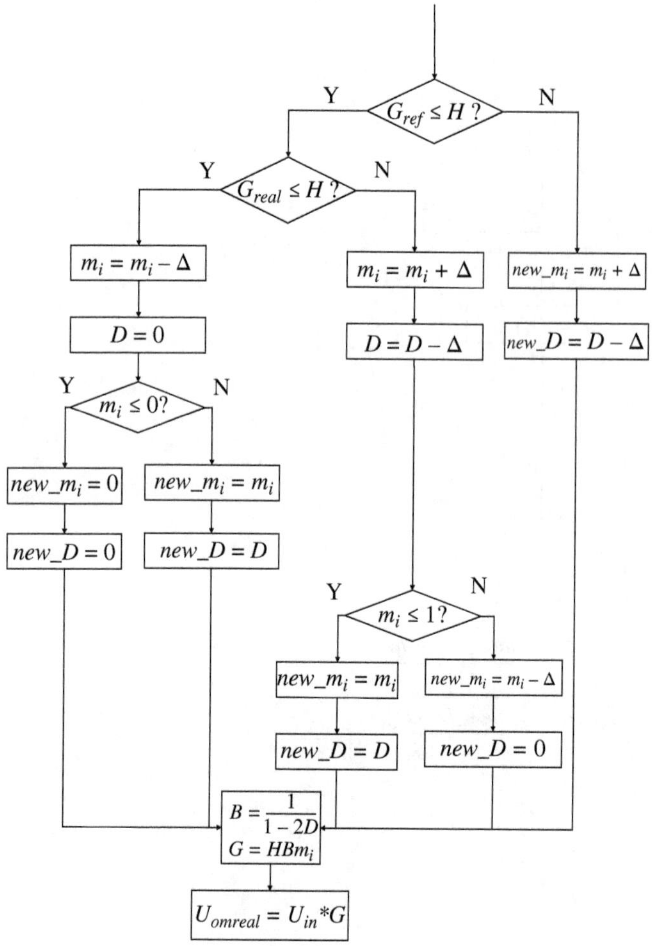

Figure 9.8 Flow chart of buck mode.

Table 9.2 QZS-IMC parameters.

Amplitude of input phase voltage, U_{in}	163 V
Frequency of input phase voltage, f_{in}	50 Hz
Switching frequency	10 kHz
Inverter modulation index, m_o	1
Load resistance	10 Ω
QZS capacitance, C_1 and C_2	100 μF
QZS inductance, L_1 and L_2	5 mH

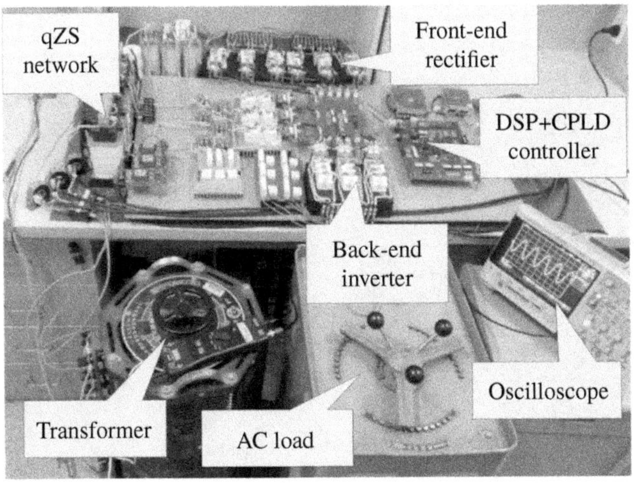

Figure 9.9 General test bench for the QZS-IMC.

Lastly, additional insight from Figure 9.10d demonstrates that the output power factor of the QZS network remains notably close to 1, with minor fluctuations occurring during the transition phases.

Additional facets of the system performance are explored in Figures 9.10e, 9.10f, 9.11, and 9.12, which collectively display both simulated and real-world results for various electrical parameters. Specifically, Figure 9.10e,f focuses on the AC source voltage and current, revealing a noticeable surge in the input current when the system is in boost mode.

Turning our attention to Figure 9.11, it delivers an in-depth view of the optimal operational behavior of the system, in terms of both voltage and current output. Employing FFT for analytical rigor, this figure demonstrates that the output voltage amplitude experiences a smooth escalation, transitioning from an initial 102 V to a more robust 176 V.

Through these visual representations, one can observe that the QZS-IMC achieves desired performance levels efficiently, especially during its operation in boost mode.

Further insights into the system's performance can be extracted from Figures 9.11d,e, and 9.12, which provide both simulated and actual measurements for the DC bus voltage as well as the output voltages from the QZS network. Intriguingly, at approximately 0.13 seconds into the operation, a pivotal event occurs: the system gain surpasses the predefined threshold, denoted as H. Consequently, the QZS network initiates the process of voltage amplification.

Upon delving into the data represented in Figures 9.11d and 9.12, one cannot help but notice a compelling feature: the output voltage propagated by the QZS network consistently follows a PWM contour. This modulation, intrinsic in its form, offers a captivating visual experience. Amplifying the intrigue is the gradual yet distinct rise in the waveform's amplitude, an eventuality that aligns seamlessly with theoretical paradigms discussed earlier.

As we spread our analytical lens across Figures 9.10–9.12, treating them as interconnected pieces of a grander puzzle, a mosaic of insights begins to take shape. Firstly, one of the key takeaways is

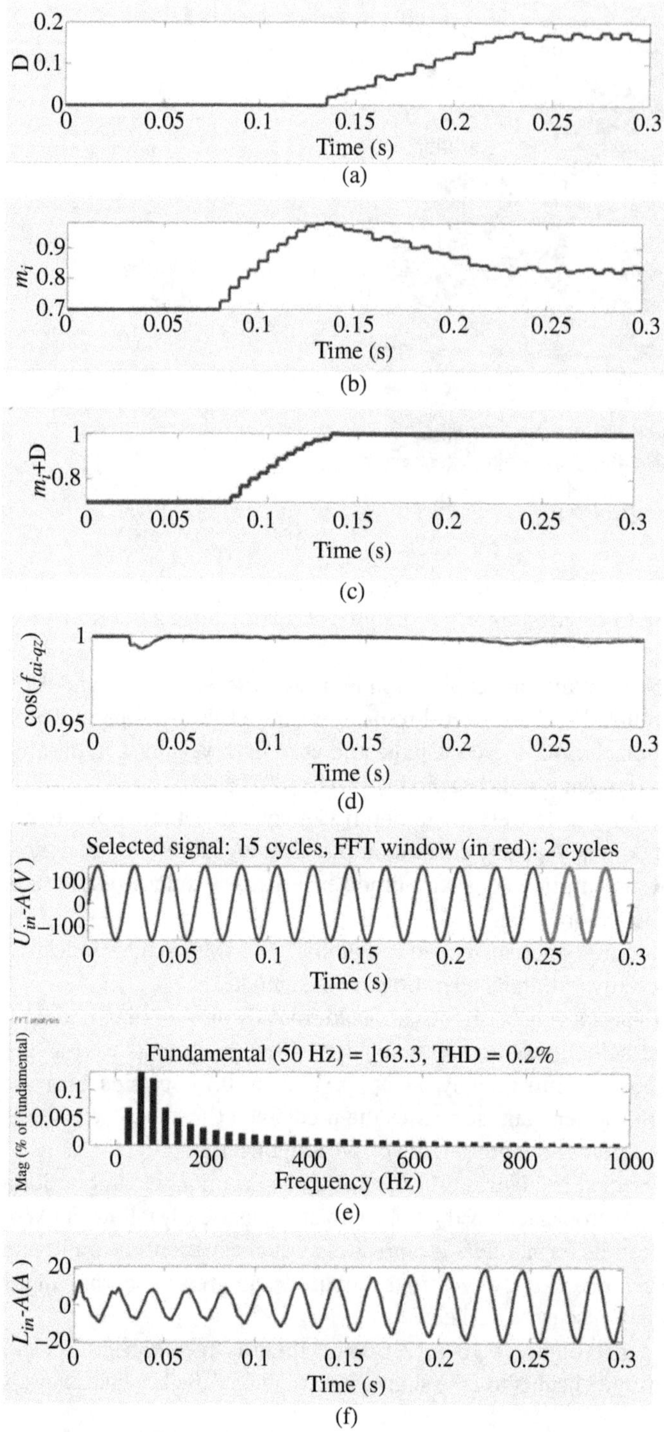

Figure 9.10 Data overview at an output frequency of 200 Hz. (a) Depiction of D's operational trajectory; (b) trace of m_i's performance; (c) the combined performance curve of both D and m_i; (d) value of $\cos \varphi_{qz}$; (e) an FFT breakdown pertaining to the input voltage; (f) measurement of the current input.

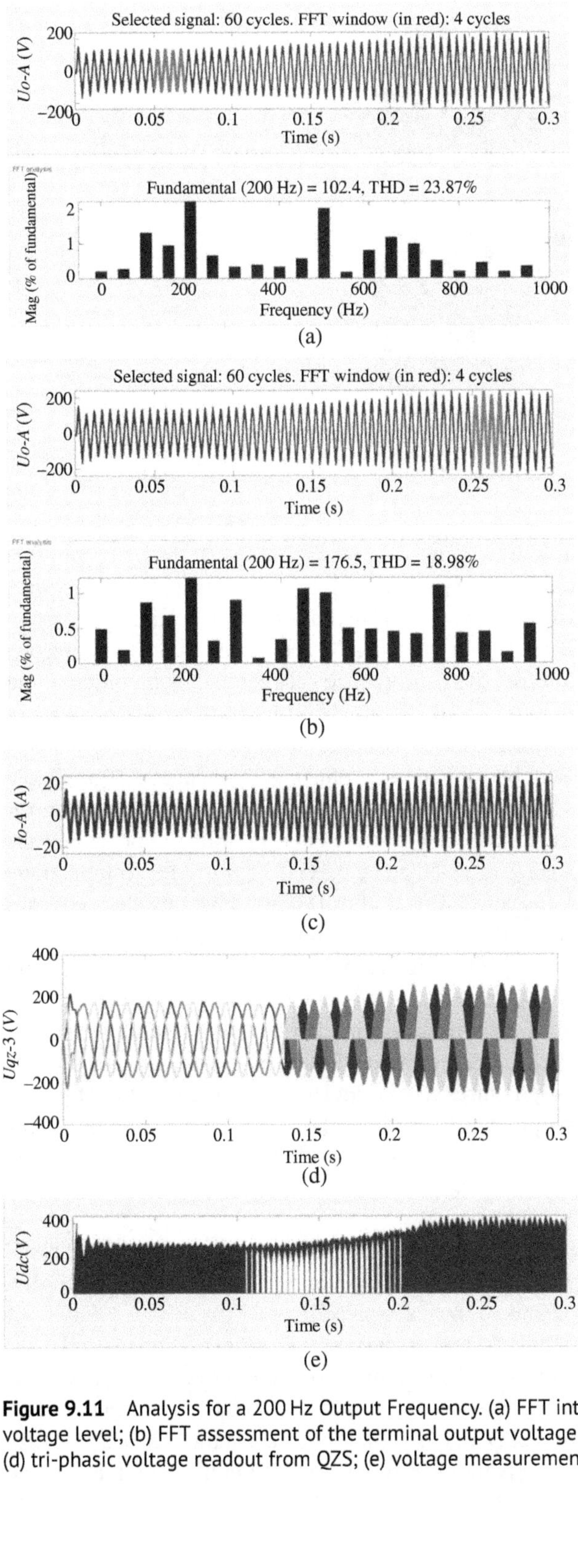

Figure 9.11 Analysis for a 200 Hz Output Frequency. (a) FFT interpretation of the beginning output voltage level; (b) FFT assessment of the terminal output voltage reading; (c) recorded outflow of current; (d) tri-phasic voltage readout from QZS; (e) voltage measurements from the DC bus.

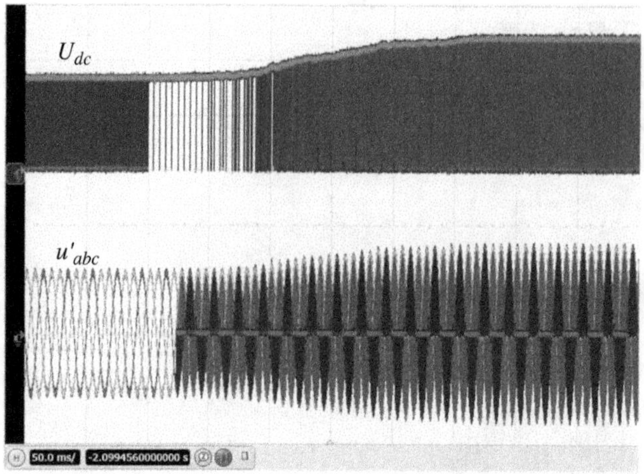

Figure 9.12 DC bus voltage and QZS output voltages (150 V/div) in boost mode at 200 Hz output frequency.

the evident sensitivity of the PWM pattern's duty cycle, as showcased at the DC bus, to alterations in the "m_i" parameter. Each minute tweak in this parameter brings forth palpable oscillations, painting a vivid empirical picture. In a parallel vein, the "D" parameter, upon adjustments, ushers in proportional shifts in the amplitude of the DC bus voltage. This behavior not only underscores the system's responsiveness but also reinforces the convergence between hands-on observations and time-honored theoretical constructs.

Within Figure 9.13a–c, the dynamic adjustments in variables D and m_i during the system's operation at an output frequency of 50 Hz in boost mode are delineated. Specifically, Figure 9.13a illustrates that D incrementally ascends from an initial value of zero to reach 0.1761 when the modulation index, m_o, is set to 1. In Figure 9.13b, one observes that m_i initially increases from 0.697 to a peak value of 1 before decreasing to stabilize at 0.8239. Figure 9.13c demonstrates that the aggregate sum of D and m_i remains constant at 1 after D has attained its elevated state. Furthermore, Figure 9.13d offers insights into the $\cos \varphi_{qz}$ values during this operation.

Within Figures 9.13f and 9.15a, an observed gradual uptick in the input current during the boost adjustment phase is highlighted. Turning to Figures 9.14a–c and 9.15a, these depict the system's performance concerning output voltage and current under optimal operating conditions. During this phase, the amplitude of the output voltage experiences a steady elevation, ranging from an initial 103 V and culminating at 179 V. Lastly, Figures 9.14d,e, and 9.15b delve into the fluctuations in DC bus voltage as well as the varying output voltages from the QZS network throughout the operation.

9.4.1.2 Case 2: Buck Mode From Line-B to Line-A

For the present study, the initial output phase voltage is set at 160 V on line-B and concludes at 80 V on line-A. Figures 9.16–9.18, and 9.19–9.21 depict the system waveforms under optimal control conditions at output frequencies of 200 Hz and 50 Hz, respectively.

Figure 9.16a–c illustrates how parameters D and m_i behave during operation. Specifically, Figure 9.16a reveals that the value of D diminishes from an initial 0.094 to zero when m_o is set at 1. Concurrently, as Figure 9.16b shows, m_i ascends from 0.906 to a peak at 1 before declining to 0.556. Figure 9.16c points out that the summation of D and m_i remains at a constant 1 as D is adjusted, and begins to drop once D reaches zero. Observations from Figure 9.16d indicate that during the

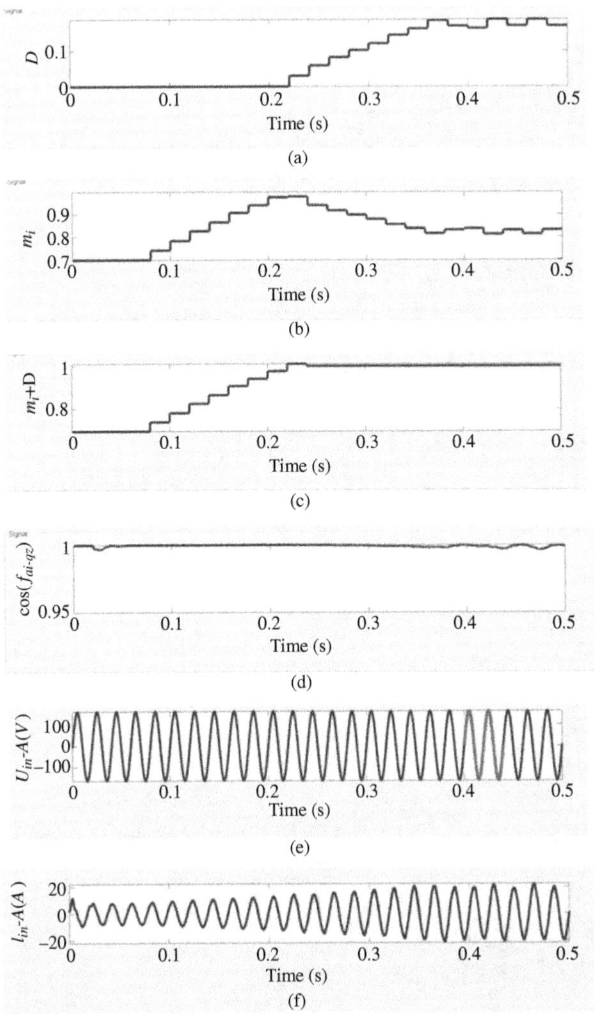

Figure 9.13 Observations for an Output of 50 Hz Frequency. (a) Performance trajectory associated with D; (b) analytical curve linked to m_i; (c) combined functional curve for ($D + m_i$); (d) value of cos φ_{qz}; (e) captured input voltage levels; (f) documented current upon input.

operation, the output of the QZS network maintains a cos φ_{qz} value of 1. In addition, Figure 9.16f illustrates a gradual decrease in the input current in the buck mode operation. Figure 9.17a–c exhibits the system's output voltage and current under optimal conditions, where FFT analysis reveals a smooth drop in the output voltage from 156.1 V to 82.7 V.

In a closer analysis of Figures 9.17d,e, and 9.18, they collectively present both simulated and empirically derived data, capturing the nuances of DC bus voltage and dynamic QZS output voltages. A noteworthy observation can be pinpointed at approximately 0.09 seconds into the operation where there's a discernible decline in the system gain, sinking beneath H, and leading D to assume a value of zero. Intriguingly, under these conditions, the PWM waveform is absent in the QZS output voltages. Expounding further, Figures 9.17e and 9.18 shed light on the intricacies of DC bus voltage modulation, especially when operating in buck mode with an output frequency set at 200 Hz. Modifying "m_i" elicits a direct transformation in the duty cycle of the PWM waveform specific to

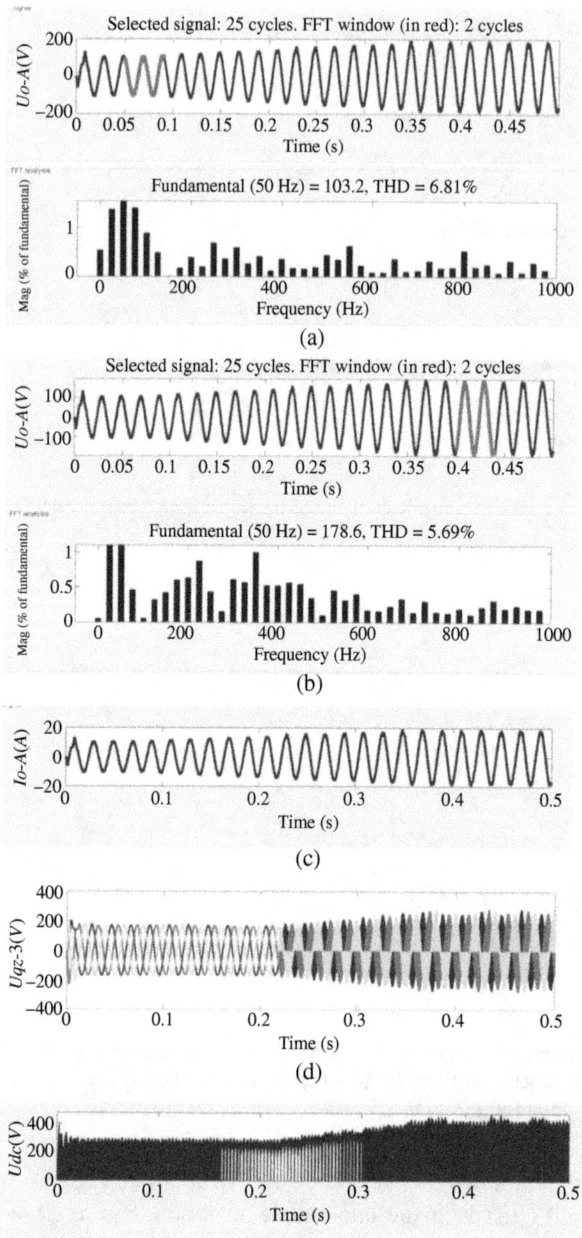

Figure 9.14 Observations at an Output Frequency of 50 Hz. (a) Initial FFT examination related to the output voltage; (b) concluding FFT study of the output voltage; (c) noted output current measurements; (d) tri-phasic QZS voltage outputs; (e) documented DC bus voltage levels.

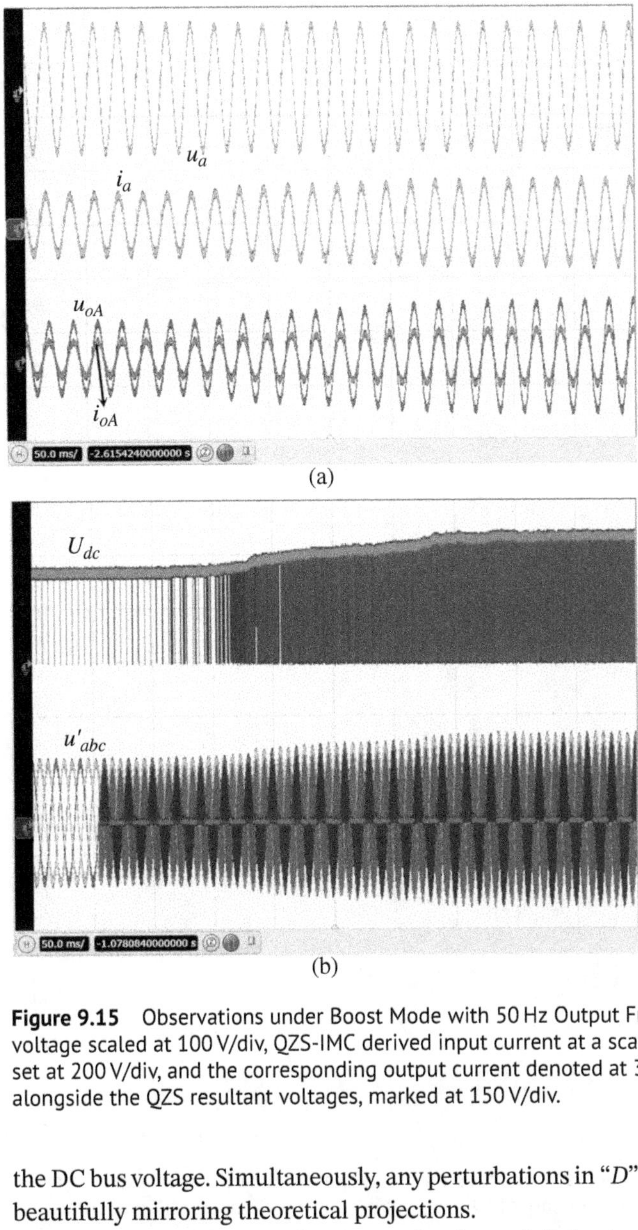

Figure 9.15 Observations under Boost Mode with 50 Hz Output Frequency. (a) Presented AC source voltage scaled at 100 V/div, QZS-IMC derived input current at a scale of 20 A/div, illustrated output voltage set at 200 V/div, and the corresponding output current denoted at 30 A/div; (b) displayed DC bus voltage alongside the QZS resultant voltages, marked at 150 V/div.

the DC bus voltage. Simultaneously, any perturbations in "D" resonate through shifts in amplitude, beautifully mirroring theoretical projections.

In the study, Figure 9.19 focuses on the dynamic behavior of parameters D and m_i during operations at a 50 Hz output frequency. As demonstrated in Figure 9.19a, the parameter D descends in value, moving from an initial 0.094 to zero when m_o is set to 1. Concurrently, Figure 9.19b shows that m_i takes an upward trajectory from its starting point of 0.906, peaking at 1, before eventually settling at 0.5527. Notably, Figure 9.19c underlines that the combined value of D and m_i remains constant at 1 as D is being reduced and m_i is elevating, but begins to diminish as soon as D hits zero. Figure 9.19d indicates that the cos φ_{qz} remains at 1 throughout the buck mode operation. Turning to Figures 9.19e,f, and 9.21a, these charts portray the behavior of the AC source voltage and the QZS-IMC input current in the buck mode at 50 Hz. The input current shows a decremental trend as the operation progresses.

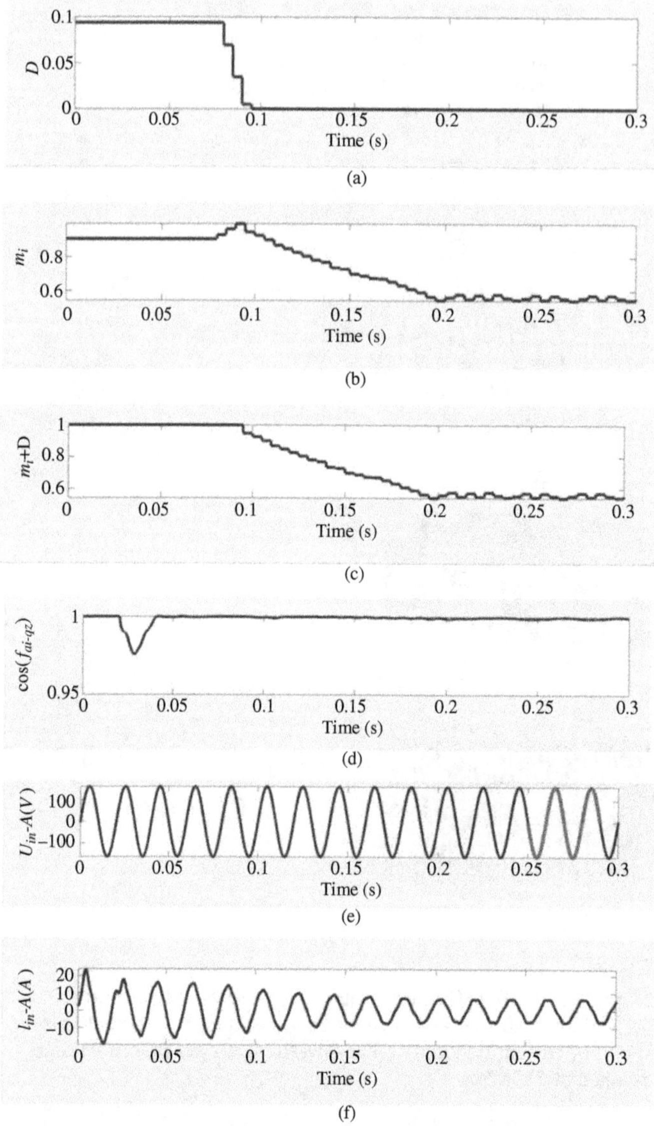

Figure 9.16 Observational data at 200 Hz output frequency. (a) Behavioral trend of D; (b) characteristic progression of m_i; (c) combined operational trajectory of $(D + m_i)$; (d) evaluation of $\cos \varphi_{qz}$; (e) presented input voltage; (f) captured input current.

Within the spectrum of research outcomes, a focused examination of Figures 9.20a–c and 9.21a unravels the nuanced behavior of the system in terms of its output voltage and current dynamics. Drawing insights from the FFT analytical data, there's a systematic attenuation in the output voltage magnitude, transitioning adeptly from an initial 156 V down to 81 V. A subsequent analysis, anchored in Figures 9.20d,e, and 9.21b, unveils the modulatory patterns influencing the DC bus voltage and the respective output voltages of the QSZ network, particularly when navigating the

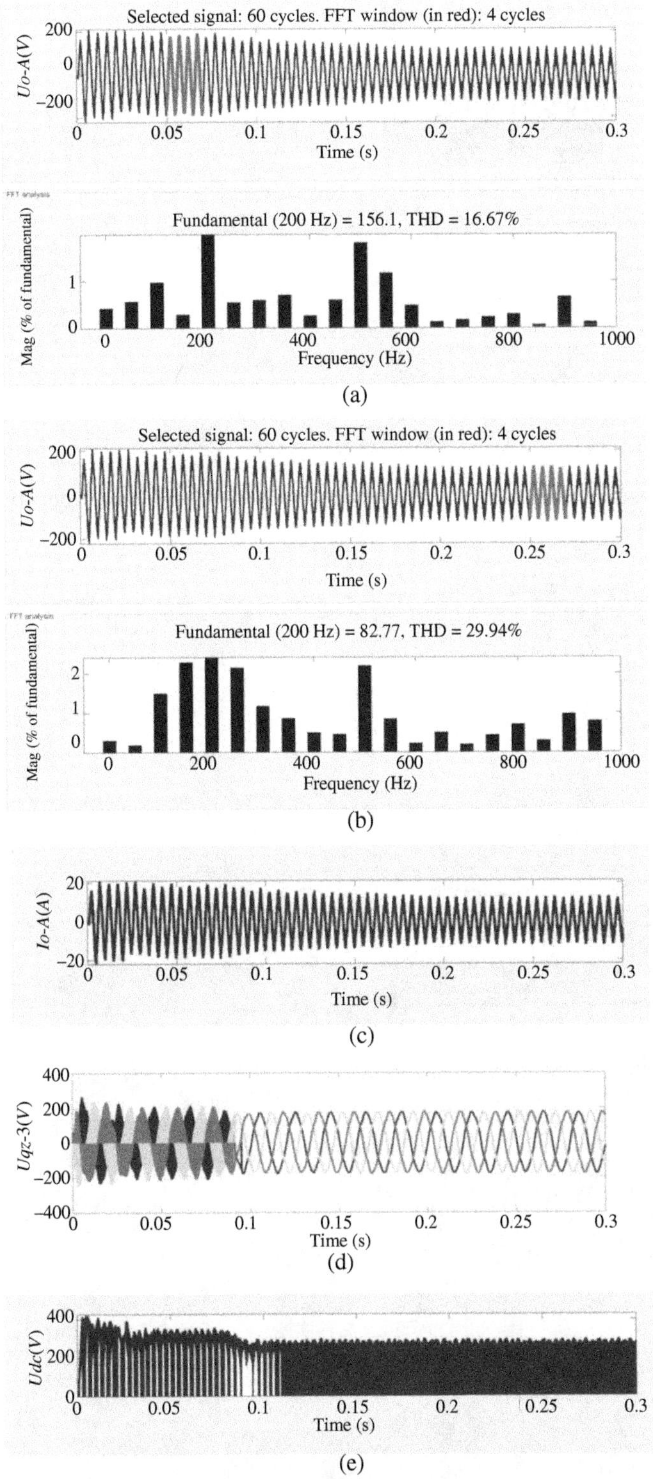

Figure 9.17 Observations at 200 Hz output frequency. (a) Initial output voltage subjected to FFT scrutiny; (b) FFT exploration of the terminal output voltage; (c) current emerging from the output; (d) three-phase voltage outputs from the QZS; (e) voltage levels within the DC bus.

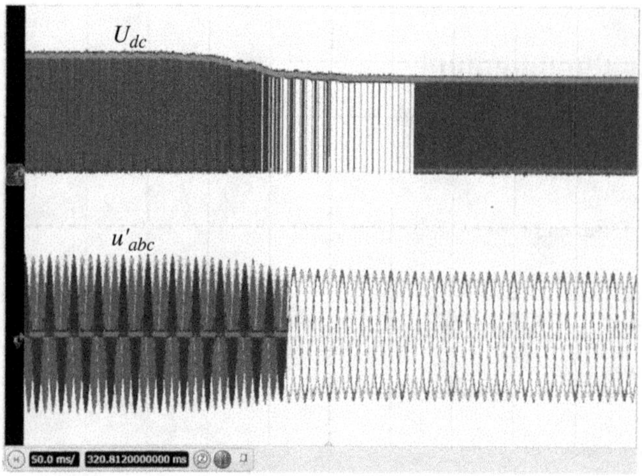

Figure 9.18 DC bus voltage and QZS output voltages (150 V/div) in buck mode at 200 Hz output frequency.

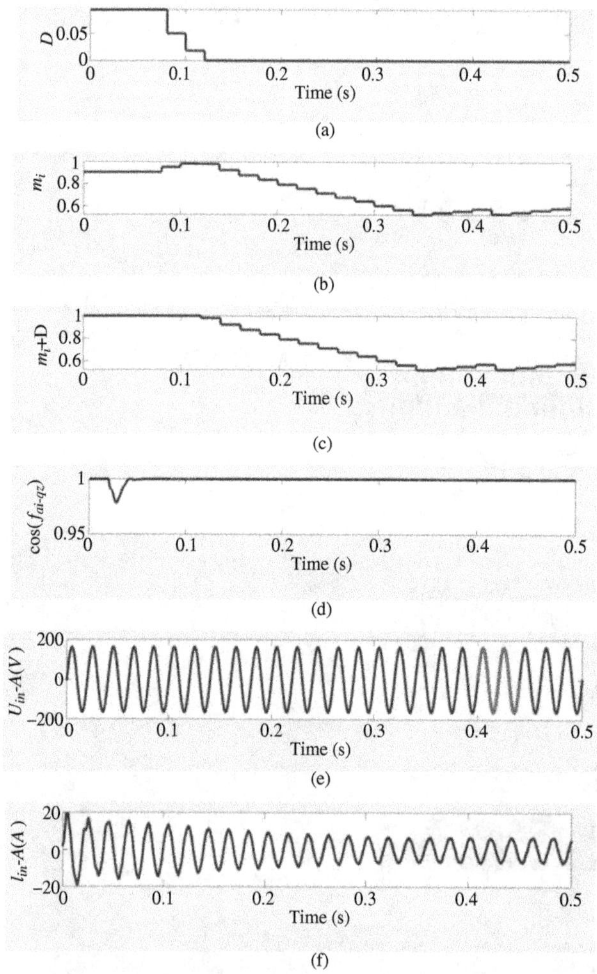

Figure 9.19 Analysis at 50 Hz output frequency. (a) Illustrative curve representing D's operation; (b) graphical representation of m_i's functionality; (c) combined operation curve for $(D + m_i)$; (d) evaluation of $\cos \varphi_{qz}$; (e) depiction of the input voltage level; (f) display of the current input.

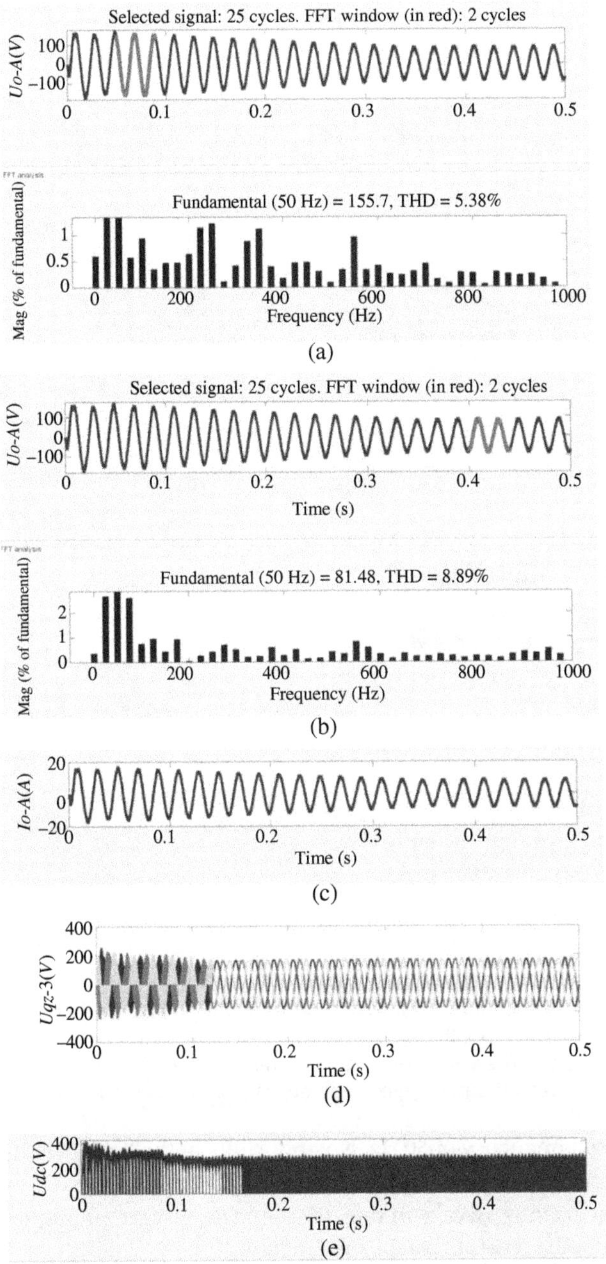

Figure 9.20 Performance Analysis at a 50 Hz output frequency. (a) Provides the FFT scrutiny of the initial output voltage; (b) delivers an FFT inspection of the terminal output voltage; (c) illustrates the value of $\cos \varphi_{qz}$; (d) depicts the QZS three-phase output voltages; (e) exhibits the voltage across the DC bus.

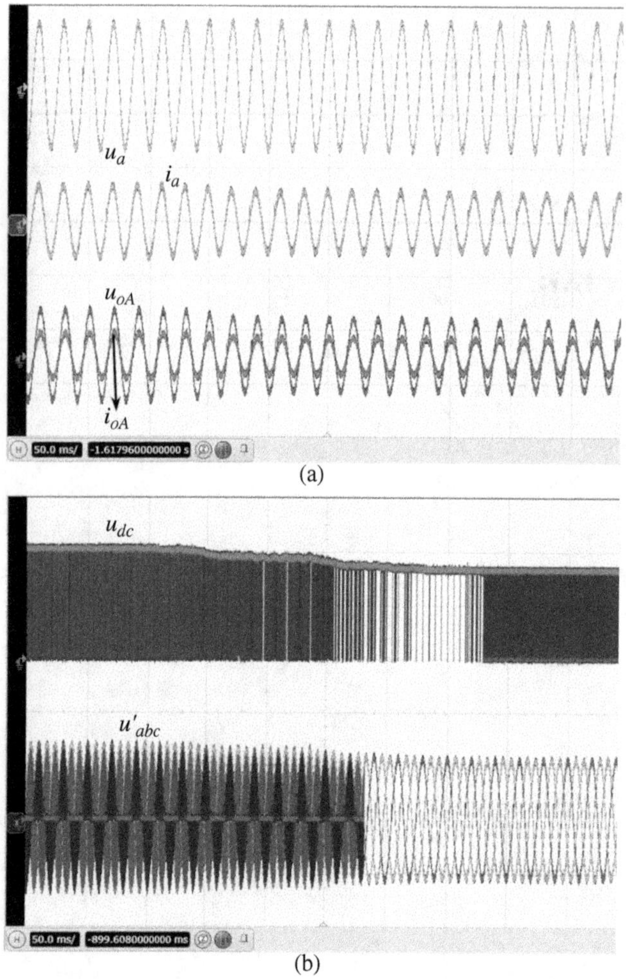

Figure 9.21 Representation of system performance in Buck mode with a 50 Hz output frequency. Part (a) showcases the AC source voltage at a scale of 100 V/div, the QZS-IMC input current measured at 20 A/div, the resultant output voltage scaled at 200 V/div, and the associated output current gauged at 30 A/div. Part (b) delineates the DC bus voltage alongside the QZS output voltages, each depicted at a scale of 150 V/div.

system in the buck mode at an operational frequency of 50 Hz. Notably, shifts in the "m_i" parameter precipitate discernible alterations in the duty cycle, which in turn, sculpt the PWM waveform associated with the DC bus voltage. Concurrently, tweaks in the "D" parameter trigger amplitude deviations in the PWM waveform, echoing the tenets of established theory.

9.4.2 Power Loss Comparison

9.4.2.1 Parameters

In the realm of this study, Table 9.3 delineates the specifics of various parameter combinations that yield an equivalent QZS-IMC voltage gain. For the foundational parameters, one can refer back to the details laid out in Table 9.2. It's worth noting that the configurations presented in cases 1 and 2 operate under non-optimized conditions, while the setup described in case 3 emerges from an optimization process.

Table 9.3 Examples to achieve the same voltage gain.

	D	m_i	B	G	$m_i + D$
Case 1	0.25	0.61	2.00	1.05	0.86
Case 2	0.21	0.70	1.72	1.05	0.91
Case 3	0.15	0.85	1.43	1.05	1.00

Table 9.4 Loss comparison in experiments.

	U_{o-A} (V)	$\cos\varphi_{qz}$	P_{in} (W)	P_o (W)	Loss (W)
Case 1	170.5	0.987	4857	4470	387
Case 2	170.3	0.994	4790	4463	327
Case 3	170.1	0.998	4682	4457	225

9.4.2.2 Measured Powers and Losses in Experiments

In Table 9.4, several specific terms are introduced for clarification. "U_{0_A}" refers to the phase A output voltage in the QZS-IMC system. Meanwhile, "P_{in}" is the term used to specify the active power input into the QZS-IMC system. Additionally, "P_0" designates the active power output emanating from the QZS-IMC system.

9.4.2.3 Analysis and Summary

a) A close examination of Table 9.4 reveals that the empirical voltage data labeled as U_{0_A} precisely aligns with the theoretical values denoted by U_0. This congruence attests to the achievement of identical system voltage gain across the three tested scenarios.
b) Concurrently, the $\cos\varphi_{qz}$ values for all three cases exhibit a remarkable proximity to unity, further validating the system's efficacy.
c) Through comparative analysis, it's evident that the optimized scenario incurs the least amount of power loss, reinforcing its superiority in terms of energy efficiency.

9.5 Conclusion

In this segment, we delineated a comprehensive approach for streamlining the functionality of QZS-IMC by anchoring it to the tenets of constrained optimization. Probing the intricacies of voltage gain within the QZS-IMC environment, we successfully articulated precise formulae for two pivotal components: the QZS shoot-through duty cycle (D) and the rectifier's modulation index (m_i). The overarching objective was to harmonize and fine-tune these parameters D and m_i, thus achieving a consistent voltage gain. We further delved into the QZS-IMC system's dynamism across buck and boost configurations, examining them through the lens of the chosen control paradigm. Crucially, the practical outcomes resonated profoundly with our theoretical postulations, underpinning the robustness of our suggested optimization matrix. Such concurrence augments the prospects of attaining pinnacle efficiency and curtailing power wastage in the QZS-IMC configurations.

References

1 P. W. Wheeler, J. Rodriguez, J. C. Clare, L. Empringham, and A. Weinstein, "Matrix converters: A technology review," *IEEE Trans. Ind. Electron.*, vol. 49, no. 2, pp. 276–288, Aug. 2002.

2 L. Empringham, J. W. Kolar, J. Rodriguez, P. W. Wheeler, and J. C. Clare, "Technological issues and industrial application of matrix converters: A review," *IEEE Trans. Ind. Electron.*, vol. 60, no. 10, pp. 4260–4271, Oct. 2013.

3 S. Safari, A. Castellazz, and P. W. Wheeler, "Experimental and analytical performance evaluation of SiC power devices in the matrix converter," *IEEE Trans. Power Electron.*, vol. 29, no. 5, pp. 2584–2596, May 2014.

4 C. F. Garcia, M. E. Rivera, J. R. Rodríguez, P. W. Wheeler, and R. S. Peña, "Predictive current control with instantaneous reactive power minimization for a four-leg indirect matrix converter," *IEEE Trans. Ind. Electron.*, vol. 64, no. 2, pp. 922–929, Feb. 2017.

5 T. Friedli and J. W. Kolar, "Milestones in matrix converter research," *IEEJ J. Ind. Appl.*, vol. 1, no. 1, pp. 2–14, 2012.

6 A. Trentin *et al.*, "Experimental efficiency comparison between a direct matrix converter and an indirect matrix converter using both Si IGBTs and SiC MOSFETs," *IEEE Trans. Ind. Appl.*, vol. 52, no. 5, pp. 4135–4145, May. 2016.

7 J.-I. Itoh, I. Sato, A. Odaka, H. Ohguchi, H. Kodachi, and N. Eguchi, "A novel approach to practical matrix converter motor drive system with reverse blocking IGBT," *IEEE Trans. Power Electron.*, vol. 20, no. 6, pp. 1356–1363, Nov. 2005.

8 T. D. Nguyen and H.-H. Lee, "A new SVM method for an indirect matrix converter with common-mode voltage reduction," *IEEE Trans. Ind. Informat.*, vol. 10, no. 1, pp. 61–72, Feb. 2014.

9 J. Riedemann, J. C. Clare, P. W. Wheeler, R. Blasco-Gimenez, M. Rivera, and R. Peña, "Open-end winding induction machine fed by a dual-output indirect matrix converter," *IEEE Trans. Ind. Electron.*, vol. 63, no. 7, pp. 4118–4128, Jul. 2016.

10 M. Hamouda, H. F. Blanchette, and K. Al-Haddad, "A hybrid modulation scheme for dual-output five-leg indirect matrix converter," *IEEE Trans. Ind. Electron.*, vol. 63, no. 12, pp. 7299–7309, Dec. 2016.

11 Y. Liu, H. Abu-Rub, B. Ge, F. Blaabjerg, O. Ellabban, and P. C. Loh, *Impedance Source Power Electronic Converters*. Hoboken, NJ: John Wiley & Sons Ltd., Oct. 2016.

12 S. Liu, B. Ge, X. Jiang, H. Abu-Rub, and F. Z. Peng, "A simplified quasi-Z-source matrix converter," *Int.J. Circ. Theory Appl.*, vol. 43, no. 11, pp. 1775–1793, Nov. 2015.

13 G. T. Chiang and J. I. Itoh, "Comparison of two overmodulation strategies in an indirect matrix converter," *IEEE Trans. Ind. Electron.*, vol. 60, no. 1, pp. 43–53, Jan. 2013.

14 M. Chai, D. Xiao, R. Dutta, and J. E. Fletcher, "Space vector PWM techniques for three-to-five-phase indirect matrix converter in the overmodulation region," *IEEE Trans. Ind. Electron.*, vol. 63, no. 1, pp. 550–561, Jan. 2016.

15 B. Jan, L. Jiri, F. Stanislav, and L. Libor, "Contribution to the matrix converter overmodulation strategies based on the virtual dc-link concept," in *Proc. 14th Int. Power Electronics and Motion Control Conf. EPE-PEMC*, 2010, pp. T3-46–T3-52.

16 G. T. Chiang and J. I. Itoh, "Performance optimization of a square wave operation in an indirect matrix converter with a reactor free DC boost converter," in *Proc. 14th European Conf. Power Electronics and Applications*, 2011, pp. 1–10.

17 G. T. Chiang and J. I. Itoh, "DC/DC boost converter functionality in a three-phase indirect matrix converter," *IEEE Trans. Power Electron.*, vol. 26, no. 5, pp. 1599–1607, May 2011.

18 F. Z. Peng, "Z-source inverter," *IEEE Trans. Ind. Appl.*, vol. 39, no. 2, pp. 504–510, Mar/Apr. 2003.

19 J. Anderson and F. Z. Peng, "Four quasi-Z-Source inverters," in *Proc. PESC '08-39th IEEE Annu. Power Electronics Specialists Conf.*, Jun. 15–19, 2008, Rhodes, Greece, pp. 2743–2749.

20 Y. P. Siwakoti, F. Z. Peng, F. Blaabjerg, P. C. Loh, and G. E. Town, "Impedance-source networks for electric power conversion part I: A topological review," *IEEE Trans. Power Electron.*, vol. 30, no. 2, pp. 699–716, Feb. 2015.

21 B. Ge, Q. Lei, W. Qian, and F. Z. Peng, "A family of Z-source matrix converters," *IEEE Trans. Ind. Electron.*, vol. 59, no. 1, pp. 35–45. Jan. 2012.

22 O. Ellabban, H. Abu-Rub, and S. Bayhan, "Z-source matrix converter: an overview," *IEEE Trans. Ind. Electron.*, vol. 31, no. 11, pp. 7436–7450, Aug. 2015.

23 X. Liu, P. C. Loh, P. Wang, and X. Han, "Improved modulation schemes for indirect Z-source matrix converter with sinusoidal input and output waveforms," *IEEE Trans. Power Electron.*, vol. 27, no. 9, pp. 4039–4050, Sept. 2012.

24 S. Liu, B. Ge, H. Abu-Rub, and F. Z. Peng. "Modeling, analysis, and motor drive application of quasi-Z-source indirect matrix converter," *Int. J. Comput.Math. Electr. Electron. Eng.*, vol. 33 no. 1/2, pp. 28–28, 2013.

25 S. Liu, B. Ge, X. Y. You, X. Jiang, H. Abu-Rub, and F. Z. Peng, "A novel quasi-Z-source indirect matrix converter," *Int. J. Circ. Theory Appl.*, vol. 43, no. 4 pp. 438–454, Apr. 2015.

26 S. Liu, B. Ge, X. Jiang, H. Abu-Rub, and F. Z. Peng, "Comparative evaluation of three Z-source/quasi-Z-Source indirect matrix converters," *IEEE Trans. Ind. Electron.*, vol. 62, no. 2, pp. 692–701, Feb. 2015.

27 M. Guo, Y. Liu, B. Ge, H. Abu-Rub, "Optimum boost control of quasi-Z source indirect matrix converter," *IEEE Trans. Ind. Electron.*, vol. 65, no. 10, pp. 8393–8404, Oct. 2018.

28 S. Liu, B. Ge, Y. Liu, H. Abu-Rub, R. S. Balog, and H. Sun, "Modeling, analysis, and parameters design of LC-filter-integrated quasi-Z-source indirect matrix converter," *IEEE Trans. Power Electron.*, vol. 31, no. 11, pp. 7544–7555, Nov. 2016.

29 X. You, B. Ge, S. Liu, N. Nie, X. Jiang, and H. Abu-Rub, "Common mode voltage reduction of quasi-Z source indirect matrix converter," *Int. J. Circ. Theory Appl.*, vol. 44, no. 1 pp. 162–184, Jan. 2016.

30 E. Karaman, M. Farasat, and A. M. Trzynadlowski, "A comparative study of series and cascaded Z-source matrix converters," *IEEE Trans. Ind. Electron.*, vol. 61, no. 10, pp. 5164–5173, Oct. 2014.

31 K. Park, K. B. Lee, and F. Blaabjerg, "Improving output performance of a Z-source sparse matrix converter under unbalanced input-voltage conditions," *IEEE Trans. Power Electron.*, vol. 27, no. 4, pp. 2043–2054, Apr. 2012.

32 O. Ellabban, H. Abu-Rub, and B. Ge, "A quasi-Z-source direct matrix converter feeding a vector controlled induction motor drive," *IEEE J. Emerg. Select. Top. Power Electron.*, vol. 3, no. 2, pp. 339–348, Jun. 2015.

33 S. Vidhya and T. Venkatesan, "Quasi-Z-source indirect matrix converter fed induction motor drive for flow control of dye in paper mill," *IEEE Trans. Power Electron.*, vol. 33, no. 2, pp. 1476–1486, Feb. 2018.

34 C.A.C. Coello, D.A. Veldhuizen, and G. B. Lamont, *Evolutionary Algorithms for Solving Multi Objective Problems*. Boston, MA: Kluwer Academic Publishers, 2003.

35 M. S. Bittermann and Ö. Ciftcioglu, "Precision constrained optimization by exponential ranking," in *IEEE Congress on Evolutionary Computation (CEC)*, Vancouver, BC, 2016, pp. 2296–2305.

10

Applications in Motor Drives

10.1 Introduction

In the evolving field of power conversion, the application of Matrix Converter (MC) technology has taken a pivotal role, especially in motor drive systems. This rise can be attributed to several defining features, including the absence of unwieldy DC-link capacitors, a design that optimizes space, and the simplicity it brings to maintenance tasks [1–3]. When we dissect the architecture of the MC, it unveils two primary configurations: the often-discussed Direct Matrix Converter (DMC) and its less-known counterpart, the Indirect Matrix Converter (IMC). While the IMC manages to echo the inherent characteristics of the DMC, it goes a step further by embedding the beneficial function of zero-current commutation. This added layer of sophistication augments its alignment with AC motor drive systems [4]. Over the years, academic researchers and industry experts have tirelessly worked on devising an array of control methodologies specifically for motor systems powered by the MCs. Their efforts have birthed techniques that span from the well-regarded vector control to the innovative direct torque control, primarily suited for induction motors [5]. In the context of synchronous motors, a bouquet of avant garde strategies has emerged. This includes innovations like sensorless position tracking, meticulous regulation of common-mode voltage, and the established vector control. All of these have been methodically curated by making precise adjustments to the modulation benchmarks of the rectifier and inverter phases [6–10]. However, the journey of MC technology isn't without its bumps. A few inherent challenges have been identified, notably the convoluted commutation protocol, limitations on voltage augmentation, and an amplified risk from external perturbations. But as with every technological hurdle, solutions are on the horizon. There have been insightful proposals aiming to elevate modulation techniques and also to weave in an additional boost mechanism, demonstrating the community's commitment to surmount these challenges and elevate performance [11–13].

When combining the quasi-Z-source (QZS) network with the IMC, a synergistic effect arises that mitigates the inherent challenges of both IMC and ZS/QZS-MC models, as evidenced by studies [14–21]. By melding these technologies, a series of noteworthy benefits emerges: The QZS network performs dual functions, acting as both a filter and voltage amplifier, thereby eliminating the necessity of an additional input filter. This union yields an appreciable enhancement in voltage gain and brings forth a design that is particularly resistant to external disturbances with an improved low voltage ride-through feature. This fusion also refines the two-stage current commutation, simplifying control mechanisms while maintaining the full-silicon traits. A comprehensive assessment contrasting the filtering and voltage magnification qualities of the LC filter integrated QZS-IMC with other ZS/QZS-MC designs is presented in [14, 15]. A novel method to curb the common-mode voltage emerged in [16]. Of importance is the triad of control variables on which

Impedance Source Matrix Converters and Control, First Edition.
Yushan Liu, Xiao Li, Baoming Ge, Haitham Abu-Rub, and Frede Blaabjerg.

the LC filter integrated QZS-IMC operates: namely, the shoot-through duty cycle D, modulation index m_i for the rectifier, and modulation index m_o for the inverter. Even though various combinations of D, m_i, and m_o can yield similar voltage boosts, a method that precisely coordinates these parameters, drawing inspiration from constrained optimization theories, is detailed in [19]. Additionally, a more refined vector control methodology suited for motor drives employing the LC filter integrated QZS-IMC has been clarified in [17, 18]. The crux of the matter remains in pinpointing an optimal coordination approach for D, m_i, and m_o within the QZS-IMC driven motor framework to further enhance its functional efficiency.

In this chapter, we explore an innovative control mechanism, precisely tailored for motor drives integrated with an LC filter and QZS-IMC [20]. This approach marks a paradigm shift from traditional methodologies, offering several distinct benefits. Firstly, a salient feature of this strategy is the rigorous maintenance of the minimum value of D. By adhering to this principle, the system effectively diminishes energy wastage occurring during the rectifier operation, ensuring enhanced efficiency. Secondly, this system doesn't rely on static values for modulation as seen in conventional methods. Instead, it introduces a pioneering method of dynamically modulating the DC-link voltage through the rectifier stage modulation index. This dynamic modulation provides the flexibility for intricate voltage adjustments, optimizing the performance of the system. In doing so, the DC-link voltage is consistently set to operate at its lowest threshold, eliminating any redundant operations, and effectively curtailing power loss during the inverter operations. A third major advancement is the strategy's unwavering focus on maintaining the inverter stage modulation index at its apex. Such an approach not only emphasizes the pivotal role of minimizing the output voltage total harmonic distortion (THD) but also leads to consequential reductions in motor energy wastage. Additionally, the system introduces an auto-tuning mechanism for D and m_i. This not only assures optimized vector control for the motor but also provides a robust countermeasure against potential fluctuations in the power supply voltage, thereby guaranteeing a stable output voltage delivery. Beyond these advantages, the strategy proposes a meticulously crafted coordination mechanism for D, m_i, and m_o, ensuring seamless operation. Furthermore, this mechanism stands out when juxtaposed against traditional control systems; it bypasses the need for cumbersome DC bus voltage measurements and eliminates superfluous PI regulators for D. This simplification ensures a more streamlined and efficient control process, minimizing chances of error. The efficacy of this novel strategy isn't just theoretical. It has undergone rigorous testing, both in simulation environments and real-world empirical setups, showcasing its robustness and adaptability. This chapter also provides a comprehensive breakdown, starting with an elucidative introduction to the integration of the LC filter with QZS-IMC, followed by a deep dive into the motor drive system's control intricacies. It further presents a detailed analysis of the results obtained from both simulations and hands-on tests, culminating in a synthesis of the critical insights garnered throughout the chapter.

10.2 LC Filter-Integrated QZS-IMC

Figure 10.1 provides a comprehensive layout of the motor drive system, seamlessly integrating both the LC filter and QZS-IMC. Drawing energy from a tri-phase AC source, the system is an assembly of key components: the QZS network, a front-end rectifier, a succeeding inverter, and the motor apparatus. Taking a closer look at the QZS network, there's a distinctive pairing of inductors: L_{a1} teams up with L_{a2}, L_{b1} collaborates with L_{b2}, and L_{c1} partners with L_{c2}. Furthermore, capacitors across each phase are demarcated meticulously: C_{a1} and C_{a2} find their place in phase a; C_{b1} and C_{b2} are assigned to phase b, and C_{c1} and C_{c2} represent phase c. The switches, specifically S_a, S_b, and

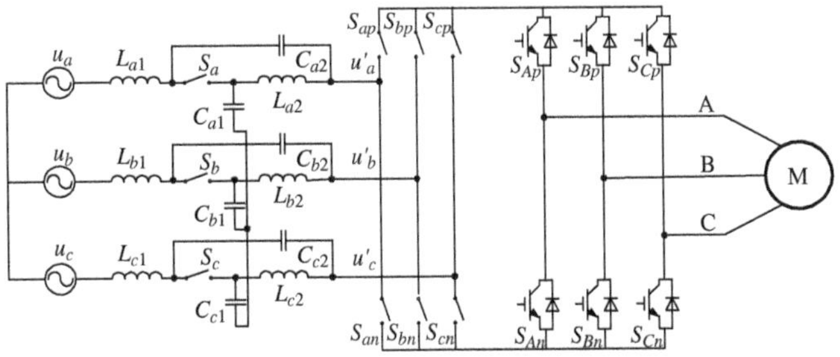

Figure 10.1 LC filter-integrated QZS-IMC-based motor drive.

S_c, operate in unison, governed by a solitary gate signal. Delving into the intricate IMC domain, the bi-directional switches of the rectifier are split into two: S_{Ap}, S_{Bp}, S_{Cp} for the positive spectrum and their counterparts S_{An}, S_{Bn}, S_{Cn} on the negative end. Similarly, the inverter's switching mechanism is bifurcated with S_{Ap}, S_{Bp}, S_{Cp} dominating the positive side while S_{An}, S_{Bn}, S_{Cn} define the negative frontier.

The operational framework of the system comprises two distinct states: the state with shoot-through and the state devoid of it. Let's use D as a representation for the shoot-through duty cycle, determined by the ratio T_{st} to T_s, where T_s symbolizes the overarching control cycle, and T_{st} stands for the designated shoot-through duration.

In assessing the voltage boost proportion, B, specific to the QZS network, one can refer to:

$$B = \frac{U_{im}}{U_{in}} = \frac{1}{1-2D} \tag{10.1}$$

In our analysis, U_{in} represents the amplitude of the incoming phase voltage. On the other hand, U_{im} denotes the amplitude of the voltage emerging from the QZS network. Notably, this same voltage serves as the introductory phase voltage when fed into the IMC rectifier. By adopting the advanced dual space vector modulation (SVM) method, we can extract the average DC-link voltage, as illustrated in the subsequent equation.

$$\overline{U}_{dc} = \frac{3}{2} m_i U_{im} \cos \varphi_{qz} \tag{10.2}$$

As delineated in references [14, 15], within this framework, the parameter m_i is used to symbolize the modulation index associated with the rectifier stage. Concurrently, cos φ_{qz} denotes the input power factor pertinent to that specific stage.

The QZS-IMC's voltage gain, as depicted in the following equation, showcases:

$$G = \frac{U_{om}}{U_{in}} = H m_o m_i B \cos \varphi_{qz} \tag{10.3}$$

factors in H as 0.866. U_{om} is the peak of the QZS-IMC's phase voltage output, and m_o characterizes the inverter stage's modulation index.

Integrating equation (10.1) into (10.3) leads us to:

$$G = \frac{U_{om}}{U_{in}} = H m_o m_i \cos \varphi_{qz} \frac{1}{1-2D} \tag{10.4}$$

The complexity of the voltage gain is undeniably a function of multiple parameters. For enhancing this gain and ensuring efficient use of the DC-link voltage, it's prudent to set the power factor at cos $\varphi_{qz} = 1$. Furthermore, by stabilizing the inverter modulation index, m_o, to unity, we can effectively mitigate power losses at the inverter phase. Using equation (10.4) as a reference, the following derivation can be made:

$$G = Hm_i \frac{1}{1-2D} \tag{10.5}$$

Conclusively, as indicated by (10.5), the QZS-IMC amplifies the system's gain to $H/(1-2D)$, marking a noticeable shift from the conventional IMC's fixed value of H.

10.3 QZS-IMC Induction Motor Drive Control

As delineated in Figure 10.2, we present an evolved methodology for overseeing the QZS-IMC within the context of the induction motor drive framework. Building upon the inherent dual-loop vector control of the induction motor, our strategy offers a more expansive approach. The focal point is the nuanced adjustment of the modulation index, m_i, paired with the careful calibration of the shoot-through duty cycle D in the rectifier stage. Meanwhile, we've ensured the modulation index m_o of the inverter phase remains unwavering at unity. To meet our control targets, we've harnessed the dual SVM methodology of QZS-IMC, as expounded in reference [19].

10.3.1 Dual Closed-Loop Vector Control of Induction Motor

Figure 10.2 illustrates the integration of both speed and current closed loops, as cited in [17, 18, 22]. The outer loop employs the *ASR* regulator, ensuring the motor's speed N remains consistent with the reference speed N^*. The desired q-axis current reference, i^*_{sq}, is derived from:

$$\text{Formula A}: i^*_{sq} = \frac{2}{3}\frac{1}{n_p}\frac{T^*_e L_r}{L_m \psi_r} \tag{10.6}$$

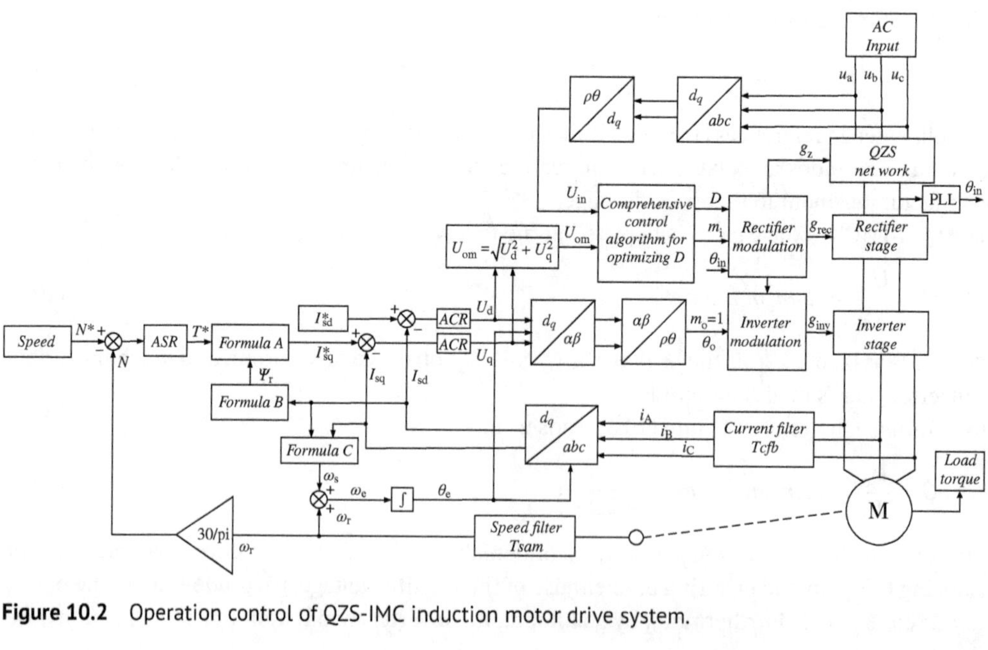

Figure 10.2 Operation control of QZS-IMC induction motor drive system.

$$\text{Formula B}: \quad \psi_r = \frac{i_{sd}L_m}{1 + T_r p} \tag{10.7}$$

In the provided equations, the variable n_p designates the number of motor pole pairs. The term L_m characterizes the mutual inductance, while T^*_e serves as a reflection of the reference torque. It's also noteworthy to mention other relevant parameters: L_r, indicative of rotor inductance; Ψ_r, representing rotor flux linkage; T_r, which stands for rotor time constant; the Laplace operator depicted as p; and i_{sd}, the d-axis current. From these relationships, we can discern the derivation of rotor speed slip.

$$\text{Formula C}: \omega_s = L_m i_{sq}/(T_r \psi_r) \tag{10.8}$$

Within the current loop, *ACR* regulators are tasked with controlling the actual d-axis and q-axis currents, ensuring they align with their references i_{sq} and i_{sd}. The outcomes of these loops produce stator voltage components, U_d and U_q, for the d-axis and q-axis. These components aid in determining the phase angle θ_o of the stator voltage. In achieving inverter stage control, both the phase angle θ_o and the unity modulation index play crucial roles. Concurrently, the values of U_d and U_q inform the calculations for D and the modulation index m_i in the rectifier stage, courtesy of the overarching control algorithm. The input phase angle, θ_{in}, is ascertained by the phase-locked loop (PLL).

10.3.2 Comprehensive Control Algorithm

Initially, the components U_d and U_q, derived from the dual closed-loop vector control, serve to deduce the amplitude U_{om} pertaining to the stator phase voltage. Using the amplitude U_{in} of the input phase voltage in conjunction with (10.3), the target voltage gain G is determined as $G = U_{om}/U_{in}$. Proceeding with the guidelines set by (10.5):

1) For scenarios where G remains less than or equal to H, voltage boosting isn't necessitated. Under these circumstances, the QZS-IMC mirrors the functionality of a conventional IMC. In essence, the QZS network functions purely as an input filter. This behavior is articulated by the control model represented as:

$$D = 0, m_i = \frac{G}{H} \tag{10.9}$$

2) However, when G surpasses H, the QZS network transitions beyond mere input filtration. It assumes the role of managing voltage elevation through the modulation of the shoot-through duty cycle D. To optimize the input voltage and decrease power losses in the QZS network and the rectifier stage, the control strategy prioritizes achieving the highest m_i while reducing D to its minimum for the rectifier stage. This strategy is constrained by:

$$m_i + D = 1 \tag{10.10}$$

Building on (10.5), the voltage gain can be expressed as:

$$G = H\frac{m_i}{2m_i - 1} \tag{10.11}$$

Subsequently, the ensuing control model is delineated as:

$$D = \frac{H - G}{H - 2G}, m_i = 1 - D \tag{10.12}$$

In the context of the QZS-IMC motor drive system, it's observed that the D value, as represented in control model (10.12), undergoes fluctuations. Such variations can be attributed to disturbances in feedback from the integrated vector control loops and discrepancies from non-modeled errors. As a result, the system may experience diminished performance, evident from heightened harmonic interference in input currents and observable control discrepancies. To mitigate these challenges, an intricate control algorithm has been introduced. Elucidated in Figures 10.3–10.5 are flowcharts detailing this approach. In these figures: D_{min} represents the theoretical optimum of D, as deduced

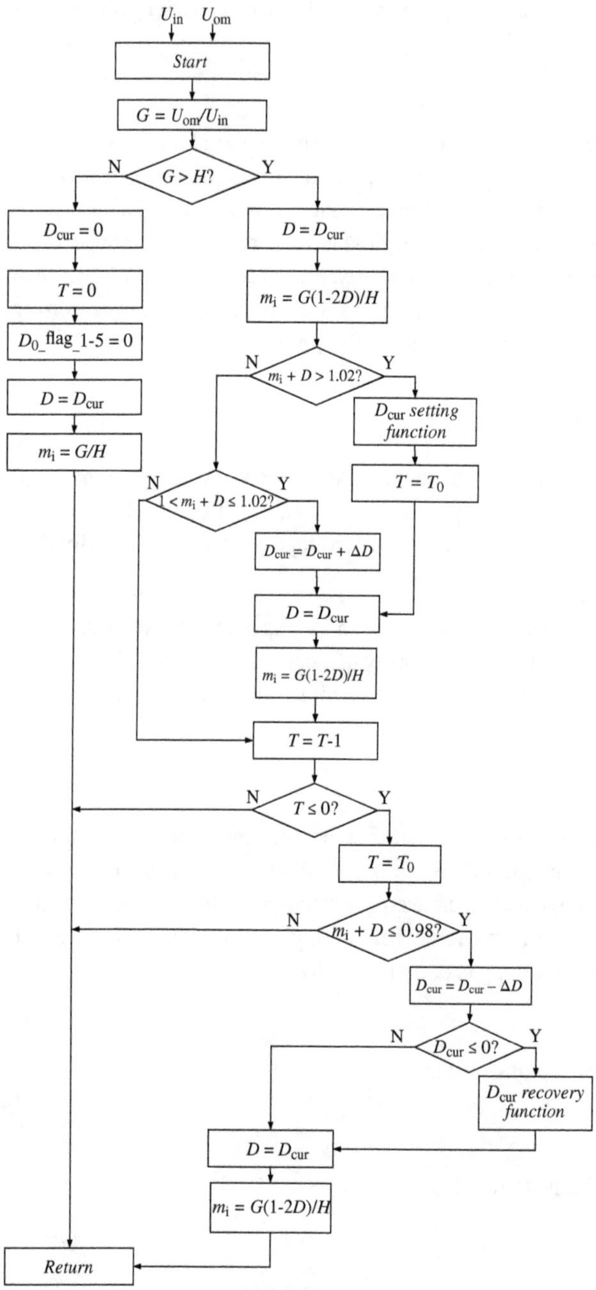

Figure 10.3 Main flow chart.

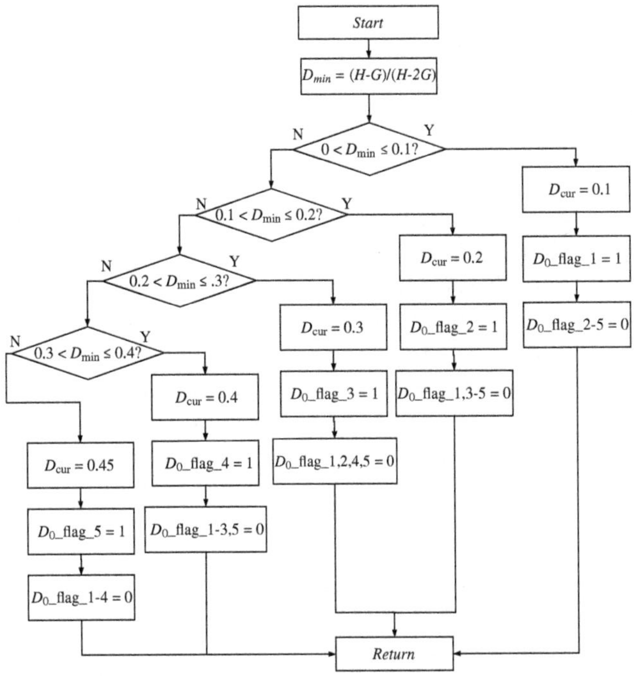

Figure 10.4 Flow chart of D setting function.

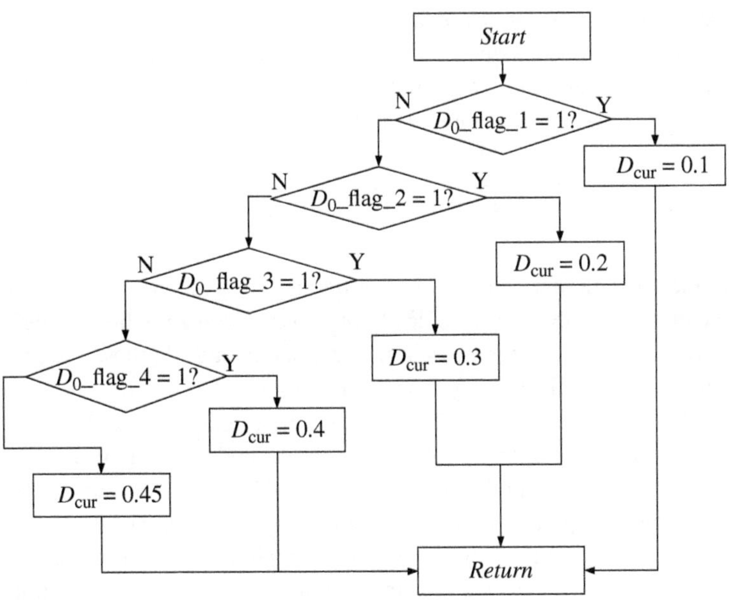

Figure 10.5 Flow chart of D recovery function.

from (10.9) or (10.12). D_{cur} refers to D's prevailing value. ΔD stands for the incremental change in D. Flags ranging from D_0_flag_1 to D_0_flag_5 indicate the initialization phase of D_{cur} across varying stages. T acts as an interval counter, assessing if the combined value of m_i and D adheres to set benchmarks. T_0 specifies the foundational timing parameter for the counter T.

Figure 10.3 delineates the primary flowchart essential for executing the intricate control algorithm. Two crucial input parameters are considered: the amplitude U_{in}, which signifies the input phase voltage, and U_{om}, reflecting the desired motor output voltage. Upon initialization, the system gauges the relationship between G and H, bifurcating into two dominant pathways. In scenarios where G is less than or equal to H, the system adheres to the conventional MC modus operandi, maintaining D at zero, consistent with the interpretation in (10.9).

However, when G surpasses H, the QZS-IMC amplifies the voltage, diverging from $D = 0$. It begins with establishing an initial value for D and subsequently determining m_i using (10.5). From previous discussions, it's clear that the collective value of D and m_i might be permeated with high-frequency noise, causing a divergence from (10.10). To account for this, a fluctuation margin of $\pm 2\%$ is allocated to the sum of D and m_i:

1) Optimal D is attained when the summation lies between 0.98 and 1.
2) For values exceeding 1.02, refer to Figure 10.4 for necessary D adjustments.
3) If $D + m_i$ is greater than 1 but remains under 1.02, an incremental step (ΔD) is added to D.
4) Should $D + m_i$ dip below 0.98, D is decreased by a step (ΔD).
5) For negative D values, the preceding value of D is reinstated, as outlined in Figure 10.5.

Initially, D is set to a significantly high value. Figure 10.4 elucidates the methodology for determining the initial value of D, categorizing it into five distinct zones. $D_{\min}$, derived from equation (10.12), is juxtaposed against these zones to find its maximal margin. Concurrently, flags corresponding to different regions are activated for use in the restoration function displayed in Figure 10.5. For instance, should the evaluated $D_{\min}$ fall below 0.1, D_0 is adjusted to 0.1, the D_0_flag_1 is activated, and flags for the remaining zones are deactivated.

10.3.3 Pulse-Width Modulation

By employing our advanced and meticulously crafted control design, the system management adeptly directs the ascertained modulation indices. Specifically, m_o is associated with the inverter's function, while m_i is inherently linked to the operations of the rectifier. Alongside these integral values, the shoot-through duty cycle, which is denoted as D, becomes a key component as it seamlessly integrates into the dual SVM method of QZS-IMC, a method eloquently expounded upon in [19]. In the subsequent stages, the controller, with its robust capabilities, engineers a set of specific PWM signals. Here, g_z corresponds to the complex operations of the QZS framework, g_{rec} is reserved for the rectifier segment's operations, and g_{inv} is attributed to the functionalities of the inverter segment. This methodical orchestration is by no means accidental. Such a systematic approach ensures not just pinpoint accuracy in speed tracking and input power factor adaptation but also meticulously fine-tunes the voltage gain associated with the inverter. Beyond this, the system promotes optimal use of the DC-link voltage, goes to great lengths to enhance the effectiveness of the applied input voltage, and makes considerable efforts to augment the value of m_i. Concurrently, there's a judicious and calculated reduction in D, all aiming to minimize potential losses occurring within the internal mechanisms of the converter.

10.4 Simulation and Experimental Verifications

To substantiate the effectiveness of the newly devised QZS-IMC motor drive control strategy, we embarked on a twofold approach: simulation-based trials and practical tests on a 4 kW induction motor drive setup. The intricacies of the QZS-IMC, along with those of the motor, have been systematically presented in Tables 10.1 and 10.2. In our endeavor to thoroughly validate the robustness of the system, we probed its behavior under certain challenging conditions, including instances of voltage drop from the primary power source, fluctuations in the load torque, and variations in motor operating speeds. In a bid to offer a broader perspective, we've juxtaposed the performance indicators of our proposed methodology with those from traditional control systems.

10.4.1 Input Voltage Sag and Load Change

In our simulation, the motor accelerates to its rated speed of 1430 r/min under no load conditions. At the 0.4 seconds mark, we introduce a load torque of 26 Nm. Initially, the input power phase voltage amplitude stands at 311 V. At the 0.6 seconds juncture, there's a 15% dip in this voltage. Furthermore, we reduce the load torque by 6 Nm at 0.8 seconds and invert it to −26 Nm by 1 seconds. Detailed outcomes of this simulation can be found in Figures 10.6–10.9. For the hands-on tests, we mimicked these conditions: introducing a 15% voltage sag and modulating load torque between 20 Nm and −26 Nm at a consistent 1430 r/min speed. The results are graphically represented in Figures 10.10 and 10.11.

Within the initial 0.6 seconds, as illustrated by Figures 10.6–10.9, the motor achieves its rated speed. This is accomplished with negligible shoot-through duty cycles and an almost perfect modulation index m_i in the rectifier stage. Between 0.6 and 0.8 seconds, as Figures 10.6–10.9 and

Table 10.1 Parameters of QZS-IMC system.

Amplitude of input phase voltage, U_{in}	311 V
Frequency of input phase voltage, f_{in}	50 Hz
Switching frequency	10 kHz
Inverter stage modulation index, m_o	1
QZS capacitance, C_1 and C_2	200 μF, 150 μF
QZS inductance, L_1 and L_2	500 μH, 250 μH

Table 10.2 Parameters of induction motor.

Rated power, P_N	4 kW
Rated speed, n_N	1430 r/min
Rated phase-phase voltage, U_N	380 V rms
Stator resistance, R_s	1.405 Ω
Stator leakage inductance, L_{ls}	5.84 mH
Rotor resistance, R_r	1.395 Ω
Rotor leakage inductance, L_{lr}	5.84 mH
Mutual inductance, L_m	172.2 mH

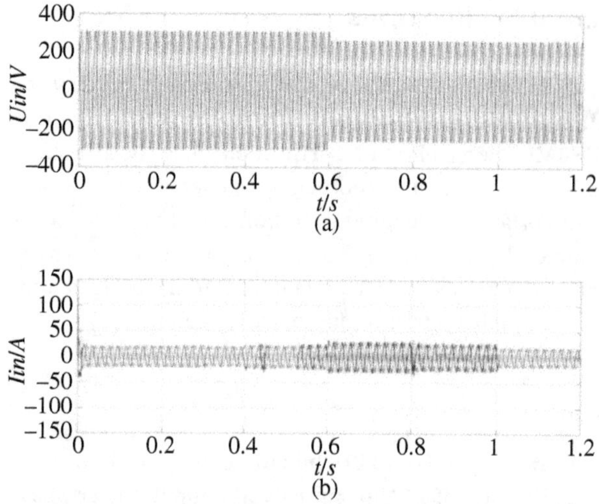

Figure 10.6 Analyzed Simulations of (a) Voltages from the input power supply and (b) Associated input electrical currents. The simulations take into account scenarios in which the supply voltage undergoes a dip at the time-stamp of 0.6 seconds, followed by modifications in the load torque at 0.8 and 1.0 seconds.

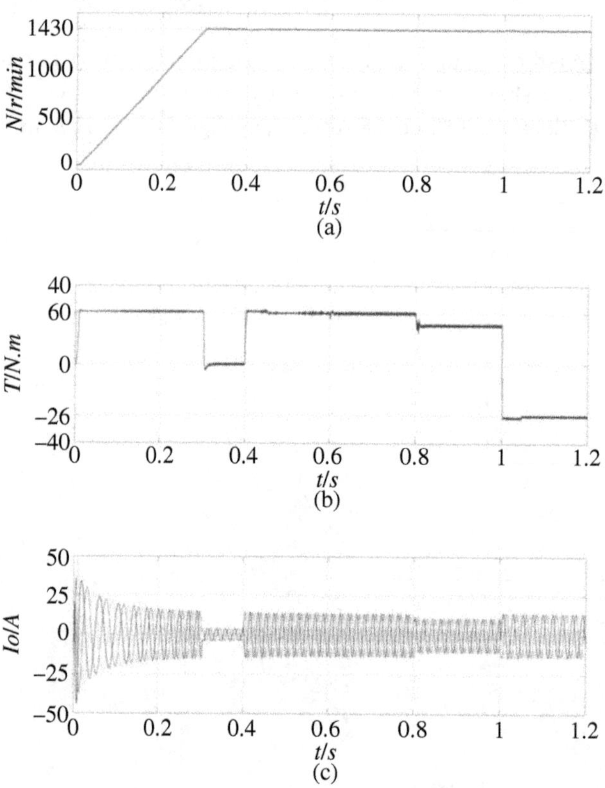

Figure 10.7 Simulation results of (a) rotational speed of the motor; (b) generated motor torque; and (c) stator electrical currents. The study focuses on the instances where the supply voltage incurs a slump at the 0.6-second mark, and alterations in torque are observed at time intervals 0.8 seconds and 1.0 seconds.

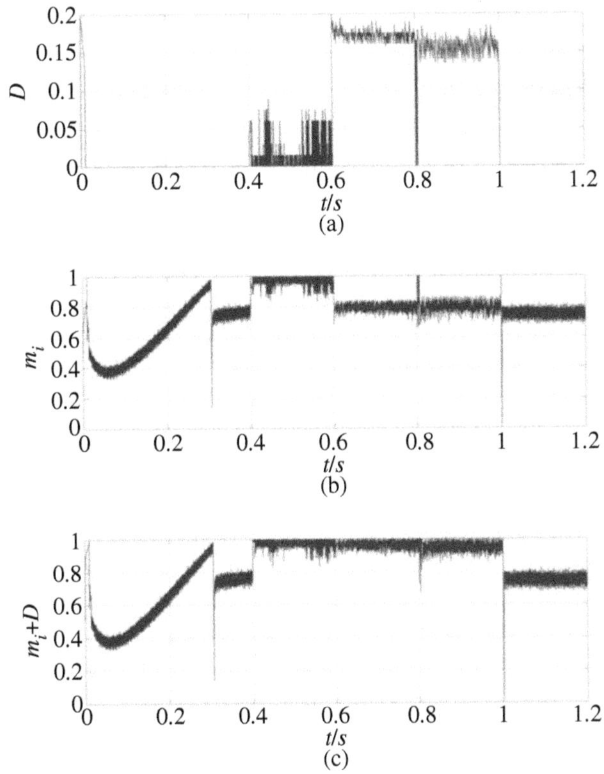

Figure 10.8 Simulated responses of control variables when power supply voltage sags at 0.6 seconds and torque changes at 0.8 seconds and 1.0 seconds. (a) Shoot-through duty cycle D; (b) modulation index m_i of rectifier stage; (c) $m_i + D$.

Figure 10.10 indicate, there's a 15% reduction in input voltage, evident from Figures 10.6a and 10.10a. In response, the duty cycle D undergoes rapid adjustment, leading to a drop in m_i, but the collective $D + m_i$ remains proximate to 0.98, as Figure 10.8 reveals. This adjustment results in an upsurge of both DC-link and QZS network output voltages, evident in Figures 10.9 and 10.10b. This ensures the consistent output of the QZS-IMC in terms of voltage and power, maintaining the motor's speed at 1430 r/min, torque at 26 Nm, and stator currents during the 0.4–0.8 seconds timeframe, as shown in Figure 10.7. However, to counterbalance the input power's voltage sag without affecting the system's power, there's a marginal increment in input currents, depicted in Figures 10.6b and 10.10a.

From the span of 0.8–1.2 seconds, as delineated in Figures 10.7–10.9 and Figure 10.11, fluctuations in the motor's phase current are directly tied to shifts in torque, remaining unaffected by sagging input voltage. Figures 10.7a,b, and 10.11a depict how the motor's currents pivot when the torque transitions from 20 Nm to a negative value of −26 Nm, causing the motor to switch from driving to power generation. During this power generation state, there's a diminished need for motor phase voltage. This leads to the deactivation of the boost function, settling D at zero, as

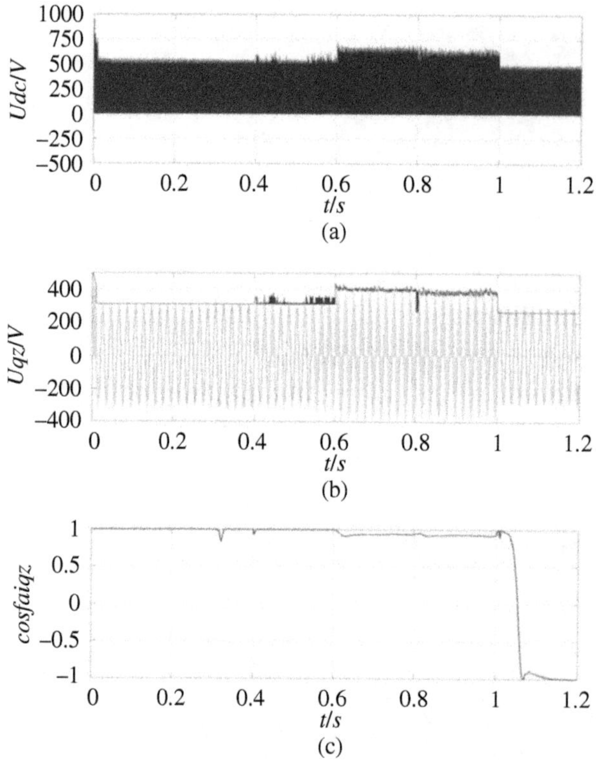

Figure 10.9 Simulation results of (a) the DC bus voltage, denoted as U_{dc}, (b) voltage outputs from the QZS network, and (c) the power factor cos φ_{qz} in scenarios where supply voltage experiences a dip at 0.6 seconds and load torque undergoes adjustments at timestamps 0.8 seconds and 1.0 seconds.

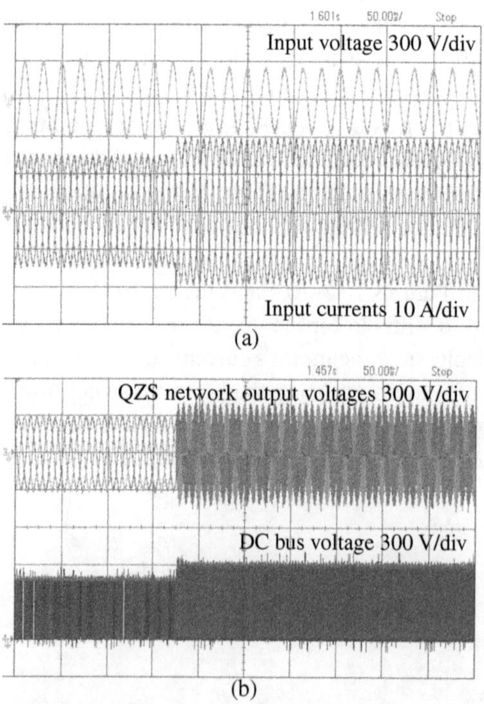

Figure 10.10 Experimental results of QZS-IMC induction motor drive system under conditions of a 15% input voltage reduction. Panel (a) highlights fluctuations in the power input parameters, specifically input voltage and currents, scaled at 300 V/div and 10 A/div, respectively. Panel (b) focuses on observed deviations in QZS network output voltage as well as the DC bus voltage, each delineated at a 300 V/div scale.

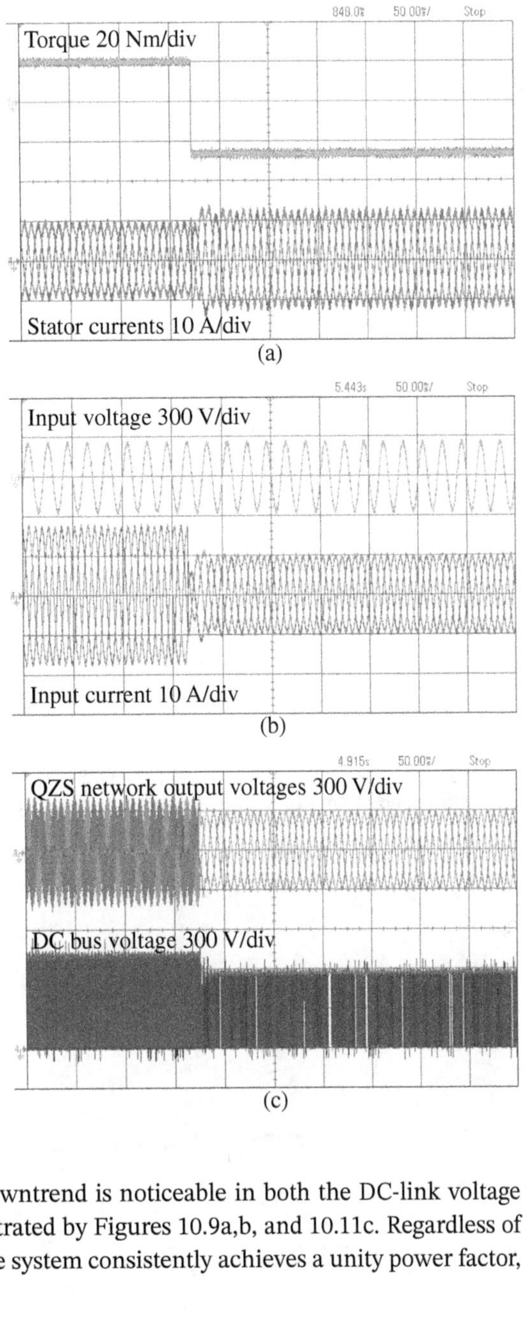

Figure 10.11 Experimental results on the behavior of QZS-IMC induction motor drive during torque fluctuations from 20 Nm to −26 Nm under a 15% three-phase voltage dip. Panel (a) delves into torque and stator current changes, scaled at 20 Nm/div and 10 A/div, correspondingly. Panel (b) sheds light on modifications in input electrical parameters, with voltage and current each presented at a scale of 300 V/div and 10 A/div, respectively. Panel (c) observes alterations in the output metrics of the QZS network and DC bus voltage, both expressed at a scale of 300 V/div.

visualized in Figure 10.8a. Consequently, a downtrend is noticeable in both the DC-link voltage and the QZS network's output voltage, as illustrated by Figures 10.9a,b, and 10.11c. Regardless of the mode—whether driving or generating—the system consistently achieves a unity power factor, as portrayed in Figure 10.9c.

10.4.2 Rotor Speed Change at Input Voltage Sags

In the simulation, the reference speed is adjusted from 1430 r/min to 800 r/min at the 0.8 seconds mark, a procedure replicated in the experimental evaluation. Throughout this phase, the voltage of the input power supply consistently displays a 15% sag, while the motor's load torque remains fixed at 26 Nm. The outcomes of both simulation and experiment are depicted in Figures 10.12–10.16.

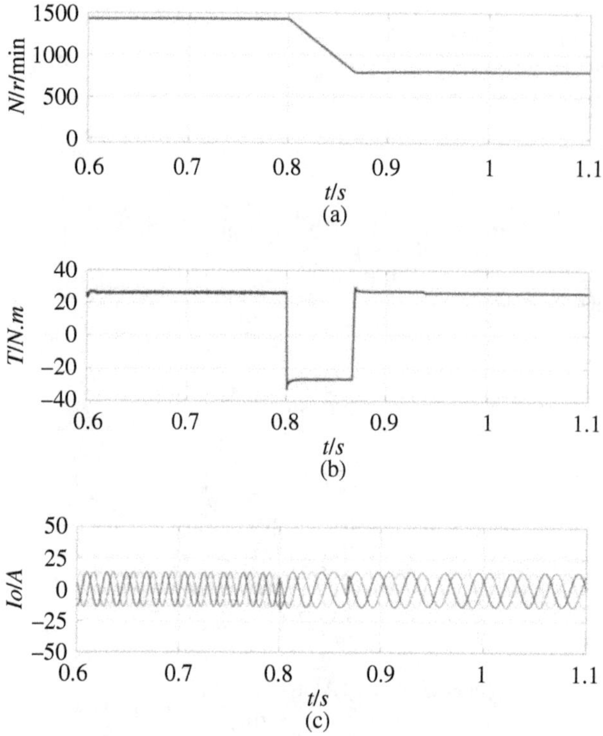

Figure 10.12 Simulation responses of the QZS-IMC motor drive when motor speed varies at 0.8 seconds. (a) motor speed; (b) motor torque; (c) motor stator currents.

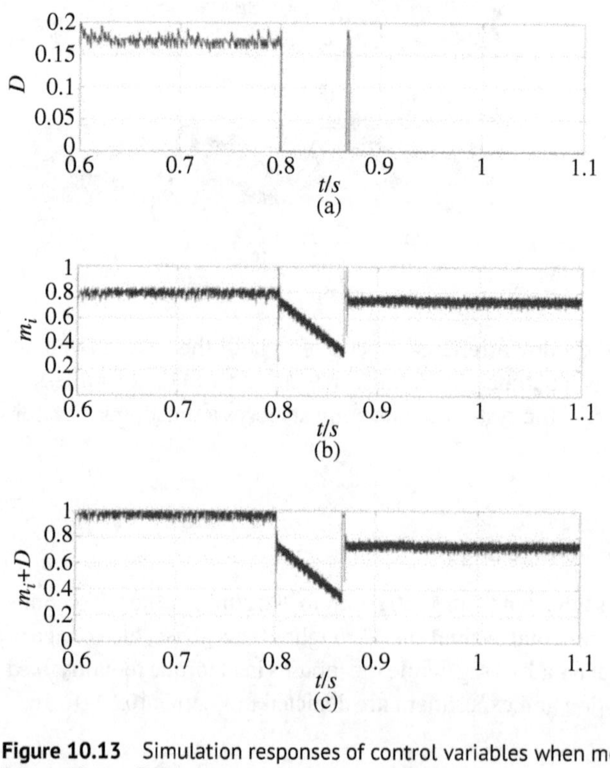

Figure 10.13 Simulation responses of control variables when motor speed varies at 0.8 seconds. (a) D; (b) m_i; (c) m_i +D.

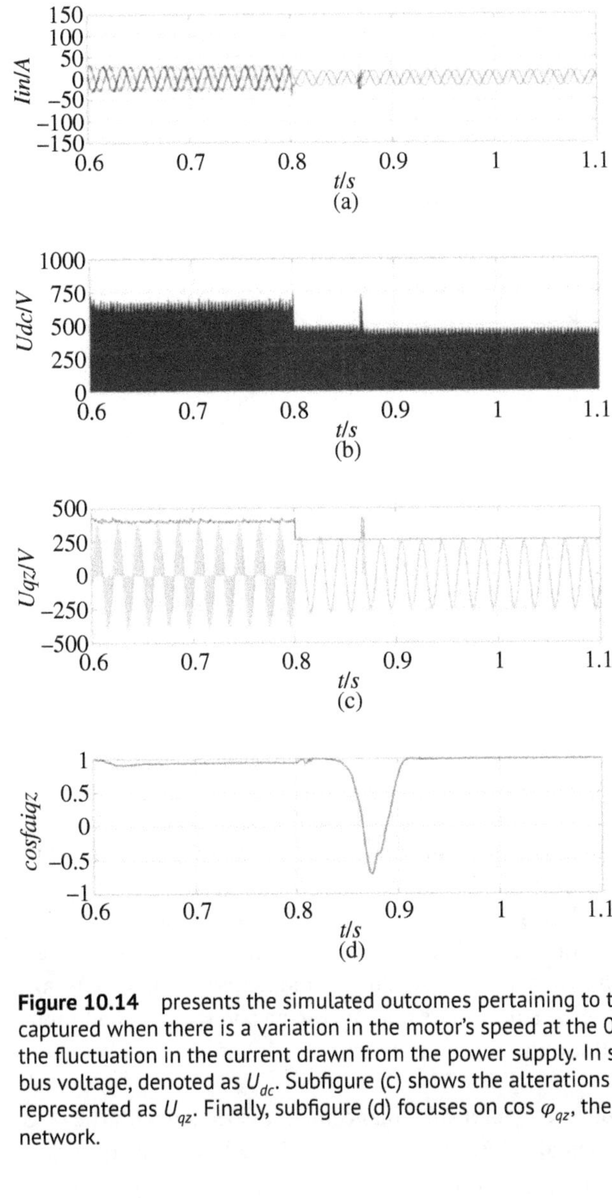

Figure 10.14 presents the simulated outcomes pertaining to the DC bus and QZS network configurations, captured when there is a variation in the motor's speed at the 0.8-second mark. The subfigure (a) depicts the fluctuation in the current drawn from the power supply. In subfigure (b), we observe changes in the DC bus voltage, denoted as U_{dc}. Subfigure (c) shows the alterations in the output voltage from the QZS network, represented as U_{qz}. Finally, subfigure (d) focuses on $\cos \varphi_{qz}$, the power factor associated with the QZS network.

Analyzing the data, one observes that the decrement in speed results in a diminished current on the input side due to the lessened motor power, a phenomenon evident in Figures 10.14a and 10.16a. With the torque remaining steady, the stator current's magnitude is unchanged, as reflected in Figures 10.12b,c and 10.15b. Moreover, when operating at 1430 r/min, the sag in input voltage necessitates a voltage augmentation at $D = 0.16$ to supply the motor's voltage demand, a detail captured in Figure 10.13.

In situations distinguished by reduced operational velocities and torque values that dip into the negatives, it becomes apparent that the essential voltage gain dips beneath a threshold of 0.866.

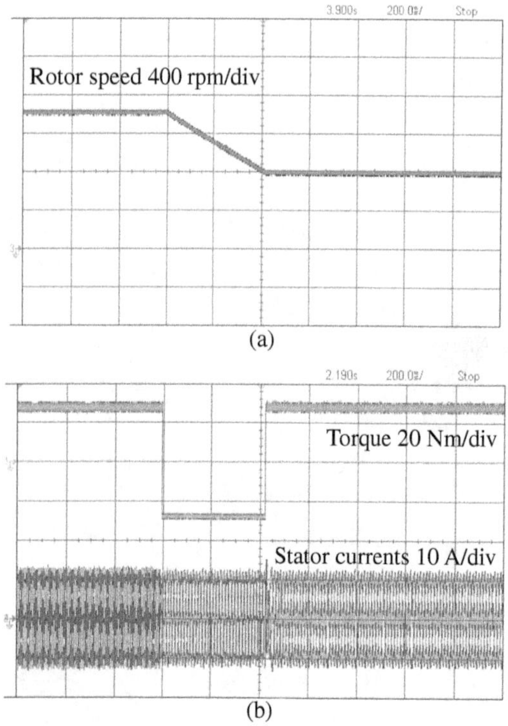

Figure 10.15 Showcases the outcomes from experiments focusing on various performance parameters of the QZS-IMC induction motor drive system. Specifically, part (a) reveals the variations in motor speed, scaled at 400 revolutions per minute per division, and torque, scaled at 20 Newton-meters per division. Part (b) highlights the changes in stator currents, scaled at 10 A/div.

This observation triggers the decision to shut off the boost functionality, achieved by configuring D to a null value. A direct visual representation of this dynamic can be observed in Figures 10.14 and 10.16b. Here, an upward trajectory in speeds is intricately linked with a concomitant surge in both DC-link voltages and the output voltages from the QZS network. This trend is primarily steered by the engagement of the boost function, which is operative during instances when D registers positive values. However, drawing a contrast, during moments when speeds are curtailed, there's an alignment between the QZS network's output voltage and the supplied input voltage, coupled with a noticeable downtrend in the DC-link voltage, attributable to D being set to zero.

As depicted in Figures 10.14c and 10.16b, the theoretical enhancement associated with the QZS network aligns closely with the observed actual values. In Figure 10.14d, we observe a consistent output power factor of the QZS network. Throughout the experimentation, the QZS-IMC motor drive consistently delivered commendable system performance, even when subjected to voltage sags in the input. Impressively, the system managed to optimally utilize both the input voltage and the DC-link voltage while maintaining the D value at a minimal level.

10.4.3 Power Loss Analysis

We undertook a comparative assessment between the conventional control approach [18] and the introduced control technique for the QZS-IMC motor drive system, focusing on power

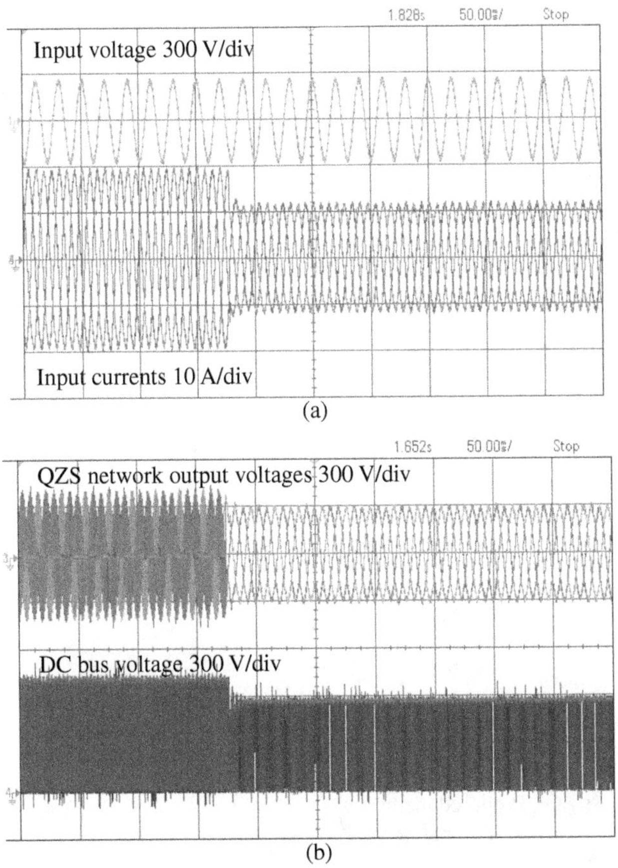

Figure 10.16 A showcase of experimental data for the QZS-IMC induction motor drive under a 15% reduction in three-phase input voltages. (a) Details the variations in input voltage (scaled at 300 V/div) alongside corresponding input currents (scaled at 10 A/div). (b) Captures the modulations in the QZS network's output voltages and the DC bus voltage (both scaled at 300 V/div), specifically when the motor's operational speed transitions from 1430 to 800 r/min.

consumption. While maintaining a constant load torque of 26 N · m, variations in power were facilitated by altering the motor speed, N. The settings for the conventional dual vector control were m_i = 0.8 and D = 0.2, with m_o being adjusted according to distinct power levels. For the newly introduced control, the comprehensive control algorithm was leveraged to implement the minimum D, optimal m_i, and unity m_o.

Table 10.3, alongside Figure 10.17, provides a visualization of the power losses and efficiency across varied motor powers. Wherein, P_e signifies the motor's power, P_{in} represents the input power sourced from the power supply, P_{QZS} is the power utilized by the QZS network, and P_o stands for output power. The difference, ΔP_{IMC}, between P_{QZS} and P_o, defines the IMC's power consumption. η represents overall system efficiency. Notably, the new control approach has proven to cut down the QZS-IMC power consumption and enhanced efficiency by 0.2 when juxtaposed with the conventional method.

Table 10.3 Power consumptions and efficiency of the two methods at different powers.

Method	N (r/min)	P_e (W)	P_{in} (W)	P_{QZS} (W)	P_o (W)	ΔP_{IMC} (W)	$\eta = P_o/P_{in}$
Presented method	200	545	2041	1385	1291	94	0.63
	600	1634	3201	2504	2383	121	0.74
	1000	2723	4319	3579	3472	107	0.8
	1400	3811	5394	4614	4563	51	0.85
Traditional method	200	545	3832	1538	1298	240	0.34
	600	1634	4999	2618	2383	235	0.48
	1000	2723	6156	3697	3467	230	0.56
	1400	3811	7310	4775	4550	225	0.62

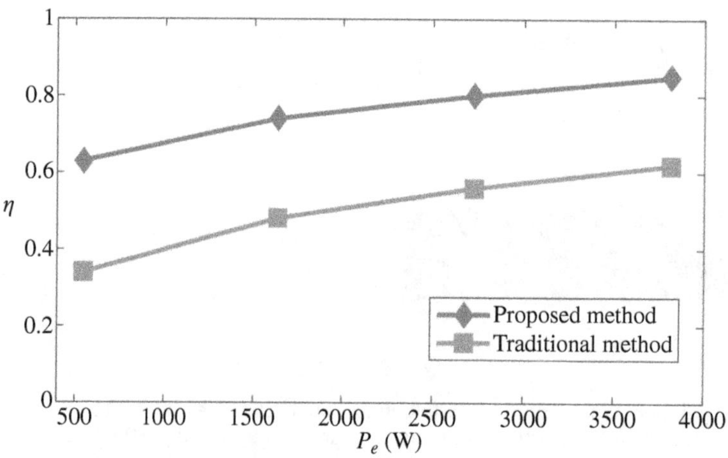

Figure 10.17 A comparative analysis highlighting the efficiency of the QZS-IMC motor drive system when operated using the newly proposed technique versus the conventional method, plotted against varying motor power levels.

10.5 Conclusions

In the ensuing discussion, we introduce a meticulously crafted control strategy, expressly designed for the QZS-IMC induction motor drive mechanism. Our method holistically integrates the actions of the QZS network with those of the rectifier and inverter, paving the way for the maximum exploitation of the DC bus voltage alongside the input voltage. A standout feature of this design is its inherent capability to adapt to fluctuations in the input voltage, obviating the necessity for additional DC-link voltage feedback mechanisms. Central to our approach is the reduction of the shoot-through duty cycle across all operational conditions, subsequently driving down the power demands of the QZS-IMC and attenuating the THD present in the motor's phase currents. The proficiency of our control paradigm for the QZS-IMC induction motor drive was ascertained both theoretically, via simulations, and practically, through real-world experiments. These assessments took into account various parameters such as voltage dips at the input, motor behavioral patterns, regenerative potential, speed variability, and dynamic shifts in load torque.

References

1 P. W. Wheeler, J. Rodriguez, J. C. Clare, L. Empringham, and A. Weinstein, "Matrix converters: A technology review," *IEEE Trans. Ind. Electron.*, vol. 49, no. 2, pp. 276–288, Aug. 2002.

2 L. Empringham, J. W. Kolar, J. Rodriguez, P. W. Wheeler, and J. C. Clare, "Technological issues and industrial application of matrix converters: A review," *IEEE Trans. Ind. Electron.*, vol. 60, no. 10, pp. 4260–4271, Sep. 2012.

3 H. She *et al.*, "Implementation of voltage-based commutation in space-vector-modulated matrix converter," *IEEE Trans. Ind. Electron.*, vol. 59, no. 1, pp. 154–166, Mar. 2011.

4 T. D. Nguyen and H.-H. Lee, "A new SVM method for an indirectmatrix converter with common-mode voltage reduction," *IEEE Trans. Industr. Inform.*, vol. 10, no. 1, pp. 61–72, Mar. 2013.

5 T. F. Podlesak *et al.*, "A 150-kVA vector-controlled matrix converter induction motor drive," *IEEE Trans. Ind. Appl.*, vol. 41, no. 3, pp. 841–847, May. 2005.

6 K.-B. Lee and F. Blaabjerg, "An improved DTC-SVM method for sensorless matrix converter drives using an over-modulation strategy and a simple nonlinearity compensation," *IEEE Trans. Ind. Electron.*, vol. 54, no. 6, pp. 3155–3166, Nov. 2007.

7 S. Bouchiker, G.-A. Capolino, and M. Poloujadoff, "Vector control of a permanent-magnet synchronous motor using AC-AC matrix converter," *IEEE Trans. Power Electron.,* vol. 13, no. 6, pp. 1089–1099, Nov. 1998.

8 Y. Yan, J. Zhao, C. Xia, and T. Shi, "Direct torque control of matrix converter-fed permanent magnet synchronous motor drives based on master and slave vectors," *IET Power Electron.*, vol. 8, no. 2, pp. 288–296, Feb. 2015.

9 T.-H. Liu, S.-H. Chen, and D.-F. Chen, "Design and implementation of a matrix converter PMSM drive without a shaft sensor," *IEEE Trans. Aerosp. Electron. Syst.*, vol. 39, no. 1, pp. 228–243, Mar. 2003.

10 C. Ortega, A. Arias, C. Caruana, and M. Apap, "Reducing the common mode voltage in a DTC-PMSM drive using matrix converters," in *IEEE International Symposium on Industrial Electronics*, Cambridge, UK, 2008, pp. 526–531.

11 M. Chai, D. Xiao, R. Dutta, and J. E. Fletcher, "Space vector PWM techniques for three-to-five-phase indirect matrix converter in the over-modulation region," *IEEE Trans. Ind. Electron.*, vol. 63, no. 1, pp. 550–561, Jan. 2016.

12 Y. Xia, X. Zhang, M. Qiao, F. Yu, Y. Wei, and P. Zhu, "Research on a new indirect space-vector over-modulation strategy in matrix converter," *IEEE Trans. Ind. Electron.*, vol. 63, no. 2, pp. 1130–1141, Feb. 2016.

13 G. T. Chiang and J. I. Itoh, "DC/DC boost converter functionality in a three-phase indirect matrix converter," *IEEE Trans. Power Electron.*, vol. 26, no. 5, pp. 1599–1607, May 2011.

14 S. Liu, B. Ge, X. Jiang, H. Abu-Rub, and F. Z. Peng, "Comparative evaluation of three Z-source/quasi-Z-Source indirect matrix converters," *IEEE Trans. Ind. Electron.*, vol. 62, no. 2, pp. 692–701, Feb. 2015.

15 S. Liu, B. Ge, Y. Liu, H. Abu-Rub, R. S. Balog, and H. Sun, "Modeling, analysis, and parameters design of LC-filter-integrated quasi-z-source indirect matrix converter," *IEEE Trans. Power Electron.*, vol. 31, no. 11 pp. 7544–7555, Apr. 2016.

16 X. You, B. Ge, S. Liu, N. Nie, X. Jiang, and H. Abu-Rub, "Common mode voltage reduction of quasi-Z source indirect matrix converter," *Int. J. Circ. Theory Appl.,* vol. 44, no. 1 pp. 162–184, Jan. 2016.

17 S. Liu, B. Ge, X. Y. You, X. Jiang, H. Abu-Rub, and F. Z. Peng, "A novel quasi-Z-source indirect matrix converter," *Int. J. Circ. Theory Appl.*, vol. 43, no. 4 pp. 438–454, Apr. 2015.

18 S. Liu, B. Ge, H. Abu-Rub, and F. Z. Peng, "Modeling, analysis, and motor drive application of quasi-Z-source indirect matrix converter," *Int. J. Comput. Math. Electric. Electron. Eng.*, vol. 33 no. 1/2, pp. 28–28, 2013.

19 M. Guo, Y. Liu, B. Ge, and H. Abu-Rub, "Optimum boost control of Quasi-Z source indirect matrix converter," *IEEE Trans. Ind. Electron.*, vol. 65, no. 10, pp. 8393–8404, Oct. 2018.

20 M. Guo *et al.*, "A quasi-Z source indirect matrix converter-fed induction motor drive," *IET Electr. Power Appl.*, vol. 14, no. 5, pp. 797–808, May 2020.

21 B. Ge, Q. Lei, W. Qian, and F. Z. Peng, "A family of Z-source matrix converters," *IEEE Trans. Ind. Electron.*, vol. 59, no. 1, pp. 35–45. Jan. 2012.

22 Y. Kawabata, T. Kawakami, Y. Sasakura, E. C. Ejiogu, and T. Kawabata, "New design method of decoupling control system for vector controlled induction motor," *IEEE Trans. Power Electron.*, vol. 19, no. 1, pp. 1–9, Jan. 2004.

11

Future Trends

11.1 General Expectation

In modern industrial settings, the architecture frequently employed for variable-speed drives (VSDs) is the AC/DC/AC converter topology. This particular configuration seamlessly blends three pivotal components into a unified system: a rectifier for the initial conversion of AC to DC, a DC bus that serves as a transitional hub for energy storage, and a two-level voltage-source inverter (2L-VSI). The latter is sometimes colloquially termed the DC/AC converter, as it transitions the direct current back to alternating current to drive the motor. While prevalent and functionally effective, this AC/DC/AC setup has its drawbacks. A primary concern is the obligation to include large capacitors in the DC bus for energy storage, which engenders issues tied to thermal dissipation and the overall reliability of the system. On the other hand, matrix converters (MCs) offer a different paradigm. They act as a direct conduit for AC/AC conversion, thus forgoing the need for any DC-link capacitors. The absence of these capacitors renders the MC design substantially simpler, thereby reducing the complexity of the overall drive system. The minimized need for periodic maintenance becomes an additional benefit, especially when you place this architecture in comparison with its more conventional AC/DC/AC counterpart. Scholarly contributions in this field present a broad range of perspectives on the merits of using matrix converters to power various types of motors, including induction motors and permanent magnet synchronous motors (PMSMs). These viewpoints are encapsulated in the literature, specifically within references [1–7]. For those interested in diving deeper into direct torque control techniques employed for both PMSMs and induction motors (IMs), fueled by a matrix converter-based supply, landmark studies are available in references [8–10]. Furthermore, the academic literature expands on an assortment of modulation schemes and control mechanisms tailor-made for matrix converters, as detailed in references [11–14].

The Direct Matrix Converter (DMC) offers a unique mode of operation, entirely forgoing the necessity for a Direct Current (DC) link. This is achieved by meticulously manipulating the input Alternating Current (AC) voltages to produce the precise output voltages. It's a design approach that eliminates an entire stage traditionally used in power conversion systems. In contrast, the Two-Level Indirect Matrix Converter (2L-IMC) is crafted with a distinct architectural layout that includes an intermediate stage. This stage essentially serves as a capacitor-free DC link, and acts as the crucial bridging element between the AC/DC rectifier, which functions as the system's input stage, and the DC/AC inverter, which operates as the output stage. One particularly remarkable advantage of employing this segmented two-stage design is its inherent ability to enable zero-current commutation, a complex electrical phenomenon that has been thoroughly elaborated in the academic paper referenced as [4].

Impedance Source Matrix Converters and Control, First Edition.
Yushan Liu, Xiao Li, Baoming Ge, Haitham Abu-Rub, and Frede Blaabjerg.

MCs are generally characterized by a range of inherent limitations that warrant careful consideration for optimal performance. One salient restriction is that they operate in a configuration commonly known as buck-type. This specific configuration imposes a ceiling on the voltage gain, limiting it to no more than 0.866. Additionally, the performance efficiency of MCs is notably vulnerable to a variety of disturbances. This sensitivity arises at both the input and output levels, a consequence of the absence of intermediary capacitors that are often used in other configurations for buffering energy fluctuations. In light of these challenges, there has been a surge of scholarly inquiry, documented in studies referenced as [15–29], that aims to circumvent these limitations through the exploration of alternate modulation techniques as well as the incorporation of supplementary circuit elements. One such promising approach that has garnered attention is the integration of a quasi-Z-source (QZS) network into the input AC stage of a two-level indirect matrix converter, known in technical parlance as 2L-QZS-IMC. Incorporating this QZS network offers several pivotal advantages. Notably, it allows for an extension of the voltage gain threshold beyond the conventional 0.866 limit. Additionally, this modification also enhances and amplifies the overall voltage profile of the converter, providing it with a measure of resilience against input–output discrepancies. As a bonus, the integration preserves the structural integrity of a converter architecture that is built entirely of silicon-based components.

Vector control emerges as a widespread method for managing the operation of three-phase squirrel-cage induction motors (SCIMs). While it easily aligns with Indirect Matrix Converter (IMC)-based Variable-Speed Drives (VSDs), it is not without its limitations—chiefly in terms of achieving the most favorable ratio between output and input fundamental voltages. When it comes to boosting the performance of a PDS, particularly during operations that require higher speeds and thus more voltage to counteract field weakening phenomena (which result in torque drops and higher rotor losses for SCIMs), having an augmented voltage capacity on the DC side of Quasi-Z-Source (QZS)-IMC-based VSDs becomes pivotal. To accomplish this enhancement, there are generally three core strategies that can be employed.

Executing open-loop control is relatively simple but comes with the downside of lacking dynamic adaptability when there are AC grid voltage sags affecting the DC-link voltage. On the flip side, closed-loop control effectively maintains a stable DC-link voltage even when subjected to fluctuations in the AC grid, regardless of what speed the motor is operating at. However, this form of control might be a double-edged sword; it can lead to an uptick in switching losses in the converter, compromise motor efficiency, and also give rise to supplementary losses in the QZS network. There's also an alternative approach tethered to the motor's speed that modulates the DC-link voltage. When the motor is running at lower speeds, this technique aims to cut down on both DC-link voltage and converter energy losses. In contrast, at higher speeds, it ramps up the DC-link voltage and compensates for any sag in the AC grid voltage.

By addressing the shortcomings inherent to the prior techniques, this particular method guarantees that the inverter's amplitude modulation index hovers close to unity. Achieving this benchmark serves a dual purpose: it both elevates the amplitude of the principal output component to its maximum and concurrently lowers the harmonic distortion levels. To make this happen, the strategy fine-tunes the DC-link voltage in real time by making adaptive changes to the amplitude modulation index of the rectifier and also the shoot-through duty cycle.

The existing research predominantly centers on the topology and modulation schemes of 2L-QZS-IMC, where the output voltage is constrained by the voltage ratings of the switches or transistors employed in IMC inverters. For instance, reference [29] showcases an open-loop control-based algorithm that was successfully applied to a 2L-QZS-IMC system for modulating the DC-link voltage. This was achieved by varying both the rectifier's amplitude modulation index and the shoot-through duty cycle. However, the approach necessitated a rather intricate process for open-loop adjustments to determine these parameters.

Boosting the output voltage of QZS-IMC while operating within a given supply voltage and without changing the voltage rating of the power switches remains a subject of ongoing research. Within the scope of this chapter, several pivotal perspectives of QZS matrix converters are made:

1) This chapter provides a prospected future research and application for a dual-three-level (D-3L)-QZS-IMC architecture [30]. This newly formulated design holds the remarkable ability to amplify the output voltage gain by a striking factor of two, thereby eclipsing the performance of the traditional 2L-QZS-IMC model that has been elaborated upon in reference [29].
2) To govern the behavior and performance of this D-3L-QZS-IMC architecture, an intricate modulation technique has been meticulously engineered. This is fundamentally different from the modulation approach elaborated for a regular 2L-IMC in existing literature, specifically in reference [4]. Our sophisticated approach not only optimizes the levels of output voltage but also judiciously curtails common-mode voltage (CMV). Moreover, this technique incorporates an enhanced array of switching states, including dual indirect matrix converters (IMCs) and shoot-through states, thereby departing markedly from the more limited scope of the conventional IMC system as described in [4].
3) Alongside these innovations, the chapter delves deeply into a state-of-the-art PDS. This system ingeniously amalgamates the features and functionalities of the D-3L-QZS-IMC with an Open-Ended Winding Squirrel-Cage Induction Motor (OEW-SCIM). At the heart of this system is a closed-loop control architecture capable of managing both variable voltage and frequency. This is achieved via specialized rectifiers and inverters. Significantly, the strategy embedded within this control system permits automatic adjustments to the DC-link voltage, thereby eliminating any need to alter the modulation index in the inverters. This has the potential benefit of reducing converter losses, all while maintaining robust motor control objectives.
4) To add empirical credence to the theoretical constructs presented herein, data have been collected through rigorous testing on a 4-kW experimental setup, effectively validating the viability and benefits of the proposed architecture and methodologies.

The chapter is organized for easy navigation. Section 11.2 takes a deep dive into the D-3L-QZS-IMC PDS, detailing its topology, underlying mechanisms, and the set of modulation rules governing its operation. Section 11.3 brings to the fore various control methodologies that have been employed. Section 11.4 lays out the empirical results, which serve to substantiate the efficacy of the chosen topology, modulation strategies, and control frameworks. In Section 11.5, we engage in a rigorous comparative analysis with traditional systems to highlight the distinct advantages of the approach. Finally, Section 11.6 offers a summary of the key findings and insights.

11.2 Dual-Three-Level QZS-IMC-Based Power Drive System

11.2.1 Topology

As showcased in Figure 11.1, we present a detailed view of the three-phase D-3L-QZS-IMC-based Power Drive System under examination. This system is specifically engineered around a Variable-Speed Drive (VSD) that employs a novel D-3L-QZS-IMC architecture. This is further integrated with an OEW-SCIM.

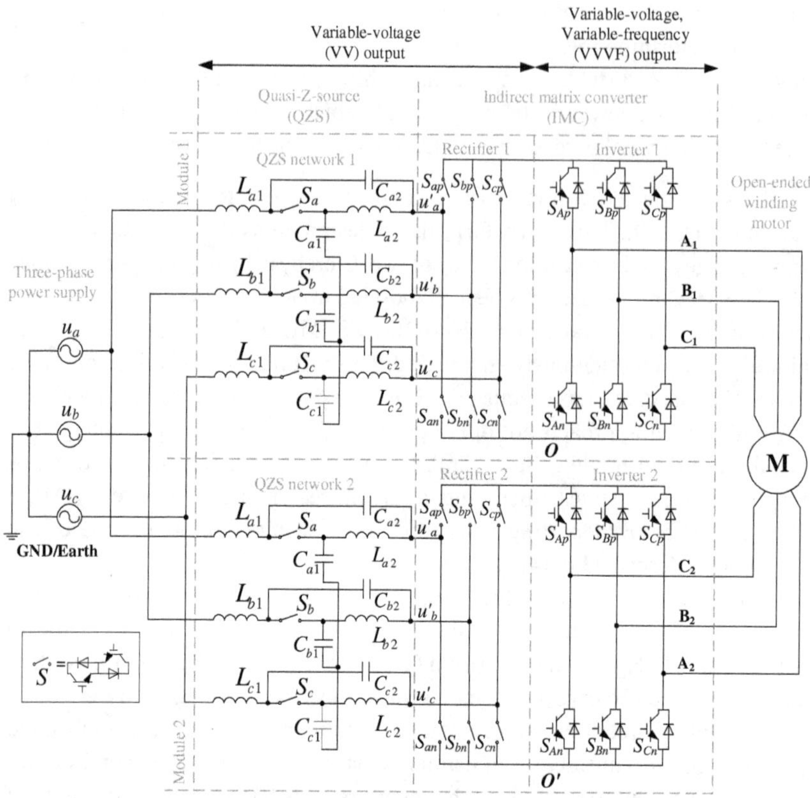

Figure 11.1 Discussed PDS integrating the D-QZS-3L-IMC and an OEW-SCIM.

The core of the VSD, the D-3L-QZS-IMC, merges two distinct modules from the traditional three-phase 2L-QZS-IMC. Each of these modules interfaces directly with the three-phase power grid via their input terminals and connects to the six terminations—namely A_1, B_1, C_1, C_2, B_2, A_2—of the motor's open-ended windings. This design thereby forms three H-bridge configurations, enabling a 3-level voltage distribution to the motor's phase windings. This results in several operational benefits, such as an enhancement in the system's voltage gain, a reduction in the mechanical stress exerted on the power switches, and improvements in Total Harmonic Distortion (THD) of the voltage output. The design also ensures a moderated rate of voltage change, often referred to as dv/dt.

Furthermore, the D-3L-QZS-IMC can furnish a variable-voltage, variable-frequency (VVVF) control for the induction motor. The frequency modulation is orchestrated through the inverter stage, while the voltage modulation can be managed via either the rectifier or inverter units. In contrast, conventional VSDs, which usually consist of a three-phase diode rectifier, a DC bus, and a three-phase voltage-source inverter (VSI), provide similar VVVF output but with a constant DC bus voltage.

In this work, an exhaustive closed-loop control schema is introduced. The control strategy efficiently adjusts the DC output voltages originating from both rectifier units and also regulates the AC frequency output from the inverter stages. Importantly, the system maintains a unity modulation index, leading to a suppression of low-order harmonic distortions in the AC voltage output and a reduction in converter losses. The minimized dv/dt also helps to mitigate voltage reflections in the power cables that link the inverters and the motor, thereby reducing the risk of voltage spikes at the motor windings.

Each of the three-phase QZS-IMC modules incorporates three primary elements:

- The QZS network features a dual inductor pair for every phase—namely L_{a1} and L_{a2} for phase A, L_{b1} and L_{b2} for phase B, and L_{c1} and L_{c2} for phase C. Each phase is also furnished with two capacitors, specifically C_{a1} and C_{a2} for phase A, C_{b1} and C_{b2} for phase B, and C_{c1} and C_{c2} for phase C. To facilitate bi-directional current flow, each phase employs a singular power switch, designated as S_a, S_b, and S_c.
- The rectification stage is equipped with a hexad of bi-directional power switches. These switches are labeled as S_{ap}, S_{bp}, and S_{cp} for the positive lines and S_{an}, S_{bn}, and S_{cn} for the negative lines.
- Lastly, the inverting segment employs an array of six unidirectional power switches—S_{Ap}, S_{Bp}, and S_{Cp} are responsible for the positive current direction, whereas S_{An}, S_{Bn}, and S_{Cn} govern the negative current path.

11.2.2 Operating Principle

As explicitly indicated in the graphical schematic shown in Figure 11.1, the intricate electrical interconnection between our primary power source and the advanced OEW-SCIM is smoothly orchestrated via a specialized converter, known as the Dual-Three-Level Quasi-Z-Source Indirect Matrix Converter (D-3L-QZS-IMC). This D-3L-QZS-IMC configuration is notable for its operational versatility, which manifests through two distinctly different operational modes or states: these are referred to as the Shoot-Through State (STS) and its contrasting operational condition, the Non-Shoot-Through State (NSTS).During periods of NSTS operation, the essential power switches of the Quasi-Z-Source (QZS) network—identified as S_a, S_b, and S_c—are actively engaged or turned on. This state of activation allows for a sequential series of operations to occur; specifically, it permits the upper-level power switches of the rectifier segment (distinctly labeled as S_{ap}, S_{bp}, and S_{cp}) to be fully operational when the system transitions into the STS mode. The transition to STS occurs when the aforementioned switches S_a, S_b, and S_c are deliberately turned off or deactivated.

The voltage amplification ratio, symbolized as B, for an individual QZS network is expressed in equation (11.1):

$$B = \frac{1}{1-2D} = \frac{1}{1-2\frac{T_0}{T_s}} = \frac{T_s}{T_s - 2T_0}, \tag{11.1}$$

Here, D symbolizes the duty cycle of STS, T_s stands for the complete switching cycle, and T_0 refers to the STS duration.

The gain factor of each QZS-IMC module, denoted by G, can be computed as in equation 11.2:

$$G = \frac{U_{om,m}}{U_{in}} = H \cdot m_o \cdot m_i \cdot B \cdot \cos \varphi_{qz}, \tag{11.2}$$

In the equation under consideration, the term $U_{om,m}$ serves as an indicator for the voltage phase output of the individual module. Meanwhile, U_{in} signifies the phase voltage entering the system. The constant H is set at approximately 0.866, while m_i and m_o act as the respective

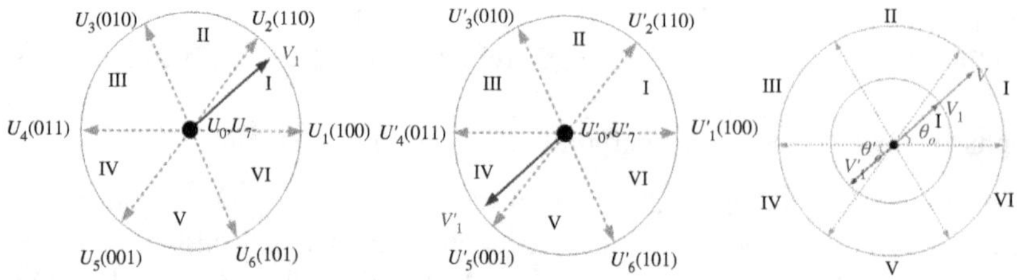

Figure 11.2 Output voltage vectors of the presented PDS: (left) Module 1 ($\underline{V}_1$); (middle) Module 2 ($\underline{V}'_1$); (right) motor winding phase terminals ($\underline{V}$).

modulation indexes for the rectifying and inverting stages. Furthermore, $\cos\varphi_{qz}$ denotes the rectifier's power factor.

Figure 11.2 illustrates the voltage vectors produced by the PDS under discussion. With respect to points O and O', the output voltage phasors of both QZS-IMC modules, $\underline{V}_1$ and $\underline{V}'_1$, are offset by 180° to maximize the output voltage of the D-3L-QZS-IMC. Consequently, the motor winding phase voltage, $\underline{V} = \underline{V}_1 - \underline{V}'_1$, then $|\underline{V}| = 2|\underline{V}_1| = 2|\underline{V}'_1|$, is doubled when compared to each module's output, restricted only by the voltage ratings of the individual power switches.

11.2.3 Modulation Method

As delineated in Table 11.1, the rectifier functions in two principal states: the active and the zero vector states, where "1" indicates the ON state and "0" signifies the OFF state. During the STS, voltage undergoes a surge to an elevated level.

Table 11.1 DC-link voltage and rectifier switching states.

State	U_{dc}	S_{ap}, S_{bp}, S_{cp}	S_{an}, S_{bn}, S_{cn}	S_x	Vector	Description
1	u_{ac}	1, 0, 0	0, 0, 1	1	$\underline{I}_1$	Active Vector
2	u_{bc}	0, 1, 0	0, 0, 1	1	$\underline{I}_2$	Active Vector
3	$-u_{ab}$	0, 1, 0	1, 0, 0	1	$\underline{I}_3$	Active Vector
4	$-u_{ac}$	0, 0, 1	1, 0, 0	1	$\underline{I}_4$	Active Vector
5	$-u_{bc}$	0, 0, 1	0, 1, 0	1	$\underline{I}_5$	Active Vector
6	u_{ab}	1, 0, 0	0, 1, 0	1	$\underline{I}_6$	Active Vector
7	0	1, 0, 0	1, 0, 0	1	$\underline{I}_7$	Zero Vector
8	0	0, 1, 0	0, 1, 0	1	$\underline{I}_8$	Zero Vector
9	0	0, 0, 1	0, 0, 1	1	$\underline{I}_9$	Zero Vector
10	0	1, 1, 1	1, 0, 0	0	$\underline{I}_{10}$	Shoot-Through Zero Vector
		1, 1, 1	0, 1, 0	0	$\underline{I}_{11}$	
		1, 1, 1	0, 0, 1	0	$\underline{I}_{12}$	
11	0	1, 0, 0	1, 1, 1	0	$\underline{I}_{13}$	Shoot-Through Zero Vector
		0, 1, 0	1, 1, 1	0	$\underline{I}_{14}$	
		0, 0, 1	1, 1, 1	0	$\underline{I}_{15}$	

Table 11.2 Output phase voltage and inverter switching states.

State	Vector	S_{Ap}, S_{Bp}, S_{Cp}	S_{An}, S_{Bn}, S_{Cn}	U_{A1}, U_{A2}	U_{B1}, U_{B2}	U_{C1}, U_{C2}	U_{01}, U_{02}
1	$\underline{U}_1$	1, 0, 0	0, 1, 1	U_{dc}	0	0	$\frac{1}{3}U_{dc}$
2	$\underline{U}_2$	1, 1, 0	0, 0, 1	U_{dc}	U_{dc}	0	$\frac{2}{3}U_{dc}$
3	$\underline{U}_3$	0, 1, 0	1, 0, 1	0	U_{dc}	0	$\frac{1}{3}U_{dc}$
4	$\underline{U}_4$	0, 1, 1	1, 0, 0	0	U_{dc}	U_{dc}	$\frac{2}{3}U_{dc}$
5	$\underline{U}_5$	0, 0, 1	1, 1, 0	0	0	U_{dc}	$\frac{1}{3}U_{dc}$
6	$\underline{U}_6$	1, 0, 1	0, 1, 0	U_{dc}	0	U_{dc}	$\frac{2}{3}U_{dc}$
7	$\underline{U}_7$	1, 1, 1	0, 0, 0	U_{dc}	U_{dc}	U_{dc}	U_{dc}
8	$\underline{U}_8$	0, 0, 0	1, 1, 1	0	0	0	0

Table 11.2 outlines the eight distinct states for Inverters 1 and 2. The zero-voltage levels, relative to the voltages at reference points O and O', are formulated as:

$$\underline{U}_{01} = \frac{1}{3}(\underline{U}_{A1} + \underline{U}_{B1} + \underline{U}_{C1}), \tag{11.3}$$

$$\underline{U}_{02} = \frac{1}{3}(\underline{U}_{A2} + \underline{U}_{B2} + \underline{U}_{C2}). \tag{11.4}$$

Duty cycles for the rectifier and inverter stages are harmonized via the ensuing equations:

$$\begin{cases} d_\alpha = m_i \sin(60° - \theta_i) \\ d_\beta = m_i \sin\theta_i \\ d_{0r} = 1 - d_\alpha - d_\beta - d_{st} \\ d_{st} = D = const. \end{cases} \tag{11.5}$$

$$\begin{cases} d_\mu = m_o \sin(60° - \theta_o) \\ d_\nu = m_o \sin\theta_o \\ d_{0i} = 1 - d_\mu - d_\nu \end{cases} \tag{11.6}$$

$$\begin{cases} d_{\alpha\mu} = d_\alpha d_\mu \\ d_{\beta\mu} = d_\beta d_\mu \\ d_{\alpha\nu} = d_\alpha d_\nu \\ d_{\beta\nu} = d_\beta d_\nu \\ d_{\alpha 0i} = d_\alpha d_{0i} \\ d_{\beta 0i} = d_\beta d_{0i} \end{cases} \tag{11.7}$$

In this context, the angle θ_i corresponds to the vector orientation of the incoming current at the rectifier stage, whereas θ_o signifies the vector direction of the output voltage emerging from the inverter stage. As for the duty cycles within one complete switching period, the values d_α, d_β, d_{st}, and d_{0r} specifically pertain to the current vectors. On the flip side, d_μ, d_ν, and d_{0i} are the duty cycles associated with the active and zero output voltage vectors, respectively.

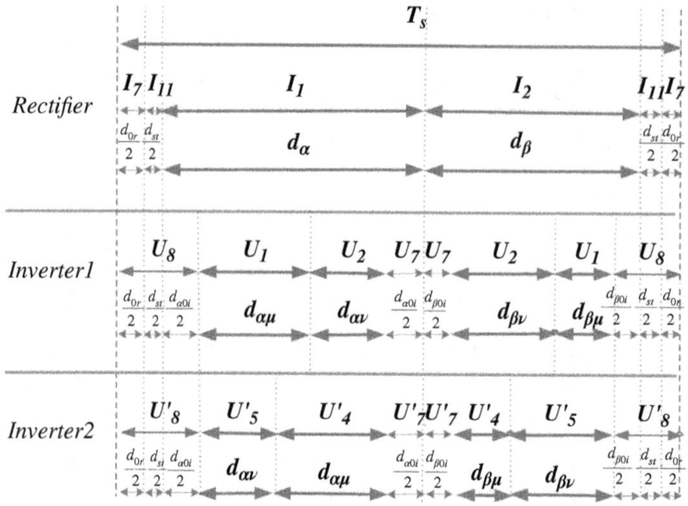

Figure 11.3 Assignment of vectors and duty cycles of D-3L-QZS-IMC.

As depicted in Figure 11.3, the vectorial representation and corresponding duty cycles for the D-3L-QZS-IMC are presented. The CMV generated between the phase terminals of the motor winding is mathematically defined as:

$$\underline{U}_{\text{CMV}} = \underline{U}_{01} - \underline{U}_{02}. \tag{11.8}$$

As Figure 11.4 graphically portrays, the resulting CMV aligns with the specifications laid out in both Figure 11.3 and Table 11.2. A key observation is the success of the suggested modulation techniques in minimizing the CMV levels at the phase winding terminals, denoted as $\underline{U}_{\text{CMV}}$. This achievement becomes particularly clear when set against the CMV measured between the phase winding terminals and the negative poles of both DC links, represented as $\underline{U}_{01}$ or $\underline{U}_{02}$. Although CMV doesn't contribute to torque generation, it does generate common-mode phase currents, leading to added losses in the winding. Therefore, managing this parameter is of paramount importance. Furthermore, during the rectification stage, the duty cycles of zero vectors are allocated in a balanced manner. On the other hand, the inverter stage utilizes a six-action switching scheme, as opposed to the five actions in the rectification stage.

The phase-to-ground CMVs, $\underline{U}_{\text{A1G}}$ and $\underline{U}_{\text{A2G}}$, exhibit the same waveforms as $\underline{U}_{01}$ and $\underline{U}_{02}$, albeit with the ground/earth as the reference, not points O and O', as illustrated in Figure 11.4 (right axes). They can be calculated using the following equations:

$$\begin{cases} \underline{U}_{\text{A1G}} = \underline{U}_{01} + \underline{U}_{\text{OG}} = \underline{U}_{01} - \dfrac{1}{2}U_{\text{dc1}} \\ \underline{U}_{\text{A2G}} = \underline{U}_{02} + \underline{U}_{\text{O'G}} = \underline{U}_{02} - \dfrac{1}{2}U_{\text{dc2}} \end{cases}, \tag{11.9}$$

In contrast to traditional 2-Level Voltage Source Inverters (2L-VSIs), the proposed system is not anticipated to provide any distinct advantages when it comes to mitigating either the phase-to-ground CMV or the rate of voltage change over time, commonly expressed as dv/dt. In more specific terms, the system does not offer foreseeable benefits in terms of reducing high-frequency currents that could affect the motor bearings or in minimizing the electrical stress that could be experienced by the phase-to-ground insulation of the motor windings. Utilizing the technique of Space Vector Modulation (SVM), as elaborated in the scholarly reference [4],

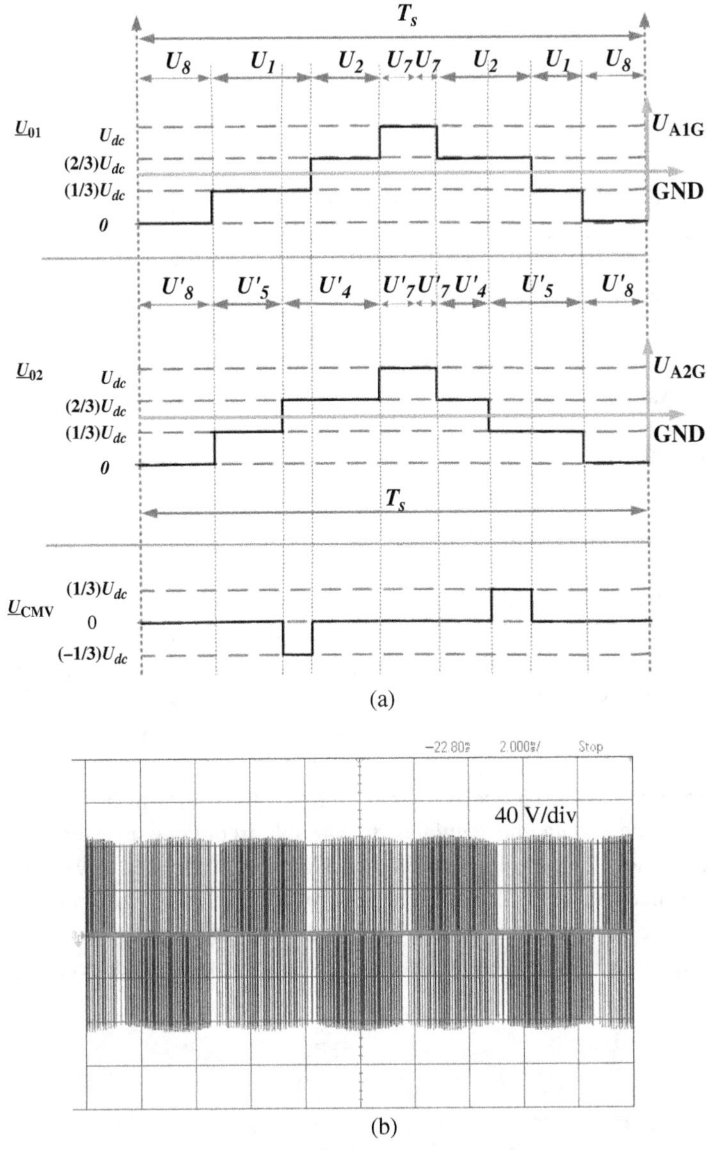

Figure 11.4 CMV produced by D-3L-QZS-IMC: (a) Graphical representation of zero voltage ($\underline{U}_{01}$ and $\underline{U}_{02}$) and CMV ($\underline{U}_{CMV}$); (b) experimental CMV($\underline{U}_{CMV}$).

a 2L-IMC can achieve a maximum CMV that is equal to $1/\sqrt{3}$ of the amplitude of the sinusoidal input phase voltage. This specific outcome is obtained through the implementation of a near-state Pulse-Width Modulation (PWM) scheme, further fine-tuned by introducing a $\pi/6$ phase shift in the inverter stage. However, when it comes to the D-3L-QZS-IMC, its modulation strategy diverges significantly from the strategy discussed in [4]. As a result, and as corroborated by the data presented in Figures 11.3 and 11.4, the peak value of phase CMV in the D-3L-QZS-IMC configuration equates to a third of the DC-link voltage.

11.3 Motor Control Strategy

11.3.1 General Description

In this segment, the focus is on elucidating a control strategy engineered to deliver Variable Voltage Variable Frequency (VVVF) output within the PDS framework under discussion. As depicted in Figure 11.5, the scheme employs a closed-loop control mechanism for motor speed, wherein a Proportional-Integral (PI) regulator is utilized to synthesize the target torque, T^*_e. This is accomplished while maintaining the motor's rotational speed N in alignment with the reference speed N^*, as cited in [17, 18, 21]. The corresponding q-axis current setpoint, I^*_{sq}, along with rotor flux linkage (ψ_r) and motor slip angular speed (ω_s), are derived within the blocks labelled as "Formula A," "Formula B," and "Formula C," respectively. These blocks are correlated with the ensuing equations 11.10–11.12, respectively.

$$I^*_{sq} = \frac{2}{3} \cdot T^*_e \cdot L_r \cdot p^{-1} \cdot L_m^{-1} \cdot \psi_r^{-1}, \tag{11.10}$$

$$\psi_r = I_{sd} \cdot L_m \cdot (1 + T_r \cdot p)^{-1}, \tag{11.11}$$

$$\omega_s = L_m \cdot I_{sq} \cdot T_r^{-1} \cdot \psi_r^{-1}, \tag{11.12}$$

In this analysis, key parameters such as the number of pole pairs (p), mutual inductance (L_m), rotor inductance (L_r), rotor time constant (T_r), and d-axis current component (I_{sd}) are factored in. The control strategy employs dual PI control loops designed to manage the actual d-axis and q-axis current components. These loops are responsible for faithfully tracking their respective reference values, I^*_{sq} and I^*_{sd}, and subsequently generating the stator voltage components for the d-axis (U_d) and q-axis (U_q).

These derived voltage components then serve two primary functions: they inform the voltage angle θ_o, which incorporates rotor-speed data and regulates the variable-frequency inverter, and the amplitude U_{om}, which governs the DC-link voltage through modulation of the STS duty cycle.

Moreover, the control strategy incorporates a Phase-Locked Loop (PLL) specifically aimed at acquiring the phase angle θ_{in} of the rectifier's input voltage. The modulation index (m_i) and angle θ_{in} work in tandem to secure a unity power factor, denoted as $\cos\varphi_{qz} = 1$, by controlling the rectifier.

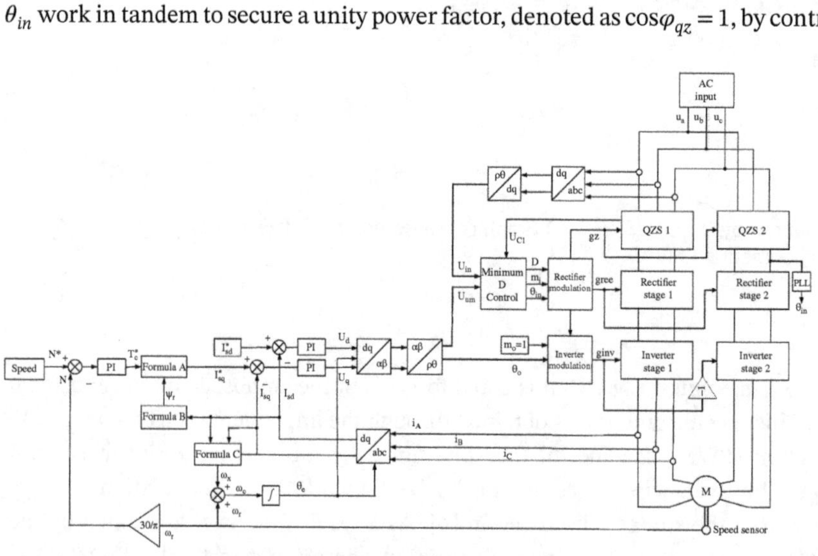

Figure 11.5 Control scheme diagram of the presented PDS.

11.3.2 Control Variables

The voltage gain for each module within the QZS-IMC system can be calculated using the following expression:

$$G = \frac{1}{2}\frac{U_{om}}{U_{in}} \tag{11.13}$$

Here, U_{om} represents the output phase voltage specific to D-3L-QZS-IMC.

Two operational modes are associated with each QZS-IMC module, governed by the following control variables:

$$\begin{cases} \text{Mode 1,} \quad G \le H: D = 0, m_i = G/H, m_o = 1 \\ \text{Mode 2,} \quad G > H: m_i = 1 - D, m_o = 1 \end{cases}. \tag{11.14}$$

In the first mode, the QZS network serves exclusively as a filter without amplification capability. On the other hand, the second mode enables both voltage amplification and filtering functions within the QZS network. In this context, the boost factor B from equation 11.2 must exceed unity, with B minimized and $m_i \times m_o$ maximized.

Based on equations 11.1–11.13, and considering $\cos\varphi_{qz} = 1$, the gain G is expressed as:

$$G = H\frac{m_i}{2m_i - 1} \tag{11.15}$$

As depicted in Figure 11.6a, the correlation between voltage gain (G) and the modulation index (m_i) for each module can be observed. When the gain value (G) is less than or equal to H (0.866), the behavior of the module falls along line-A on the graph. However, if the gain (G) surpasses the threshold of H, the module follows the trajectory defined by line-B. On a different note, Figure 11.6b

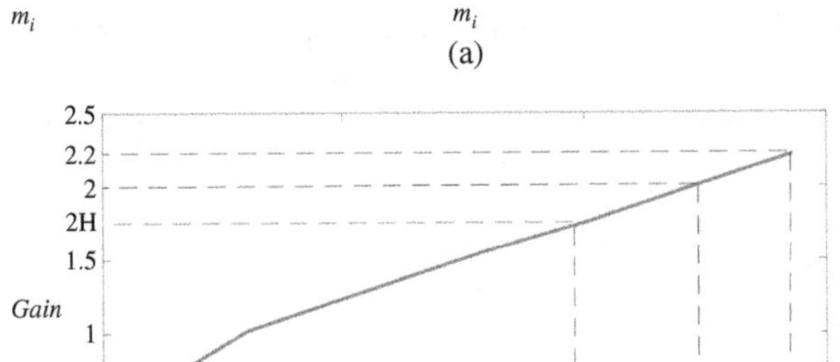

(a)

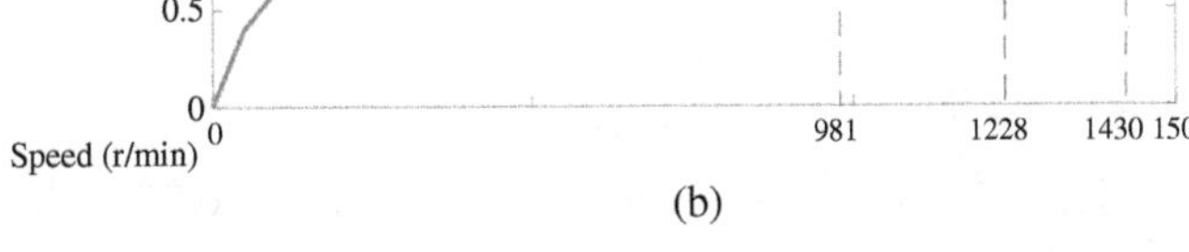

(b)

Figure 11.6 Gain curves of D-3L-QZS-IMC when operating with the presented control method: (a) gain as a function of m_i for each module; (b) gain as a function of motor speed for the D-3L-QZS-IMC.

delves into how voltage gain varies in relation to the rotational speed of the motor within the context of a D-3L-QZS-IMC setup. If the gain falls below $2H$, then Mode 1 gets triggered, which primarily employs the QZS network for filtration purposes. Alternatively, when the motor speed crosses 981 revolutions per minute, the system transitions into Mode 2. In this mode, voltage amplification is performed alongside filtration within the QZS network.

11.3.3 Boost Controller Design

In scenarios where $G > H$, an output voltage control mechanism can be formulated to stabilize U_{om}, the output voltage of the converter. The mean DC-link voltage can be represented as:

$$\overline{U}_{dc} = \frac{3}{2} m_i U_{qz} \cos \varphi_{qz} \tag{11.16}$$

In this context, U_{qz} represents the apex voltage emanating from the output phase of the QZS network. As for each of the individual inverters in the system, labeled as U_{o1} and U_{o2}, they conform to a specific mathematical equation to determine their voltages.

$$U_{o1} = U_{o2} = \frac{1}{\sqrt{3}} m_o \overline{U}_{dc} \tag{11.17}$$

Consequently, the voltage across the capacitor C_1 can be described by:

$$U_{C1} = \frac{2}{\sqrt{3}} U_{o1} = \frac{1}{\sqrt{3}} U_{om}, \tag{11.18}$$

This formulation indicates that the control of U_{om} is achievable via U_{C1}.

As illustrated in Figure 11.1, a comprehensive state-space representation for each module in the QZS-IMC architecture can be established, denoted as equation 11.19.

$$\begin{bmatrix} L & 0 & 0 \\ 0 & C_1 & 0 \\ 0 & 0 & C_2 \end{bmatrix} p \begin{bmatrix} i_L \\ u_{C1} \\ u_{C2} \end{bmatrix} = \begin{bmatrix} -(R_C + R_L) & D-1 & D \\ 1-2D & 0 & 0 \\ 1-2D & 0 & 0 \end{bmatrix} \begin{bmatrix} i_L \\ u_{C1} \\ u_{C2} \end{bmatrix} + \begin{bmatrix} 1 & R_C(1-D) \\ 0 & D-1 \\ 0 & D-1 \end{bmatrix} \begin{bmatrix} u \\ i' \end{bmatrix} \tag{11.19}$$

This equation further integrates a specific set of conditions, detailed by variables

$$i' = \frac{1-2D}{1-D} i_i$$

The Equivalent Series Resistance (ESR) for both the inductive and capacitive components are symbolized by R_L and R_C, respectively.

For analytical purposes, a small-signal model is subsequently generated, which we reference as equation 11.20.

$$\begin{bmatrix} L & 0 & 0 \\ 0 & C_1 & 0 \\ 0 & 0 & C_2 \end{bmatrix} p \begin{bmatrix} \hat{i}_L \\ \hat{u}_{C1} \\ \hat{u}_{C2} \end{bmatrix} = \begin{bmatrix} -(R_C + R_L) & D-1 & D \\ 1-2D & 0 & 0 \\ 1-2D & 0 & 0 \end{bmatrix} \begin{bmatrix} \hat{i}_L \\ \hat{u}_{C1} \\ \hat{u}_{C2} \end{bmatrix} + \begin{bmatrix} 1 & R_C(1-D) & u_{C1} + u_{C2} - R_C i' \\ 0 & D-1 & i' - 2i_L \\ 0 & D-1 & i' - 2i_L \end{bmatrix} \begin{bmatrix} \hat{u} \\ \hat{i}' \\ d \end{bmatrix} \tag{11.20}$$

Delving into the specifics of capacitor C_1, its voltage transfer function is codified in equation 11.21.

$$\hat{u}_{C1}(s) = G_{C1ui}(s)\hat{u}_i(s) + G_{C1ii'}(s) \cdot \hat{i}'(s) + G_{C1d}(s) \cdot \hat{d}(s) \tag{11.21}$$

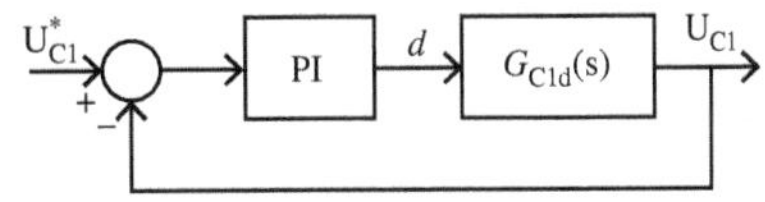

Figure 11.7 Capacitor C_1 voltage control system.

Variables related to this function are further described by

$$G_{C1ui}(s) = \frac{sC_2(1-2D)}{s^3C_2C_1L + s^2(R_C+R_L)C_2C_1 + s((1-D)(1-2D)C_2 - DC_1(1-2D))}$$

$$G_{C1ii'}(s) = \frac{s(1-2D)(1-D)C_2R_C + s(D-1)C_2(sL+R_C+R_L)}{s^3C_2C_1L + s^2(R_C+R_L)C_2C_1 + s((1-D)(1-2D)C_2 - DC_1(1-2D))}$$

$$G_{C1d}(s) = \frac{s(1-2D)C_2(u_{C1}+u_{C2}-R_Ci') + sC_2(sL+R_C+R_L)(i'-2i_L)}{s^3C_2C_1L + s^2(R_C+R_L)C_2C_1 + s((1-D)(1-2D)C_2 - DC_1(1-2D))}$$

Utilizing this framework, U_{C1} can be efficiently regulated through the implementation of a closed-loop control paradigm as depicted in Figure 11.7.

The PI regulator/control equation is:

$$d = \left(k_p + k_i\frac{1}{s}\right)\left(U_{C1}^* - U_{C1}\right) \tag{11.22}$$

As illustrated in Figure 11.8, the Bode plot corresponding to $G_{C1d}(s)$ indicates a cutoff angular frequency occurring at 1.01×10^3 rad/s. While tuning the crossover angular frequencies associated with both $G_{PI}(s)$ and the product of $G_{C1d}(s)$ and $G_{PI}(s)$, the PI regulator's parameters are configured with values $k_p = 3.51 \times 10^{-5}$ and $k_i = 0.0353$. A detailed examination of the Bode diagrams pertinent to $G_{PI}(s)$ and the combined $G_{C1d}(s) \times G_{PI}(s)$ system can be found in Figure 11.9. For additional insights into the closed-loop control scheme specific to U_{C1}, one may refer to Figure 11.10.

Upon juxtaposing Figures 11.8 and 11.9, one discerns noticeable improvements in the system's characteristics post-compensation. Specifically, in the low angular frequency domain of $G_{C1d}(s) \times G_{PI}(s)$, a magnitude slope of −20–20 dB/dec and a phase shift of −90°–90° manifest, signifying gains in both tracking precision and response velocity. Additionally, phase and magnitude

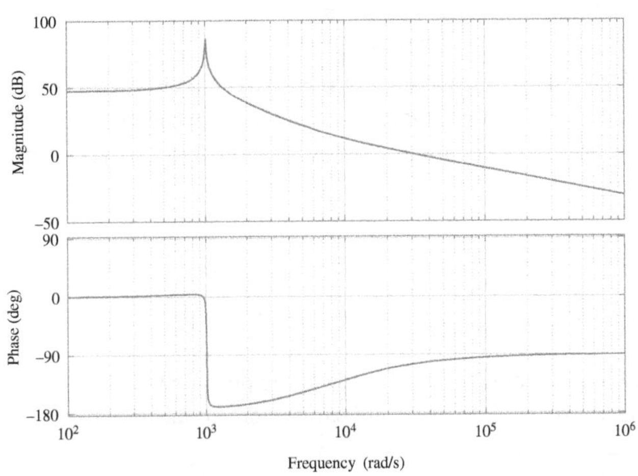

Figure 11.8 Bode plot of $G_{C1d}(s)$.

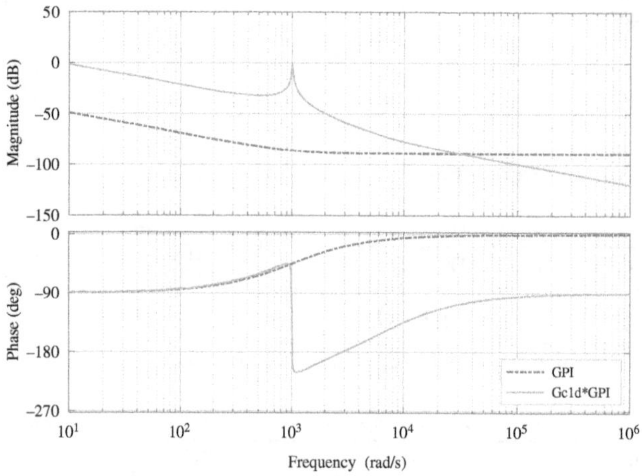

Figure 11.9 Bode plots of $G_{PI}(s)$ and $G_{C1d}(s)G_{PI}(s)$.

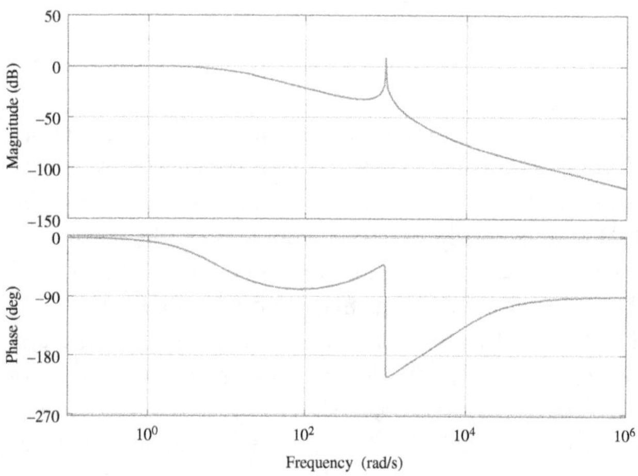

Figure 11.10 Bode plots of U_{C1} closed-loop control system.

margins of 52.3° and 57.4 dB, respectively, at the crossover and cutoff frequencies, underscore advancements in system stability and dynamic performance. Figure 11.10 subsequently corroborates these enhancements by illustrating the closed-loop system's sustained stability in both magnitude and phase domains.

11.4 Experimental Verifications

An empirical evaluation was conducted to authenticate the efficacy of the proposed PDS architecture, modulation techniques, and motor control approaches. Vital parameters for this experimental configuration are delineated in Table 11.3.

The research team conducted a series of three rigorous tests to investigate the performance of the proposed system under different scenarios.

Table 11.3 Parameters of the experimental PDS setup.

QZS-IMC	Input phase voltage	$U_{in} = 220\,\text{V}$, rms
	Switching frequency	$f_{sw} = 10\,\text{kHz}$
	Inverter modulation index	$m_o = 1$
	QZS capacitors	$C_1 = C_2 = 500\,\mu\text{F}/480\,\text{V}$
	QZS inductors	$L_1 = L_2 = 500\,\mu\text{H}/20\,\text{A}$
OEW-SCIM	Rated power	$P_n = 4\,\text{kW}$
	Rated phase-to-phase voltage	$U_n = 380\,\text{V}$, rms
	Rated speed	$n_n = 1430\,\text{r/min}$
	Number of pole pairs	$p = 2$
	Stator resistance	$R_s = 1.4\,\Omega$
	Stator leakage inductance	$L_{s\delta} = 5.8\,\text{mH}$
	Rotor leakage inductance	$L_{r\delta} = 5.8\,\text{mH}$
	Rotor resistance	$R_r = 1.4\,\Omega$
	Mutual inductance	$L_m = 172.2\,\text{mH}$

Note: Motor inductances and resistances are referred to stator.

Experimental Setup I: In the initial experimental setup, the power supply voltage was programmed to experience a 15% reduction. Despite this voltage drop, the motor's torque and speed targets were consistently maintained at 26 N · m and 1430 r/min, respectively. Key insights derived from this experimental condition can be further explored in Figures 11.11, 11.12, and 11.15. Specifically, Figure 11.11 offers a graphical representation of the PDS input electrical variables, which include both the voltage and current parameters. Figure 11.12, on the other hand, discloses the motor's torque performance as well as its corresponding phase currents and voltage levels. Last but not least, Figure 11.15 provides an in-depth analysis of the phase-to-phase voltage attributes and the individual phase voltage within the D-3L-QZS-IMC framework.

Experimental Setup II: In the established experimental configuration, the system commenced its run from a standstill, eventually reaching and maintaining a consistent rotor speed of 143 r/min.

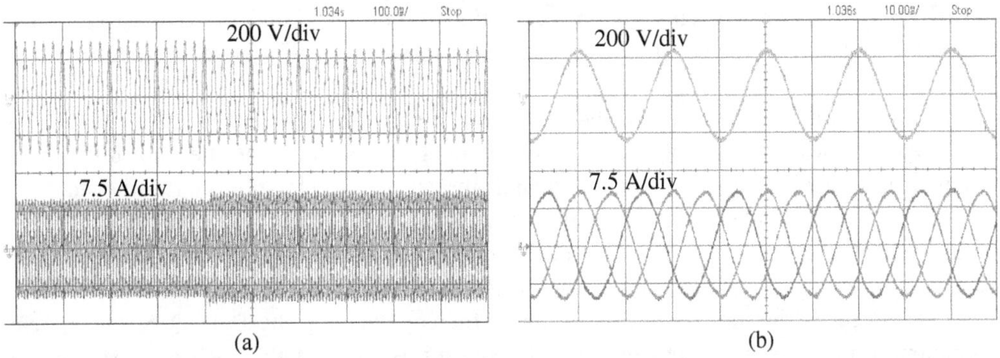

Figure 11.11 Experimental PDS input voltages and currents at 1430 r/min and 26 N · m: (a) 1000-ms time window with the transient instant of 15% voltage drop; (b) 100-ms time window after 15% voltage drop.

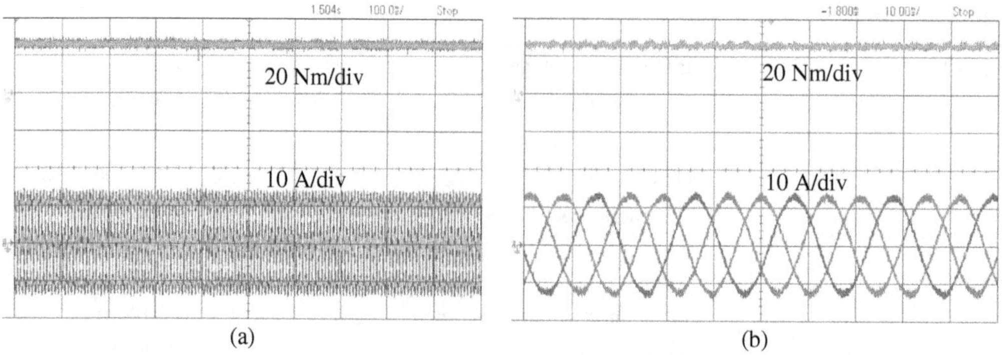

Figure 11.12 At 1430 r/min, the motor's torque and stator phase currents were analyzed amid a 15% reduction in supply voltage. The observations span: (a) a longer duration of 1000 ms; and (b) a shorter interval of 100 ms.

Figure 11.14 serves as a valuable resource for scrutinizing the motor's torque output, the accompanying phase currents, and the speed at which the rotor turns.

Experimental Setup III: In this third setup, we manipulated the rotor-speed reference, changing it from an initial value of 1430 to 800 r/min. Figure 11.13 displays the subsequent torque and phase current changes in the motor, while Figure 11.16 visualizes the changes in phase-to-phase and phase voltages for the D-3L-QZS-IMC model.

As revealed in Figures 11.11 and 11.12, the decrease in supply voltage led to a compensatory rise in input current to preserve output power levels. This occurred while maintaining constant torque and stator currents. Simply put, the outcomes demonstrated that the proposed system is capable of autonomously adjusting its voltage levels to counteract the adverse impacts of a voltage drop on the functioning of the motor.

Upon closely examining Figure 11.13, it's apparent that the motor's speed control mechanism instantaneously induced negative torque, transitioning into a braking state, to decrease the speed from its original 1430 r/min down to 800 r/min. Subsequently, once the motor stabilized at the newly set speed reference, the torque readjusted to its original value of 26 N · m. This braking phase interestingly enabled a flow of power back into the grid, thereby confirming the regenerative capabilities of the PDS. Furthermore, despite the shift in rotor speed affecting the frequency of the phase current, the torque remained consistent both pre and post the speed alteration. It's also noteworthy that the THDs of the phase currents remained exceptionally low during both speed settings, as illustrated in Figure 11.13d,e. The system thus displayed the anticipated performance metrics for motor operation.

By closely examining Figure 11.14, the high level of precision achieved in the regulation of rotor speed becomes unmistakably clear. This is particularly evident when the rotor initiates its movement from a point of zero velocity, moving through the acceleration phase to eventually stabilize at a speed of 143 r/min, all while maintaining a consistent torque of 13 N · m. This careful modulation is facilitated by a complex control system, resulting in smooth speed transition and effective performance. Operating at such a moderate speed, the back Electromotive Force (EMF) generated by the motor remains minimal, thereby enabling the D-3L-QZS-IMC system to function primarily in its designated Mode 1. Within this mode, the QZS-IMCs serve not just as basic electrical components, but take on the specialized role of acting as advanced filtering mechanisms.

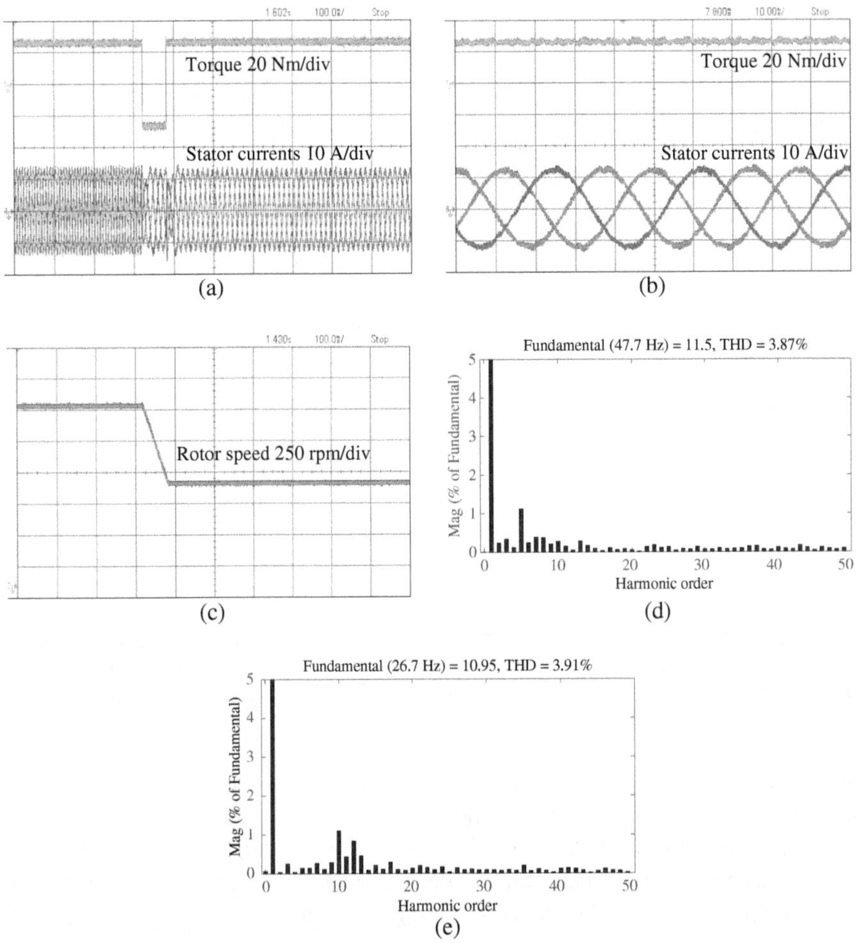

Figure 11.13 During the motor's deceleration from 1430 to 800 r/min, observations were made on its torque and stator phase currents. Specifically: (a) captures a 1000-ms period including the transient event; (b) focuses on a 100-ms interval post the deceleration; (c) details the rotor's velocity; (d) provides the THD of phase current when cruising at 1430 r/min; (e) describes the phase current's THD at 800 r/min.

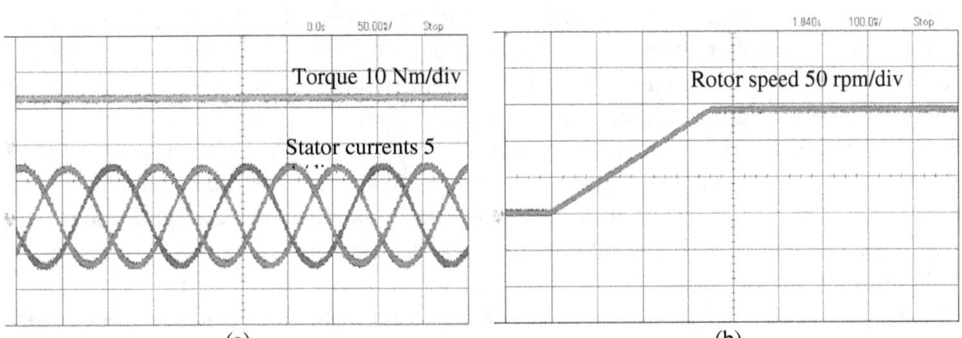

Figure 11.14 The examined system, initially at a standstill and eventually reaching 143 r/min, showcases variations in rotor speed, motor torque, and stator phase currents. Specifically: (a) illustrates the motor's torque combined with stator phase currents; (b) highlights the progression of rotor speed.

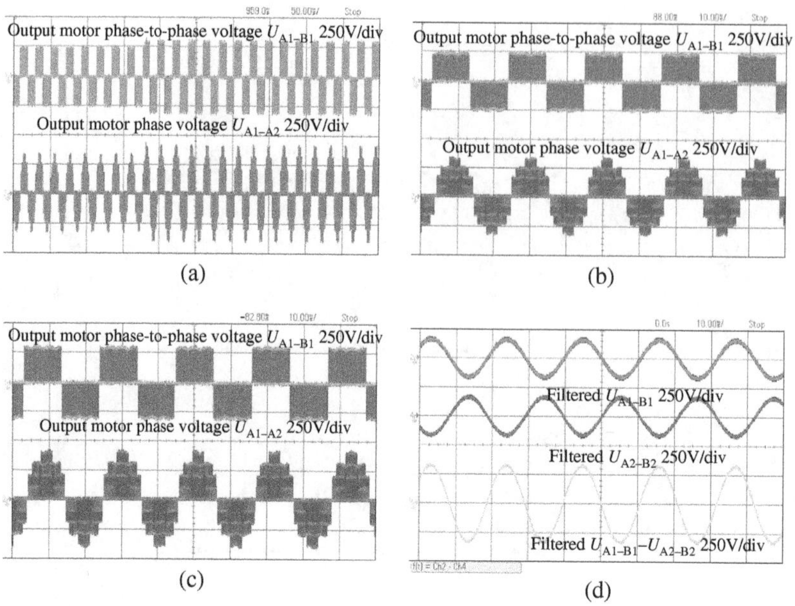

Figure 11.15 Observations highlight motor's phase-to-phase and phase voltages amid a 15% reduction in supply voltage. This is detailed as follows: (a) spanning 500 ms that includes the momentary change; (b) a 100 ms interval leading up to the transient event; (c) a subsequent 100 ms period post-transient; and (d) a juxtaposition of output voltages between 2-L-QZS-IMC and D-3L-QZS-IMC, post filtration.

Directing focus toward Figures 11.15 and 11.16, these illustrations provide comprehensive graphical representations of both the inter-phase and individual phase voltages that the motor produces. These voltages are effectively the electrical output of the intricate D-3L-QZS-IMC system. The specific voltage differential between the phases of Module 1 is represented by the parameter $U_{A1\text{-}B1}$, while the voltage applied directly to the motor's phase A winding is identified as $U_{A1\text{-}A2}$. Notably, both these voltage categories demonstrate a pulsing nature, providing an oscillatory electrical behavior. Figure 11.15d expands on this by detailing the core components of these voltages after undergoing specialized filtration, where $U_{A1\text{-}B1}-U_{A2\text{-}B2}$ is employed as a term to define the inter-phase voltage within the D-3L-QZS-IMC system.

In an even more specific dive into Figure 11.15a, it becomes apparent that there is a perceptible increase in the amplitudes of the two QZS-IMC output voltages, labeled as $U_{A1\text{-}B1}$ and $U_{A1\text{-}A2}$, when the supply voltage experiences a reduction. In order to counteract this adverse effect, the system employs a closed-loop control mechanism to adjust the shoot-through duty cycle "*D*," thereby boosting the voltage to an optimal level for sustained operation.

In an alternate situation where the motor's speed reduces to 800 r/min, the existing power supply voltage is shown to be more than sufficient for continuous motor operation, as demonstrated by Figure 11.16a. This adequacy negates the need for any additional voltage amplification, thus resulting in the nullification of the shoot-through duty cycle, marked as "*D*." As a consequence, the amplitude levels of the voltages $U_{A1\text{-}B1}$ and $U_{A1\text{-}A2}$ undergo a corresponding decline.

Empirical evidence substantiates that the system's closed-loop control mechanism is remarkably skillful at adapting the output voltage from the D-3L-QZS-IMC in a manner that aligns with the fluctuating operational conditions of the motor. For instance, during periods of diminished power supply, the system's inherent voltage-boosting feature is activated to elevate the output voltage. In contrast, this feature becomes dormant when the motor operates at slower speeds, resulting in a

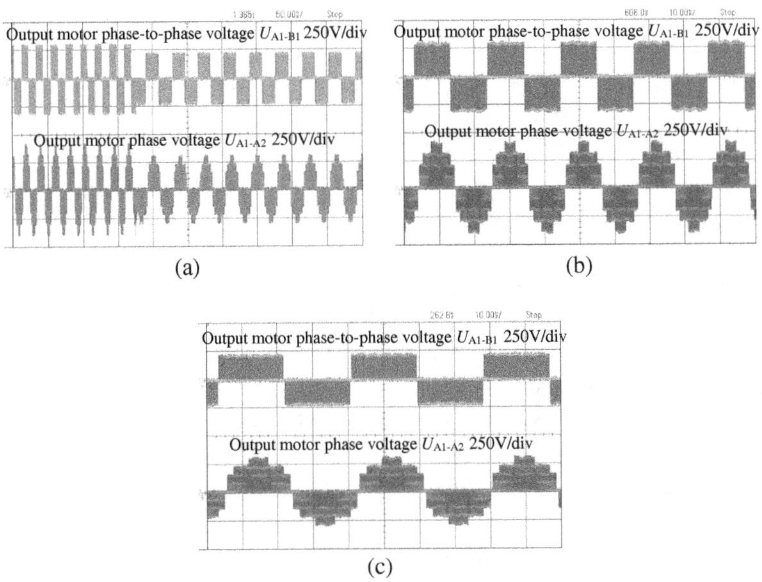

Figure 11.16 Observations of motor phase-to-phase and phase voltages coincide with a speed decline from 1430 to 800 r/min. Specifically, (a) a span of 500 ms capturing the transient moment; (b) a glimpse of 100 ms preceding the transient event; (c) another 100 ms snapshot post the transient.

consequent reduction in output voltage. Notably, the rectifier stage in the system is tasked with managing these voltage fluctuations effectively, while the inverter stage is responsible for controlling frequency variations, as clearly depicted in Figure 11.16a.

Regarding the steady-state phase voltage of the motor, referred to as $U_{A1\text{-}A2}$, Figures 11.15b,c, and 11.16b,c offer a detailed representation of a multilevel voltage waveform. This intricate waveform design is instrumental in diminishing the low-order harmonic content inherent in the motor phase current, thereby affirming both the effectiveness of the system's modulation techniques and the robustness of the overall PDS.

Moreover, a comparative analysis depicted in Figure 11.15d between the 2L-QZS-IMC and the D-3L-QZS-IMC reveals a significant disparity in voltage output. Specifically, the latter yields a voltage output that is approximately twice as high as the former. This data underscores the utility of using the novel converter design over traditional two-level converters when operating under the same voltage-rated power switches. This essentially means that while traditional converters may only sustain lower speeds, the cutting-edge converter technology can deliver the required voltage for higher motor speeds, thereby amplifying its operational range.

11.5 Discussion

When compared with traditional power drive systems that rely on matrix converters as well as those based on two-level Quasi-Z-Source Indirect Matrix Converters (2L-QZS-IMC), the D-3L-QZS-IMC put forth in this study exhibits significantly enhanced voltage gains. Specifically, these gains are as much as $2B$ and double that of the aforementioned conventional systems.

Given a consistent amplitude for the supply voltage and assuming that power switches of the same voltage rating are used, the advanced output voltage gain of our newly proposed

PDS establishes it as a particularly compelling alternative for high-power motor applications. These are applications where the demand for elevated voltage levels is often paired with a corresponding need for increased efficiency, a characteristic inherently present in such motors.

Our system introduces an adaptive closed-loop control framework that fine-tunes not only the VVVF output but accomplishes this through a coordinated strategy. It simultaneously adjusts the DC-link voltage at the end of the rectifier stage and modulates the frequency of the AC output at the inverter stage. This dual-mode operation leads to an agile and dynamic adaptation of the DC-link voltage, rather than a mere adjustment of the amplitude modulation index as is common in traditional inverter systems. Such an orchestrated approach helps to optimize the system by reducing shoot-through duty cycles and lowering the DC-link voltage, thereby achieving the desired motor control parameters without adding unnecessary layers of complexity or redundancy.

Moreover, the innovative design architecture of the presented PDS system addresses and mitigates several technical challenges commonly encountered in industrial setups. These challenges include issues such as common-mode components, the presence of low-order harmonics, and the rate of voltage changes, or dv/dt, in the motor phase currents. Cumulatively, these enhancements contribute to reducing the stress on winding insulation, especially during operations at lower speeds and voltages. They also minimize harmonic-induced losses in motors and bolster the system's resilience against voltage sags from the primary power supply—features that are critically important in industrial applications.

For a more comprehensive understanding of passive elements like inductors and capacitors, Table 11.4 provides a thorough comparative analysis across traditional MC-based PDS, 2L-QZS-IMC-based PDS, and our cutting-edge D-3L-QZS-IMC-based PDS. Those interested in further nuances of passive element design can refer to the architectural details of the 2L-QZS-IMC, as elaborated in reference [23].

It should be noted that while the proposed PDS incorporates a greater quantity of passive elements, the system's unique architecture permits superior output voltage and power when compared to traditional DMC and 2L-QZS-IMC systems. This is especially significant given that all components—inductors, capacitors, and switches—have the same voltage and current ratings, as well as the same supply voltage amplitude. When delivering equivalent output voltage and power, our newly introduced PDS requires switches and capacitors with only half the voltage ratings, and inductors with half the current ratings, in comparison to conventional setups.

Table 11.4 Comparison of amount and rating of passive elements.

Passive element		PDS converter		
		Typical MC	**2L-QZS-IMC**	**D-3L-QZS-IMC**
Inductor	Amount	3	3 (coupled)	6 (coupled)
	Rating	$I_r = 1$ p.u.	$I_r = 1$ p.u.	$I_r = 0.5$ p.u.
Capacitor	Amount	$3 \times C$	$3 \times C_1$ $3 \times C_2$	$6 \times C_1$ $6 \times C_2$
	Rating	$V_r = 1$ p.u.	C_1: $V_r = 1.5$ p.u. C_2: $V_r = 0.5$ p.u.	C_1: $V_r = 0.75$ p.u. C_2: $V_r = 0.25$ p.u.

11.6 Conclusion

This chapter provides an in-depth examination of a novel prospected Power Drive System (PDS) that seamlessly integrates a Dual Three-Level Quasi-Z-Source Indirect Matrix Converter (D-3L-QZS-IMC) with an Open-Ended Winding Squirrel Cage Induction Motor (OEW-SCIM). The viability of this unique design has been thoroughly tested by deploying extensive empirical experiments on a 4-kW prototype of the PDS. These tests were meticulously designed to assess the system's robustness, particularly its resilience against a 15% reduction in supply voltage and unexpected shifts in speed reference. The test outcomes consistently reinforced the robust characteristics of this newly devised system.

As opposed to conventional Voltage Source Inverters (VSIs), which commonly necessitate the use of high-capacity capacitors, the D-3L-QZS-IMC model functions efficiently without requiring any capacitors in the DC link. This design modification significantly mitigates the associated cost and enhances overall system reliability. Employing a 3-level voltage modulation technique on the motor phase, the architecture achieves a marked reduction in harmonic distortions, instances of peak voltages, and rapid fluctuations in voltage (dv/dt). These improvements translate to reduced motor losses due to harmonics and an extended longevity for the winding insulation. Importantly, the approach efficiently manages to lower the motor phase's common-mode voltage while preserving the phase-to-ground voltage, especially when juxtaposed with traditional VSIs.

Further enhancing this groundbreaking design is an integrated rectifier that incorporates a simplified yet effective buck–boost closed-loop control mechanism. This setup allows for dynamic optimization of the shoot-through duty cycle while ensuring a variable DC link voltage. The latter feature furnishes the PDS with a remarkable low-voltage ride-through capability, thereby tempering the impact of voltage sags on both the torque and speed of the motor. Also, the unified variable-voltage and variable-frequency control scheme is especially advantageous during lower-speed operations where the motor runs on a diminished voltage supply. This integrated control strategy surpasses the performance of conventional 2-level Indirect Matrix Converters (2L-IMCs) and VSIs in terms of minimizing switching losses and moderating voltage peaks.

Given the same amplitude for the supply voltage and assuming identical voltage ratings for power switches, our cutting-edge D-3L-QZS-IMC system has the potential to double the voltage applied to the motor phase windings. This distinct capability marks it as particularly suitable for high-power applications where motors are designed with stator windings that have higher voltage ratings.

While the system's modular nature, augmented voltage gains, and bidirectional power flow are its cornerstone advantages, it's essential to acknowledge that there are challenges. Primarily, the financial overhead can be considerable, largely because of the requirement for a large number of power switches and the intricacies of the cabling infrastructure.

References

1 R. Vargas, J. Rodriguez, C. A. Rojas, and M. Rivera, "Predictive control of an induction machine fed by a matrix converter with increased efficiency and reduced common-mode voltage," *IEEE Trans. Energy Convers.*, vol. 29, no. 2, pp. 473–485, Jun. 2014.

2 A. Olloqui *et al.*, "Model-based predictive rotor current control strategy for indirect power control of a DFIM driven by an indirect matrix converter," *IEEE Trans. Energy Convers.*, vol. 36, no. 2, pp. 1510–1516, Jun. 2021.

3 R. Sadooghi, *et al.*, "An effective non-square matrix converter based approach for active power control of multiple DGs in microgrids: Experimental implementation," *IEEE Trans. Energy Convers.*, vol. 37, no. 2, pp. 755–765. https://doi.org/10.1109/TEC.2021.3116310.

4 T. D. Nguyen and H. Lee, "A new SVM method for an indirect matrix converter with common-mode voltage reduction," *IEEE Trans. Industr. Inform.*, vol. 10, no. 1, pp. 61–72, Feb. 2014.

5 T. N. Mir, B. Singh, and A. H. Bhat, "FS-MPC-based speed sensorless control of matrix converter fed induction motor drive with zero common mode voltage," *IEEE Trans. Ind. Electron.*, vol. 68, no. 10, pp. 9185–9195, Oct. 2021.

6 L. Empringham, J. W. Kolar, J. Rodriguez, P. W. Wheeler, and J. C. Clare, "Technological issues and industrial application of matrix converters: A review," *IEEE Trans. Ind. Electron.*, vol. 60, no. 10, pp. 4260–4271, Oct. 2013.

7 T. Tayar, A. Navon, D. Shmilovitz, and Y. Levron, "Heterodyne controlled matrix converter as a direct frequency reduction method for microturbine generators (MTG)," *IEEE Trans. Energy Convers.*, vol. 37, no. 3, pp. 1533–1544, https://doi.org/10.1109/TEC.2022.3163346.

8 W. Deng, H. Li, and J. Rong, "A novel direct torque control of matrix converter-fed PMSM drives using dynamic sector boundary for common-mode voltage minimization," *IEEE Trans. Ind. Electron.*, vol. 68, no. 1, pp. 70–80, Jan. 2021.

9 J. Lei, S. Feng, B. Zhou, H.-N. Nguyen, J. Zhao, and W. Chen, "A simple modulation scheme with zero common-mode voltage and improved efficiency for direct matrix converter-fed PMSM drives," *IEEE J. Emerg. Select. Top. Power Electron.*, vol. 8, no. 4, pp. 3712–3722, Dec. 2020.

10 H. Dan, P. Zeng, W. Xiong, M. Wen, M. Su, and M. Rivera, "Model predictive control-based direct torque control for matrix converter-fed induction motor with reduced torque ripple," *CES Trans. Electric. Mach. Syst.*, vol. 5, no. 2, pp. 90–99, Jun. 2021.

11 S. Khwan-on, L. de Lillo, L. Empringham, and P. Wheeler, "Fault- tolerant matrix converter motor drives with fault detection of open switch faults," *IEEE Trans. Ind. Electron.*, vol. 59, no. 1, pp. 257–268, Jan. 2012.

12 T. N. Mir, B. Singh, and A. H. Bhat, "Delta-sigma modulation based common-mode voltage elimination in direct matrix converter," in *IEEE Trans. Industr. Inform.*, vol. 17, no. 2, pp. 1048–1057, Feb. 2021.

13 S. Mondal and D. Kastha, "Input reactive power controller with a novel active damping strategy for a matrix converter fed direct torque controlled DFIG for wind power generation," in *IEEE J. Emerg. Select. Top. Power Electron.*, vol. 8, no. 4, pp. 3700–3711, Dec. 2020.

14 R. Vargas, U. Ammann, B. Hudoffsky, J. Rodriguez, and P. Wheeler, "Predictive torque control of an induction machine fed by a matrix converter with reactive input power control," *IEEE Trans. Power Electron.*, vol. 25, no. 6, pp. 1426–1438, Jun. 2010.

15 U. R. Muduli, R. K. Behera, K. Al Hosani, and M. S. Elmoursi, "Direct torque control with constant switching frequency for three-to-five phase direct matrix converter fed five-phase induction motor drive," *IEEE Trans. Power Electron.*, vol. 37, no. 9, pp. 11019–11033, Sept. 2022. https://doi.org/10.1109/TPEL.2022.3167477.

16 Z. Jin, S. Li, W. Wang, X. Liu, Y. Liu, and B. Ji, "A novel space vector overmodulation strategy based on input current vector for indirect matrix converter," in *IEEE Energy Conversion Congr. Expo. (ECCE)*, Baltimore, MD, USA, 2019, pp. 3554–3558.

17 A. Bento, R. Luís, S. Pinto, and F. Silva, "Open-end winding synchronous reluctance sensorless drive based on indirect matrix converter with dual VSI output," in *The 10th Int. Conf. Power Electronics, Machines and Drives (PEMD 2020)*, Online Conference, 2020, pp. 995–1000.

18 E. Karaman, M. Farasat, and A. M. Trzynadlowski, "Indirect matrix converters as generator–grid interfaces for wind energy systems," in *IEEE J. Emerg. Select. Top. Power Electron.*, vol. 2, no. 4, pp. 776–783, Dec. 2014.

19 E. Karaman, M. Farasat, and A. M. Trzynadlowski, "A comparative study of series and cascaded Z-source matrix converters," *IEEE Trans. Ind. Electron.*, vol. 61, no. 10, pp. 5164–5173, Oct. 2014.

20 C. N. El-Khoury, H. Y. Kanaan, I. Mougharbel, and K. Al-Haddad, "Implementation of a series Z-source very sparse matrix converter in a PMSG-based WECS," in *IEEE Int. Conf. Industrial Technology (ICIT)*, Lyon, France, 2018, pp. 1887–1892.

21 W. Song and Y. Zhong, "A study of Z-source matrix converter with high voltage transfer ratio," in *IEEE Vehicle Power and Propulsion Conf.*, 2008, pp. 1–6.

22 W. Song, Y. Zhong, H. Zhang, X. Sun, Q. Zhang, and W. Wang, "A study of Z-source dual-bridge matrix converter immune to abnormal input voltage disturbance and with high voltage transfer ratio," *IEEE Trans. Industr. Inform.*, vol. 9, no. 2, pp. 828–838, May 2013.

23 S. Liu, B. Ge, X. Jiang, H. Abu-Rub, and F. Z. Peng, "Comparative evaluation of three Z-source/quasi-Z-source indirect matrix converters," in *IEEE Trans. Ind. Electron.*, vol. 62, no. 2, pp. 692–701, Feb. 2015.

24 K. Park, K. Lee, and F. Blaabjerg, "Improving output performance of a Z-source sparse matrix converter under unbalanced input-voltage conditions," *IEEE Trans. Power Electron.*, vol. 27, no. 4, pp. 2043–2054, Apr. 2012.

25 S. Sousa, S. Pinto, F. Silva, and J. Maia, "Extended voltage range AC drive using a Z source indirect matrix converter," *20th Int. Conf. Electrical Machines*, 2012, pp. 953–958.

26 O. Ellabban, H. Abu-Rub, and S. Bayhan, "Z-source matrix converter: an overview," *IEEE Trans. Power Electron.*, vol. 31, no. 11, pp. 7436–7450, Nov. 2016.

27 S. Liu, B. Ge, Y. Liu, H. Abu-Rub, R. S. Balog, and H. Sun, "Modeling, analysis, and parameters design of LC-filter-integrated quasi-Z -source indirect matrix converter," *IEEE Trans. Power Electron.*, vol. 31, no. 11, pp. 7544–7555, Nov. 2016.

28 O. Ellabban and H. Abu-Rub, "A quasi-Z-source direct matrix converter feeding a vector controlled induction motor drive," *IEEE J. Emerg. Select. Top. Power Electron.*, vol. 3, no. 2, pp. 339–348, Jun. 2015.

29 M. Guo, Y. Liu, B. Ge, and H. Abu-Rub, "Optimum boost control of quasi-Z source indirect matrix converter," *IEEE Trans. Ind. Electron.*, vol. 65, no. 10, pp. 8393–8404, Oct. 2018.

30 M. Guo, Y. Liu, B. Ge, X. Li, A. T. de Almeida, and F. J. T. E. Ferreira, "Dual, three-level, quasi-Z-source, indirect matrix converter for motors with open-ended windings," *IEEE Trans. Energy Convers.*, vol. 38, no. 1, pp. 64–74, Mar. 2023.

Index

Impedance Source Matrix Converters and Control, First Edition.
Yushan Liu, Xiao Li, Baoming Ge, Haitham Abu-Rub, and Frede Blaabjerg.

n

o

p

q

t

u

v

www.ingramcontent.com/pod-product-compliance
Lightning Source LLC
LaVergne TN
LVHW061154200225
803807LV00003B/224